Vector Fields with Applications to Thermodynamics and Irreversibility

Mathematics and Physics for Science and Technology
Series Editor: L.M.B.C. Campos
former Director of the Center for Aeronautical
and Space Science and Technology
Lisbon University

Volumes in the series:

TOPIC A – THEORY OF FUNCTIONS AND POTENTIAL PROBLEMS

Volume I (Book 1) – Complex Analysis with Applications to Flows and Fields
L.M.B.C. Campos

Volume II (Book 2) – Elementary Transcendentals with Applications to Solids and Fluids
L.M.B.C. Campos

Volume III (Book 3) – Generalized Calculus with Applications to Matter and Forces
L.M.B.C. Campos

TOPIC B – BOUNDARY AND INITIAL-VALUE PROBLEMS

Volume IV – Ordinary Differential Equations with Applications to Trajectories and Vibrations
L.M.B.C. Campos

Book 4 – Linear Differential Equations and Oscillators
L.M.B.C. Campos

Book 5 – Non-Linear Differential Equations and Dynamical Systems
L.M.B.C. Campos

Book 6 – Higher-Order Differential Equations and Elasticity
L.M.B.C. Campos

Book 7 – Simultaneous Differential Equations and Multi-Dimensional Vibrations
L.M.B.C. Campos

Book 8 – Singular Differential Equations and Special Functions
L.M.B.C. Campos

Book 9 – Classification and Examples of Differential Equations and their Applications
L.M.B.C. Campos

Volume V – Partial Differential Equations with Applications to Waves and Diffusion
L.M.B.C. Campos & L.A.R. Vilela

Book 10 – Vector Fields with Applications to Thermodynamics and Irreversibility
L.M.B.C. Campos & L.A.R. Vilela

Book 11 – Compressible Flow with Application to Shocks and Propulsion
L.M.B.C. Campos & L.A.R. Vilela

For more information about this series, please visit: http://www.crcpress.com/Mathematics-and-Physics-for-Science-and-Technology/book-series/CRCMATPHYSCI

Vector Fields with Applications to Thermodynamics and Irreversibility

Luis Manuel Braga da Costa Campos
and
Luís António Raio Vilela

CRC Press
Taylor & Francis Group
Boca Raton London New York

CRC Press is an imprint of the
Taylor & Francis Group, an **informa** business

First edition published 2023
by CRC Press
6000 Broken Sound Parkway NW, Suite 300, Boca Raton, FL 33487-2742

and by CRC Press
2 Park Square, Milton Park, Abingdon, Oxon, OX14 4RN

Library of Congress Cataloging-in-Publication Data
Names: Campos, Luis Manuel Braga da Costa, author. | Vilela, Luís António Raio, author.
Title: Partial differentials with applications to thermodynamics and compressible flow / Luis Manuel Braga da Costa Campos and Luís António Raio Vilela.
Description: First edition. | Boca Raton, FL : CRC Press, 2021.
| Series: Mathematics and physics for science and technology | Includes bibliographical references and index.
| Summary: "This book is part of the series "Mathematics and Physics Applied to Science and Technology." It combines rigorous mathematics with general physical principles to model practical engineering systems with a detailed derivation and interpretation of results"– Provided by publisher.
Identifiers: LCCN 2021005670 (print) | LCCN 2021005671 (ebook) | ISBN 9781032029870 (v. 1; hardback) | ISBN 9781032030838 (v. 1; paperback) | ISBN 9781032030838 (v. 2; hardback) | ISBN 9781032030814 (v. 2; paperback) | ISBN 9781003186595 (v. 1; ebook) | ISBN 9781003186588 (v. 2; ebook) Subjects: LCSH: Thermodynamics. | Fluid dynamics. | Gases, Compressed—Mathematical models. | Differential equations, Partial.
Classification: LCC QC168 .C29 2021 (print) | LCC QC168 (ebook) | DDC 515/.353—dc23
LC record available at https://lccn.loc.gov/2021005670
LC ebook record available at https://lccn.loc.gov/2021005671

ISBN: 978-1-032-02987-0 (hbk)
ISBN: 978-1-032-03083-8 (pbk)
ISBN: 978-1-003-18659-5 (ebk)

DOI: 10.1201/9781003186595

Typeset in Times
by SPi Technologies India Pvt Ltd (Straive)

To Sir Michael James Lighthill

Contents

Notes, Tables, and Diagrams

NOTES

TABLES

DIAGRAMS

Series Preface

The aim of the "Mathematics and Physics for Science and Technology" series is to describe the mathematical methods as they are applied to model natural physical phenomena and solve scientific and technological problems. The primary emphasis is on the application, including formulation of the problem, detailed solution and interpretation of results. Mathematical methods are presented in sufficient detail to justify every step of solution, and to avoid superfluous assumptions.

The main areas of physics are covered, namely:

- Mechanics of particles, rigid bodies, deformable solids and fluids;
- Electromagnetism, thermodynamics and statistical physics, as well as their classical, relativistic and quantum formulations;
- Interactions and combined effects (e.g., thermal stresses, magneto-hydrodynamics, plasmas, piezoelectricity and chemically reacting and radiating flows).

The examples and problems chosen include natural phenomena in our environment, geophysics and astrophysics; the technological implications in various branches of engineering – aerospace, mechanical, civil, electrical, chemical, computer and other mathematical models in biological, economic and social sciences.

The coverage of areas of mathematics and branches of physics is sufficient to lay the foundations of all branches of engineering, namely:

- Mechanical - including machines, engines, structures and vehicles;
- Civil - including structures and hydraulics;
- Electrical - including circuits, waves and quantum effects;
- Chemical - including transport phenomena and multiphase media;
- Computer - including analytical and numerical methods and associated algorithms.

Particular emphasis is given to interdisciplinary areas, such as electromechanics and aerospace engineering. These require combined knowledge of several areas and have an increasing importance in modern technology.

Analogies are applied in an efficient and concise way, across distinct disciplines, but also stressing the differences and aspects specific to each area, for example:

- Potential flow, electrostatics, magnetostatics, gravity field, steady heat conduction, plane elasticity and viscous flow;
- Acoustic, elastic, electromagnetic, internal and surface waves;
- Diffusion of heat, mass, electricity and momentum.

The acoustic, internal and magnetic waves appear combined as magneto-acoustic-gravity-inertial (MAGI) waves in a compressible, ionized, stratified and rotating fluid. The simplest exact solutions of the MAGI wave equation require special functions. Thus, the topic of MAGI waves combines six subjects: gravity field, fluid mechanics and electromagnetism, and uses complex analysis, differential equations and special functions. This is not such a remote subject since many astrophysical phenomena do involve a combination of several of these effects, as does the technology of controlled nuclear fusion. The latter is the main source of energy in stars and in the universe; if harnessed, it would provide a clean and inexhaustible source of energy on Earth. Closer to our everyday experience, there is a variety of electromechanical and control systems that use modern interdisciplinary technology. The ultimate aim of the present series is to build up knowledge seamlessly from undergraduate to research level, across a range of subjects, to cover contemporary or likely interdisciplinary needs. This requires a consistent treatment of all subjects so that their combination fits together as a whole.

The approach followed in the present series is a combined study of mathematics, physics and engineering, so that the practical motivation develops side by side with the theoretical concepts; the mathematical methods are applied without delay to "real" problems, not just to exercises. The electromechanical and other analogies stimulate the ability to combine different disciplines, which is the basis of much of modern interdisciplinary science and technology. Starting with the simpler mathematical methods, and the consolidating them with the detailed solutions of physical and engineering problems, gradually widens the range of topics that can be covered. The traditional method of separate monodisciplinary study remains possible, selecting mathematical disciplines (e.g., complex functions) or sets of applications (e.g., fluid mechanics). The combined multidisciplinary method of study has the advantage of connecting mathematics, physics and technology at an earlier stage. Moreover, preserving that link provides a broader view of the subject and the ability to innovate. Innovation requires an understanding of the practical aims, the physical phenomena that can implement them, and the mathematical methods that quantify the expected results. The combined interdisciplinary approach to the study of mathematics, physics and engineering is thus a direct introduction to a professional experience in scientific discovery and technological innovation.

Preface to Volume V

RELATIONSHIP WITH PREVIOUS VOLUMES OF THE SERIES

Topic A was concerned with the theory of functions: (volume I) starting with complex functions of a complex variable; (volume II) proceeding to the elementary transcendental functions in the complex plane; (volume III) extending to any dimension via the generalized functions, like the Dirac delta, derivatives and primitives. The applications of the theory of functions in topic A were mostly potential fields, that is, solutions of the Laplace and Poisson equations that arise in various domains such as: (volume I) potential flow, gravity field, electrostatics, magnetostatics and steady heat conduction in two dimensions; (volume II) still in two dimensions, solutions of the unforced and forced Laplace and biharmonic partial differential equations, including some vortical, compressible, rotating and viscous flows, plane elasticity, torsion of bars, deflection of membranes and capillarity; (volume III) extension of potential fields to three and higher dimensions, including irrotational and/or incompressible flows, electro and magnetostatics and one-dimensional elastic bodies like strings, bars and beams.

The applications of the theory of functions in topic A already include many boundary and initial-value problems; the latter are the focus of topic B, starting from the solution of ordinary and partial differential equations, which uses the theory of functions in topic A. In particular, the mechanical oscillators, electrical circuits and elastic bodies already briefly considered in topic A become the main focus of the first volume, IV, of topic B on *Ordinary Differential Equations with Applications to Trajectories and Oscillations*. The fields satisfying potential, wave or diffusion equations, partially covered in topic A, become the main focus of the second volume V of topic B on *Partial Differential Equations with Applications to Waves and Diffusion*. Thus the present volume V of the series "Mathematics and Physics Applied to Science and Technology" covers, on the mathematical side the solution of single partial differential equations or simultaneous systems, linear or non-linear, satisfying boundary and/or initial conditions ensuring existence, unicity and other properties. On the applications' side the thermodynamics of mechanical, electric, magnetic and chemical processes in solids and fluids is used to study linear and non-linear waves, including elastic, acoustic, electromagnetic, surface and internal waves and diffusion mechanisms, including electric, thermal, mass and viscosity.

OBJECTIVES AND CONTENTS OF VOLUME V

A large number and wide variety of problems in mathematics, physics, engineering and other quantitative sciences involve the solution of differential equations or simultaneous systems, that are: (i) ordinary differential equations (o.d.e.s) if there is only one independent variable (volume IV), for example time or one spatial coordinate; (ii) partial differential equations (p.d.e.s), if there are several independent variables (volume V), for example (a) several spatial coordinates or (b) time and at least one spatial coordinate. Configuration spaces with several variables also lead to p.d.e.s. The study of p.d.e.s (partial differential equations) starts with some simple classes (chapter V.1), which relate to differentials and to the principles of thermodynamics (chapter V.2). The p.d.e.s with constant or power coefficients (chapter V.3) can be solved by an extension of the methods used for o.d.e.s (ordinary differential equations) with constant and power coefficients, respectively (chapter IV.1). The singular integrals occur both for o.d.e.s (chapter IV.3) and p.d.e.s (chapter V.7); there are also methods of solution for particular classes of o.d.e.s (chapter IV.7) and p.d.e.s (chapter V.7). The most important p.d.e.s are linear of second-order (chapter V.5) and are classified into three types: (i) elliptic, like the Laplace equation, which applies to potential fields (volumes I to II); (ii) parabolic, like the heat equation, which applies to diffusion problems; (iii) hyperbolic, like the wave equation, which applies to propagation problems (volume III). The theorems of existence, unicity and regularity of solution of p.d.e.s (chapter V.9) use methods dissimilar or similar to those for o.d.e.s (chapter IV.9), as the solution of some general classes of p.d.e.s is reducible to o.d.e.s; one of the methods is similarity solutions, if the p.d.e. with several independent variables can be reduced to a set of o.d.e.s each for a single combined variable. Waves are an important class of phenomena described by p.d.e.s, such as: (i) elastic waves in solids and electromagnetic waves in vacuo and in matter (chapter V.4); (ii) sound waves in fluids, corresponding to pressure, density or velocity fluctuations (chapter V.6); (iii) waves associated with gravity, such as internal waves in a stratified fluid and water waves on the surface of a liquid (chapter V.8). When the waves are a weak (strong) perturbation of the background state they are described by linear (non-linear) p.d.e.s; if the background medium is uniform and steady (inhomogeneous or unsteady) its properties appear in the wave equation as constant (variable) coefficients.

ORGANIZATION AND PRESENTATION OF THE SUBJECT MATTER

Volume V (*Partial Differential Equation with Applications to Waves and Diffusion*) is organized like the preceding four volumes of the series "Mathematics and Physics Applied to Science and Technology": (volume IV) *Ordinary Differential Equations with Applications to Trajectories and Oscillations*; (volume III) *Generalized Calculus with Applications to Matter and Forces*; (volume II) *Elementary Transcendentals with Applications to Solids and Fluids*; and (volume I) *Complex Analysis with Applications to Flows and Fields*. Volume V consists of ten chapters: (i) the odd-numbered chapters present mathematical developments; (ii) the even-numbered chapters contain physical and engineering applications; (iii) the last chapter is a set of 20 detailed examples of (i) and (ii). The chapters are divided into sections and subsections, for example, chapter 1, section 1.1, and subsection 1.1.1. The formulas are numbered by chapters in curved brackets, for example (1.2) is equation 2 of chapter 1. When referring to volume I the symbol I is inserted at the beginning, for example: (i) chapter I.36, section I.36.1, subsection I.36.1.2; (ii) equation (I.36.33a). The final part of each chapter includes: (i) a conclusion referring to the figures as a kind of visual summary; (ii) the notes, lists, tables, diagrams and classifications as additional support. The latter (ii) appear at the end of each chapter, and are numbered within the chapter (for example, diagram D2.1, note N1.1, table T2.1); if there is more than one diagram, note or table, they are numbered sequentially (for example, notes N2.1 to N2.20). The chapter starts with an introductory preview, and related topics may be mentioned in the notes at the end. The lists of mathematical symbols and physical quantities appear before the main text, and the index of subjects and bibliography are found at the end of the book.

About the Authors

L.M.B.C. Campos was born on March 28, 1950, in Lisbon, Portugal. He graduated in 1972 as a mechanical engineer from the Instituto Superior Técnico (IST) of Lisbon Technical University. The tutorials as a student (1970) were followed by a career at the same institution (IST) through all levels: assistant (1972), assistant with tenure (1974), assistant professor (1978), associate professor (1982) and chair of Applied Mathematics and Mechanics (1985). He has served as the coordinator of undergraduate and postgraduate degrees in Aerospace Engineering since the creation of the programs in 1991. He was the coordinator of the Scientific Area of Applied and Aerospace Mechanics in the Department of Mechanical Engineering and also the director (and founder) of the Center for Aeronautical and Space Science and Technology until retirement in 2020.

In 1977, Campos received his doctorate on "waves in fluids" from the Engineering Department of Cambridge University, England. Afterward, he received a Senior Rouse Ball Scholarship to study at Trinity College, while on leave from IST. In 1984, his first sabbatical was as a Senior Visitor at the Department of Applied Mathematics and Theoretical Physics of Cambridge University, England. In 1991, he spent a second sabbatical as an Alexander von Humboldt scholar at the Max-Planck Institut für Aeronomie in Katlenburg-Lindau, Germany. Further sabbaticals abroad were excluded by major commitments at the home institution. The latter were always compatible with extensive professional travel related to participation in scientific meetings, individual or national representation in international institutions and collaborative research projects.

Campos received the von Karman medal from the Advisory Group for Aerospace Research and Development (AGARD) and Research and Technology Organization (RTO). Participation in AGARD/RTO included serving as a vice-chairman of the System Concepts and Integration Panel, and chairman of the Flight Mechanics Panel and of the Flight Vehicle Integration Panel. He was also a member of the Flight Test Techniques Working Group. Here he was involved in the creation of an independent flight test capability, active in Portugal during the last 40 years, which has been used in national and international projects, including Eurocontrol and the European Space Agency. The participation in the European Space Agency (ESA) has afforded Campos the opportunity to serve on various program boards at the levels of national representative and Council of Ministers.

His participation in activities sponsored by the European Union (EU) has included (i) 27 research projects with industry, research and academic institutions; (ii) membership of various Committees, including Vice-Chairman of the Aeronautical Science and Technology Advisory Committee; (iii) participation on the Space Advisory Panel on the future role of EU in space. Campos has been a member of the Space Science Committee of the European Science Foundation, which liaises with the Space Science Board of the National Science Foundation of the United States. He has been a member of the Committee for Peaceful Uses of Outer Space (COPUOS) of the United Nations. He has served as a consultant and advisor on behalf of these organizations and other institutions. His participation in professional societies includes member and vice-chairman of the Portuguese Academy of Engineering, fellow of the Royal Aeronautical Society, Royal Astronomical Society and Cambridge Philosophical Society, associate fellow of the American Institute of Aeronautics and Astronautics, and founding and life member of the European Astronomical Society.

Campos has published and worked on numerous books and articles. His publications include 15 books as a single author, one as an editor and one as a co-editor. He has published 166 papers (82 as the single author, including 12 reviews) in 60 journals, and 262 communications to symposia. He has served as reviewer for 40 different journals, in addition to 28 reviews published in *Mathematics Reviews*. He is or has been member of the editorial boards of several journals, including *Progress in Aerospace Sciences, International Journal of Aeroacoustics, International Journal of Sound and Vibration, Open Physics Journal* and *Air & Space Europe*.

Campos' areas of research focus on four topics: acoustics, magnetohydrodynamics, special functions and flight dynamics. His work on acoustics has concerned the generation, propagation and refraction of sound in flows with mostly aeronautical applications. His work on magnetohydrodynamics has concerned magneto-acoustic-gravity-inertial waves in solar-terrestrial and stellar physics. His developments on special functions have used differintegration operators, generalizing the ordinary derivative and primitive to complex order; they have led to the introduction of new special functions. His work on flight dynamics has concerned aircraft and rockets, including trajectory optimization, performance, stability, control and atmospheric disturbances. Other publications concern mechanics of solids, general relativity and population models.

Campos' professional activities on the technical side are balanced by other cultural and humanistic interests. Complementary non-technical interests include classical music (mostly orchestral and choral), plastic arts (painting, sculpture, architecture), social sciences (psychology and biography), history (classical, renaissance and overseas expansion) and technology (automotive, photo, audio). Campos is listed in various biographical publications, including *Who's Who in the World* since 1986, *Who's Who in Science and Technology* since 1994 and *Who's Who* in America since 2011.

L.A.R. Vilela was born on March 11, 1994, in Vila Real, Portugal. He was in his school's honours board for best academic results every year from the 5th to the 10th grade. In 2012, he started taking the first year of the program of study for the Informatics Engineering bachelor's degree at Universidade de Trás-os-Montes e Alto-Douro (UTAD), but changed, in 2013, to the Integrated Master's degree in Aerospace Engineering at Instituto Superior Técnico (IST) of Lisbon University, and is currently completing it.

Besides physical sciences, Vilela has a special interest in philosophy, psychology, anthropology and, in general, an advanced understanding of the human mind.

Acknowledgments

The fifth volume of the series justifies renewing some of the acknowledgments made in the first four volumes. Thanks are due to those who contributed more directly to the final form of this book: L. Sousa for help with manuscripts; Mr. J. Coelho for all the drawings; and at last, but not least, to my wife, my companion in preparing this work.

Acknowledgements

List of Mathematical Symbols

The mathematical symbols presented are those of more common use in the context of: (i) sets, quantifiers and logic; (ii) numbers, ordering and bounds; (iii) operations, limits and convergence; (iv) vectors, functions and the calculus. This section includes a list of functional spaces, not all of which appear in the present volume. The section where the symbol first appears may be indicated after a colon, for example: "V10.2" means section 10.2, of the present volume V; "II.1.1.4" means subsection 1.1.4 of the volume II; "N.V.9.42" means note 9.42 of the present volume V; and "NIII.9.38" means note 9.38 of the volume III.

1 SETS, QUANTIFIERS AND LOGIC

1.1 SETS

$A \equiv \{x\ldots\}$	set A whose elements x have the property…
$A \cup B$	union of sets A and B
$A \cap B$	intersection of sets A and B
$A \supset B$	set A contains set B
$A \subset B$	set A is contained in set B

1.2 QUANTIFIERS

$\forall_{x \in A}$	for all x belonging to A holds
$\exists_{x \in A}$	there exists at least one x belonging to A such that
$\exists^{1}_{x \in A}$	there exists one and only one x belonging to A such that
$\exists^{\infty}_{x \in A}$	there exist infinitely many x belonging to A such that

1.3 LOGIC

$a \wedge b$	a and b
$a \vee b$	or (inclusive): a or b or both
$a \veebar b$	or (exclusive): a or b but not both
$a \Rightarrow b$	implication: a implies b
$a \Leftrightarrow b$	equivalence: a implies b and b implies a

1.4 CONSTANTS

$$e = 2.7182\ \ 81828\ \ 45904\ \ 52353\ \ 60287$$
$$\pi = 3.1415\ \ 92653\ \ 58979\ \ 32384\ \ 62643$$
$$\gamma = 0.5772\ \ 15664\ \ 90153\ \ 28606\ \ 06512$$
$$\log 10 = 2.3025\ \ 85092\ \ 99404\ \ 56840\ \ 179915$$

2   NUMBERS, ORDERING AND OPERATIONS

2.1   Types of Numbers

$|C$	complex numbers: I.1.2
$|F$	transfinite numbers: II.9.8
$|I$	irrational numbers: real non-rational numbers: I.1.2
$|L$	rational numbers: ratios of integers: I.1.1
$|N$	natural numbers: positive integers: I.1.1
$|N_0$	non-negative integers: zero plus natural numbers: I.1.1
$|Q$	quaternions: I.1.9
$|R$	real numbers: I.1.2
$|Z$	integer numbers: I.1.1

2.2   Complex Numbers

$|...|$	modulus of complex number...: I.1.4
$\arg(...)$	argument of complex number...: I.1.4
$Re(...)$	real part of complex number...: I.1.3
$Im(...)$	imaginary part of complex number...: I.1.3
$...*$	conjugate of complex number...: I.1.6

2.3   Relations and Ordering

$a > b$	a greater than b
$a \geq b$	a greater or equal to b
$a = b$	a equal to b
$a \leq b$	a smaller or equal to b
$a < b$	a smaller than b
$\sup(...)$	supremum: smallest number larger or equal than all numbers in the set
$\max(...)$	maximum: largest number in set
$\min(...)$	minimum: smallest number in set
$\inf(...)$	infimum: largest number smaller or equal than all numbers in set

2.4   Operations between Numbers

$a + b$	sum: a plus b
$a - b$	difference
$a \times b$	product: a times b
a/b	ratio: a divided by b (alternative $a : b$)
a^b	power: a to the power b
$\sqrt[b]{a}$	root: root b of a

3   FUNCTIONS, LIMITS AND CONVERGENCE

3.1   LIMITS AND VALUES

$\lim$	limit when x tends to a: $x \to a$: I.11.2
$a \sim O(b)$	a is of order b: $\lim b/a \neq 0, \ \infty$: I.19.7
$a \sim o(b)$	b is of lower order than a: $\lim b/a = 0$: I.19.7
$f_{(n)}(a)$	residue at pole of order n at a: I.15.8
$\bar{B}$ or M	upper bound: $\left| f(z) \right| \leq \bar{B}$ for z in ...: I.39.2
$\underline{B}$ or m	lower bound: $\left| f(z) \right| \geq \underline{B}$ for z in ...: I.39.2
$f \circ g$	composition of functions f and g : N.I.38.1

3.2   ITERATED SUMS AND PRODUCTS

$\displaystyle\sum_{a}$	sum over a set
$\displaystyle\sum_{n=a}^{b}$	sum from $n = a$ to $n = b$
$\displaystyle\sum_{n,m=a}^{b}$	double sum over n, $m = a$, ..., b
$\displaystyle\prod_{a}$	product over a set
$\displaystyle\prod_{n=a}^{b}$	product from $n = a$ to $n = b$

3.3   CONVERGENCE

A.C.	absolutely convergent: I.21.2
A.D.	absolutely divergent: I.21.2
C.	convergent: I.21.2
C.C.	conditionally convergent: I.21.2
$C\lambda$	converges to class λ: II.9.6
D	divergent: I.21.1
N.C.	non-convergent: divergent or oscillatory: I.21.1
O.	oscillatory: I.21.1
T.C.	totally convergent: I.21.7
U.C.	uniformly convergent: I.21.5

applies to:

power series: I.21.1
series of fractions: I.36.6, I.1.2
infinite products: I.36.6, II.1.4
continued fractions: II.1.6

3.4 INTEGRALS

$\displaystyle\int \cdots dx$ primitive of ... with regard to x: I.13.1

$\displaystyle\int^{y} \cdots dx$ indefinite integral of ... at y: I.13.2

$\displaystyle\int_{a}^{b} \cdots dx$ definite integral of ... between a and b: I.13.2

$\displaystyle\fint_{a}^{b} \ldots dx$ principal value of integral: I.17.8

$\displaystyle\int^{(z+1)} \cdots dx$ integral along a loop around z in the positive (counterclockwise) direction: I.13.5

$\displaystyle\int^{(z-1)} \cdots dx$ idem in the negative (clockwise) direction: I.13.5

$\displaystyle\int_{L} \cdots dx$ integral along a path L: I.13.2

$\displaystyle\int_{C}^{(+)}$ integral along a closed path or loop C in the positive direction: I.13.5

$\displaystyle\int_{C}^{(-)}$ integral along a closed path or loop C in the negative direction: I.13.5

4 VECTOR, MATRICES AND TENSORS

4.1 VECTORS

$\vec{A}.\vec{B}$ inner product
$\vec{A} \wedge \vec{B}$ outer product

$\vec{A}.\left(\vec{B} \wedge \vec{C}\right)$ mixed product

$\vec{A} \wedge \left(\vec{B} \wedge \vec{C}\right)$ double outer product

$\left|\vec{A}\right|$ modulus

$ang\left(\vec{A},\vec{B}\right)$ angle of vector $\vec{B}$ with vector $\vec{A}$

4.2 MATRICES

δ_{j}^{i} identity matrix: III.5.8

$\overset{c}{A}_{j}^{i}$ matrix of co-factors: N.III.9.10

$\overset{I}{A}_{j}^{i}$ inverse matrix: N.III.9.10

$Det\left(A_{j}^{i}\right)$ determinant of matrix: N.III.9.11

$Ra\left(A_{j}^{i}\right)$ rank of matrix: N.III.5.5.1

4.3 TENSORS

$T^{i_1\ldots i_p}_{j_1\ldots j_q}$ $\overset{(\vartheta)}{}$ tensor with weight ϑ, contravariance p and covariance q: N.III.9.14

$\delta^{i_1\ldots i_N}_{j_1\ldots j_N}$ identity symbol: N.III.9.9

$e_{i_1\ldots i_N}$ covariant permutation symbol: N.III.9.9

$e^{i_1\ldots i_N}$ contravariant permutation symbol: N.III.9.9

4.4 OPERATIONS

$T_{(i_1\ldots i_p)}$ mixing: N.III.9.12

$T_{[i_1\ldots i_p]}$ alternation: N.III.9.12

T^{ij}_{ik} contraction: N.III.9.12

$T^i_{jk}\, S^j_{i\ell}$ transvection: N.III.9.12

5 DERIVATIVES AND DIFFERENTIAL OPERATORS

5.1 DIFFERENTIALS AND DERIVATIVES

$d\Phi$ differential of Φ

$d\Phi/dt$ derivative of Φ with regard to t

$\partial\Phi/\partial t \equiv \partial_t\Phi$ partial derivative of Φ with regard to t

$\partial\Phi/\partial x_i \equiv \partial_i\Phi \equiv \Phi_i$ partial derivative of Φ with regard to x_i

$\partial^n\Phi/\partial x_{i_1}\cdots\partial x_{i_n} \equiv \partial_{x_{i_1}\cdots i_n}\Phi \equiv \Phi_{i_1\cdots i_n}$ n-order partial derivative of Φ with regard to $x_{i_1},\cdots,x_{i_n}$

5.2 VECTOR OPERATORS

$\nabla\Phi = \partial_i\Phi$ gradient of a scalar: III.6.1

$\nabla\cdot\vec{A} = \partial_i A_i$ divergence of a vector: III.6.1

$\nabla\wedge\vec{A} = e_{ijk}\partial_j A_k$ curl of a vector: III.6.1, 3.9.1

$H \equiv \vec{A}\cdot\left(\nabla\wedge\vec{A}\right) = A_k e_{ijk}\partial_i A_i$ helicity: IV.3.9.2

$\nabla^2 = \partial_{ii} = \partial_i\partial_i$ Laplacian: III.6.1

$\vee^2$ modified Laplacian: III.6.2

5.3 TENSOR OPERATORS

$\partial_{[i_{M+1}}\mathcal{U}_{i_1\ldots i_n]}$ curl of a covariant M-vector: N.III.9.20

$\partial_{i_{M+1}}\mathcal{U}^{i_1\ldots i_{M+1}}$ divergence of a contravariant $(M+1)$ – vector density:N.III.9.20.

5.4 ADJOINTNESS

$\{L(\partial/\partial x_i)\}\,\Phi$ linear differential operator: III.7.6

$\left\{\bar{L}\left[\left(\partial/\partial x_i\right)\right]\right\}\Psi$ adjoint differential operator: III.7.6

$W(\Phi,\Psi)$ bilinear concomitant: III.7.6

6   FUNCTIONAL SPACES

The spaces of functions are denoted by calligraphic letters, in alphabetical order:

$...(a,b)$ — set of functions over interval from a to b. Omission of interval means the set of functions over real line $) - \infty , + \infty ($

$\mathcal{A}$ $(...)$ — analytic functions in ...: I.27.1

$\bar{\mathcal{A}}(...)$ — monogenic function in ...: I.31.1

$\mathcal{B}$ $(...)$ — bounded functions in ...: I.13.3

$\mathcal{C}$ $(...)$ — continuous functions in ...: $C \equiv C^0$: I.11.2

$\hat{\mathcal{C}}$ $(...)$ — uniformly continuous functions in...: I.13.4

$\mathcal{C}^n(...)$ — functions with continuous n-th derivative: III.3.2.1

$\mathcal{D}$ $(...)$ — differentiable functions in ...: $D \equiv D^0$: I.12.2

$\mathcal{D}^\infty(...)$ — infinitely differentiable functions or smooth in ...: I.27.1

$\mathcal{E}$ $(...)$ — Riemann integrable functions in ...: I.13.2

$\mathcal{F}$ $(...)$ — functions of bounded oscillation (or bounded fluctuation or bounded variation) in ...: I.27.9.5

$\mathcal{G}$ $(...)$ — generalized functions (or distributions) in ...: III.3.4.1

$\mathcal{H}(...)$ — harmonic functions in ...: I.11.4, II.4.6.4

$\mathcal{H}_2(...)$ — biharmonic functions in ...: II.4.6.4

$\mathcal{H}_n(...)$ — multiharmonic functions of order n in ...: II.4.6.6

$\mathcal{I}$ $(...)$ — integral functions in ...: $I \equiv I_0$: I.27.9, II.1.1.7

$\mathcal{I}_m(...)$ — rational-integral functions of degree m in ... : I.27.9, II.1.1.9

$\bar{\mathcal{K}}(...)$ — Lipschitz functions in ...: IV.9.1.5

$\mathcal{L}^1(...)$ — absolutely integrable functions in ...: N.III.3.7

$\mathcal{L}^2(...)$ — square integrable functions in ...: III.7.1.1

$\mathcal{L}^p(...)$ — functions with power p of modulus integrable in *normed* space: III.7.1.1

$\mathcal{M}^+(...)$ — monotonic increasing functions in ...: I.9.1.1

$\mathcal{M}_0^+(...)$ — monotonic non-decreasing functions in ...: I.9.1.1

$\mathcal{M}_0^-(...)$ — monotonic non-increasing functions in ...: I.9.1.2

$\mathcal{M}^-(...)$ — monotonic decreasing functions in ...: I.9.1.2

$\mathcal{O}$ $(...)$ — orthogonal systems of functions in ...: II.5.7.2

$\bar{\mathcal{O}}(...)$ — orthonormal systems of functions in ...: II.5.7.2

$\tilde{\mathcal{O}}(...)$ — complete orthogonal systems of functions in ...: II.5.7.5

$\mathcal{P}(...)$ — polynomials in ...: I.27.7m II.1.2.6

$\mathcal{P}_n(...)$ — polynomials of degree n in ...: I.27.7, II.1.1.6

$\mathcal{Q}(...)$ — rational functions in ...: I.27.7, II.1.2.6

$\mathcal{T}^0(...)$ — functions with compact support, that is which vanish outside a finite interval: III.3.3.2

$\mathcal{T}^n(...)$ — excellent functions of order n: n-times differentiable functions with compact support: III.3.3.2

$\mathcal{T}^\infty(...)$ — excellent functions: smooth or infinitely differentiable functions with compact support: II.3.3.6

$\mathcal{U}$ $(...)$ — single-valued functions in ...: I.9.1

$\tilde{\mathcal{U}}(\ldots)$	injective functions in…: I.9.1
$\bar{\mathcal{U}}(\ldots)$	surjective functions: I.9.1
$\tilde{\bar{\mathcal{U}}}(\ldots)$	bijective functions: I.9.1
$\mathcal{U}_n(\ldots)$	multivalued functions with n branches in …: I.1.6.1
$\mathcal{U}_\infty(\ldots)$	many-valued functions in …: I.6.2
$\mathcal{U}^1(\ldots)$	univalent functions, in …: I.37.4
$\mathcal{U}^m(\ldots)$	multivalent functions taking m values in …: I.37.4
$\mathcal{U}^\infty(\ldots)$	multi-valued functions in …: N.I.37.4
$\mathcal{U}_n^m(\ldots)$	many-valued multivalent function with n branches and m values in…: N.I.37.4
$\mathcal{V}^o(\ldots)$	good functions, that is with slow decay at infinity faster than some power: III.3.3.1
$\mathcal{V}^n(\ldots)$	good or slow decay functions of degree n, that is with decay at infinity faster than the inverse of a polynomial of degree n: III.3.3.1
$\mathcal{V}_0(\ldots)$	fairly good or slow growth functions, that is with growth at infinity slower than some power: III.3.3.1
$\mathcal{V}_n(\ldots)$	fairly good or slow growth functions of degree n, that is with growth at infinity slower than a polynomial of degree n: III.3.3.1
$\mathcal{V}_\infty(\ldots)$	very good or fast decay functions, that is with faster decay at infinity than any power: II.3.3.1
$\mathcal{V}_\infty^\infty(\ldots)$	optimal functions, that is smooth functions with compact support: II.3.3.6
$\mathcal{W}_q^p(\ldots)$	functions with generalized derivatives of orders up to q such that for all the power p of the modulus is integrable…*Sobolev space*
$\mathcal{X}_0(\ldots)$	self-inverse linear functions in
$\mathcal{X}_1(\ldots)$	linear functions in…: I.35.2
$\mathcal{X}_2(\ldots)$	bilinear, homographic, or Mobius functions in …: I.3.5.4
$\mathcal{X}_3(\ldots)$	self-inverse bilinear functions in …: II.37.5
$\mathcal{X}_a(\ldots)$	automorphic functions in …: II.37.5
$\mathcal{X}_m(\ldots)$	isometric mappings in …: I.35.1
$\mathcal{X}_r(\ldots)$	rotation mappings in …: II.35.1
$\mathcal{X}_t(\ldots)$	translation mappings in …: II.35.1
$\mathcal{Y}(\ldots)$	meromorphic functions in …: I.37.9, II.1.1.10
$\mathcal{Z}(\ldots)$	polymorphic functions in…: I.37.0, II.1.1.10

7 GEOMETRIES WITH *N* DIMENSIONS

$\mathfrak{A}_N$	rectilinear: N.III.9.6
$\mathfrak{W}_N$	metric: N.III.9.34
$\mathfrak{N}_N$	orthonormal: N.III.9.38
$\mathfrak{O}_N$	orthogonal: N.III.9.38
$\mathfrak{X}_N$	curvilinear: N.III.9.38
$\mathfrak{X}_N^2$	curvilinear with curvature: N.III.9.17

8 GENERALIZED FUNCTIONS

$H(x)$	Heaviside unit jump: III.1.2
$G(x;\xi)$	Green's or influence function: N1.5
$\delta(x)$	Dirac unit impulse: III.1.3
$sgn\ (x)$	sign function: I.36.4.1, III.1.7.1

9 AUXILIARY FUNCTIONS

$erf(x)$	error function: III.1.2.2
$\sin(x;k)$	elliptic sine of modulus k: I.39.9.2
$\tilde{y}(k)$	Fourier transform of $y(x)$: N.IV.1.11
$\bar{y}(s)$	Laplace transform of $y(x)$: N.IV.1.15
$W(y_1,\dots,y_N)$	Wronskian of a set of functions: IV.1.2.3
$\psi(x)$	Digamma function: I.29.5.2
$\Gamma(x)$	Gamma function: N.III.1.8.

List of Physical Quantities

The location of first appearance is indicated, for example, "2.7" means *section 2.7*; "6.8.4" means *subsection 6.8.4*; "N.8.8" means *note 8.8*; and "E10.13.1" means *example 10.13.1*.

1 SUFFIXES

a	adiabatic
b	radiative
c	chemical (Carnot cycle)
d	dissipative
e	electrical (engine)
f	centrifugal
g	gravity
h	heat
i	inertial
j	(free)
k	kinetic of translation
ℓ	longitudinal
m	magnetic
n	normal component
o	reference
p	pressure (heat pump)
q	(free)
r	kinetic of rotation (refrigerator)
s	sound (Stirling cycle)
t	transversal, tangential
u	elastic
v	viscous
w	water
0	stagnation (at rest)
$*$	critical (velocity equal to sound speed)

2 SMALL ARABIC LETTERS

$\bar{a}$	acceleration: 2.4.10.
b	dissipation velocity: N1.15
c	phase speed of waves: N1.7.
	speed of light in vacuo: 2.1.9
c_s	adiabatic sound speed: 2.3.12
c_t	isothermal sound speed: 2.3.12
c_V	specific heat at constant volume per unit mass: 2.6.5;
c_p	specific heat at constant pressure per unit mass: 2.6.5;

d_{1-12}	thermodynamic derivatives:
e_{ijk}	three-dimensional permutation symbol: 2.1.12
f	free energy per particle: 2.3.30
$\vec{f}$	force density per unit volume: 2.1.1
$\vec{f}_e$	force density of an electrostatic field: 2.1.4
$\vec{f}_g$	force density of a gravity field: 2.1.4
$\vec{f}_k$	inertia force: 2.1.3
$\vec{f}_m$	force density of a magnetostatic field: 2.1.4
$\vec{f}_p$	force density due to the pressure: 2.1.16
$\vec{f}_u$	force density due to stresses: 2.1.16
$\vec{f}_v$	friction force: 2.1.1
f_i	pyroelectric vector: 2.2.7
$\bar{g}$	free enthalpy per particle: 2.3.30
$\vec{g}$	acceleration of gravity: 2.1.5
h	enthalpy per unit mass: 2.6.4
$\bar{h}$	enthalpy per particle: 2.3.30
h_0	stagnation enthalpy: 2.6.4
h^-/h^+	enthalpy upstream/downstream of a shock wave: 2.7.3
h_i	pyromagnetic vector: 2.2.7
j	mass flux: 2.7.2
k	thermal conductivity: N1.4
	kinetic friction coefficient: 2.1.1
k_*	thermal conductivity with Thomson effect: 2.4.18
k_{ij}	thermal conductivity tensor: 2.4.1
m	mass: 2.1.20
$\dot{m}$	mass flow rate: 2.6.30
n	polytropic exponent: 2.6.13;
	number of particles: 2.1.20
p	pressure: N1.6
p_0/p_*	stagnation/critical pressure: 2.6.13
p^-/p^+	pressure upstream/downstream of a shock wave: 2.7.2
$\vec{p}$	linear momentum: 2.1.3
p_{ijk}	piezoelectric tensor: 2.2.7
q	electric charge density per unit volume: 2.1.5
q_{ijk}	piezomagnetic tensor: 2.2.3
$\vec{r}$	position vector: 1.5.2
s	rate of shear: N1.12
	entropy per unit mass: 2.6.3
$\bar{s}_h$	entropy per particle in terms of enthalpy: 2.3.30
$\bar{s}_u$	entropy per particle in terms of internal energy: 2.3.30
s^-/s^+	entropy upstream/downstream of a shock wave: 2.7.6.
t	time: N1.3
u	internal energy per unit mass: 2.6.3

$\bar{u}$	internal energy per particle: 2.3.30
u_0	stagnation internal energy: 2.6.4
u^-/u^+	internal energy upstream/downstream of a shock wave: 2.7.5
$\bar{u}$	displacement vector: 2.1.17
$\bar{v}$	velocity vector: 1.5.2
v_n	velocity normal to a surface: 2.7.1
v_t	velocity tangent to a surface: 2.7.1
w	heat rate of source/sink: N1.3
x	Cartesian coordinate: N1.7

3 CAPITAL ARABIC LETTERS

A	Avogadro number: 2.1.20
$\bar{A}$	vector potential: 1.5.12
	magnetic potential: 2.1.9
$\bar{B}$	magnetic induction vector: 2.1.9
C_V	specific heat at constant volume: N1.4
C_p	specific heat at constant pressure: 2.3.10
$C_{ijk\ell}$	elastic stiffness tensor: 2.2.6
D	thermoelectric diffusion coupling coefficient: 2.4.4
	dissipation rate: N.1.14
D_m	mass diffusion coefficient: 2.4.18
D_t	Thomson diffusion coefficient: 2.4.16
D_p	Peltier diffusion coefficient: 2.4.17
$\bar{D}$	electric displacement vector: 2.1.6
D_{ij}	displacement tensor: 2.1.17
	thermoelectric diffusion coupling tensor: 2.4.5
E	energy density: 2.6.4
E_d	elastic energy: 2.1.7
E_e	electrical energy: 2.1.5
E_{eh}	electromagnetic energy: $E_{eh} = E_e + E_h$: 2.4.3.
E_g	energy of gravity field: 2.1.4
E_h	magnetic energy: 2.1.10
E_p	kinetic energy: $E_p = E_t + E_r$:
E_k	kinetic energy of translation: N1.14
E_r	kinetic energy of rotation: 2.2.19
E	Young's modulus of elastic material: 2.2.12
$\bar{E}$	electric field vector: 2.1.5
F	free energy: 2.2.3
	thrust: 2.6.32
$\bar{F}_p$	force due to the pressure: 2.1.15
G	free enthalpy: 2.2.4
$\bar{\bar{G}}$	energy flux: 2.6.4
$\bar{\bar{G}}^c$	convective energy flux: 2.6.4
$\bar{\bar{G}}^d$	dissipative energy flux: N2.6

$\vec{G}^t$	heat flux: N1.3
$\vec{G}^{eh}$	Poynting vector: electromagnetic energy flux: 2.4.3
H	enthalpy: 2.2.3
$\bar{H}$	helicity: 1.5.2
$\vec{H}$	magnetic field vector: 2.1.9
$\vec{I}$	diffusive mass flux: 2.4.18
$\vec{J}$	electric current: 2.1.9
K_p	coefficient of thermal expansion: 2.3.10
K_T	coefficient of thermal compression: 2.3.10
K_m	gravitational constant: 2.1.4
L	length: 2.1.1
M	Mach number: 2.6.13
M_0	stagnation Mach number: 2.6.13
M_*	critical Mach number: 2.6.13
M^-/M^+	Mach number upstream/downstream of a shock wave: 2.7.8
N	mole number: 2.1.20
$\vec{N}$	normal to a surface: 2.1.15
Q	heat density per unit volume: N1.3
$\dot{Q}_e$	rate of heat release by Ohmic electrical resistance: 2.4.3
$\dot{Q}_v$	rate of heat release by viscosity: 2.4.10
$\dot{Q}$	rate of heat release by Thomson effect: 2.4.16
R	ideal gas constant: 2.3.12
$\bar{R}$	ideal gas constant per mole: 2.5.4
$\bar{\bar{R}}$	ideal gas constant per unit mass: 2.5.4
S	entropy: N1.2
S_V	non-adiabatic volume coefficient: 2.3.13
S_p	non-adiabatic pressure coefficient: 2.3.13
$\dot{S}$	rate of entropy production: 2.4.1
$\dot{S}_q$	rate of entropy production by heat conduction: 2.4.1
$\dot{S}_{qe}$	rate of entropy production by thermoelectric coupling: 2.4.5
$\dot{S}_{qev}$	rate of entropy production by thermal conduction, electrical resistance and viscosity
$\dot{S}_{qm}$	rate of entropy production by heat conduction and mass diffusion: N2.14
$\vec{S}$	area element: 2.1.9
S_{ij}	strain tensor: 2.1.19
$\dot{S}_{ij}$	rate-of-strain tensor: 2.4.9
$\dot{\bar{S}}_{ij}$	sliding rate-of-strain tensor: 2.4.9
T	temperature: N1.4
T^-/T^+	temperature upstream/downstream of a shock wave: 2.7.7
T_0/T_*	stagnation/critical temperature: 2.6.13
$\vec{T}$	stress vector: 2.1.16
T_{ij}	stress tensor: 2.1.16
U	internal energy: 2.1.2
$\bar{U}$	augmented internal energy: 2.1.22

$\tilde{U}$	modified internal energy: 2.1.23
V	volume: N1.3
W	work: 2.1.1
$\dot{W}$	power or activity: 2.1.1
W_c	chemical work: 2.1.20
W_u	work of elastic forces: 2.1.17
W_e	work of electric forces: 2.1.5
W_g	work of gravity forces. 2.1.5
W_m	work of magnetic forces: 2.1.14
W_k	work of inertia force: 2.1.3
X_n	extensive thermodynamic variables: 2.1.14
$\bar{X}_n$	diffusion gradients: 2.4.6
Y_n	intensive thermodynamic variables: 2.1.24
$\bar{Y}_n$	diffusion fluxes: 2.4.6
Z_{mn}	constitutive coefficients: 2.2.5
$\bar{Z}_{mn}$	diffusive coefficients: 2.4.6

4   SMALL GREEK LETTERS

α	Clebsch potential: 1.5.15
	thermal pressure coefficient: 2.2.13
	thermal-mass diffusion coupling coefficient: N2.13
α_{ij}	thermal stress tensor: 2.2.7
β	Clebsch potential: 1.5.15
	coefficient of isothermal expansion: 2.2.14
	thermal-mass diffusion coupling coefficient: N.2.13
δ^i_j	identity matrix: 2.1.7
ε	dielectric permittivity: 2.1.7
ε_{ij}	dielectric permittivity tensor: 2.1.7
γ	adiabatic exponent: 2.3.12
η	shear viscosity: N1.6
η_e	efficiency of engine: 2.5.20
η_p	efficiency of heat pump: 2.5.21
η_r	efficiency of refrigerator: 2.5.22
φ	angle of deflection of the velocity across a shock wave: 2.7.4
λ	Lamé modulus: 2.2.12
μ	magnetic permeability: 2.1.11
	chemical potential: 2.1.20
$\bar{\mu}$	relative chemical potential: 2.1.20
μ_{ij}	magnetic permeability tensor: 2.1.11
ν	Lamé modulus: 2.2.12
	affinity: 2.1.20
$\bar{\nu}$	relative affinity: 2.1.20
ρ	mass density per unit volume: N1.4

ρ_0/ρ_* stagnation/critical mass density: 2.6.13
ρ^-/ρ^+ mass density upstream/downstream of a shock wave: 2.7.2
σ Poisson's ratio of elastic material: 2.2.12
σ_e Ohmic electric conductivity of isotropic material: 2.4.4
σ_h Hall electrical conductivity: 2.4.14
σ_{ij} Ohmic electric conductivity tensor of anisotropic material: 2.4.4
τ_{ij} viscous stress tensor: 2.4.10
ϑ specific volume: 2.1.2
ϑ^-/ϑ^+ specific volume upstream/downstream of a shock wave: 2.7.5.
ϑ_{ij} electromagnetic coupling tensor: 2.2.7
$\bar{\varpi}$ vorticity: 2.6.6
ξ convected coordinate: N1.7
     mass fraction: 2.1.20
ζ bulk viscosity: N1.6
χ_t thermal diffusivity: N1.4
χ_D thermoelectric diffusivity: 2.4.20
χ_v total viscous diffusivity: N1.6

5 CAPITAL GREEK LETTERS AND OTHERS

Φ scalar potential: 1.5.7
Φ_e electric potential: 2.1.5
Φ_g gravity potential: 2.1.4
$\bar{\Omega}$ angular velocity: 1.5.2
Ω_{ij} rotation bivector: 2.1.19
Θ Euler potential: 1.5.9
Ξ Clebsch potential: 1.5.15
Ψ Euler potential: 1.5.8

1 Classes of Equations and Similarity Solutions

The study of ordinary differential equations (volume IV) is extended to partial differential equations (volume V), either single equations or simultaneous systems leading to an analogous classification (Section 1.1), changing from one to several independent variables. The general integral of an ordinary (partial) differential equation of order N involves an equal number N of arbitrary constants (functions) determined from boundary conditions (Section 1.2). The simplest partial differential equations of the first-order are linear or quasi-linear (Sections 1.3–1.4) and are related to differentials and vector fields (Section 1.5) providing a method of solution via ordinary differential equations. The simplest partial differential equations of order N are linear with constant coefficients and derivatives all of the same order (Section 1.6). In the unforced case, their solutions can be obtained from the roots of a characteristic polynomial. The characteristic polynomial can also be used to obtain forced solutions (Section 1.8). The unforced (forced) Laplace and biharmonic equations are particular cases of this type [Section 1.7 (1.9)]. The linear partial differential equation with constant coefficients and all derivatives of the same order have similarity solutions, involving arbitrary functions whose argument is a linear combination of the independent variables. Similarity solutions may exist also for non-linear partial differential equations, but in this case as specific functions, that are particular rather than general integrals (notes N1.1 to N1.16).

1.1 HIERARCHY OF PARTIAL DIFFERENTIAL EQUATIONS

The distinction between ordinary and partial differential equations applies both to single equations and simultaneous systems (subsection 1.1.1). A single differential equation has one dependent variable and one (several) independent variables if it is an ordinary (partial) differential equation (subsection 1.1.2). The methods of solution of a single partial differential equation, which one dependent variable and N independent variables, can often be illustrated most simply in the case $N = 2$ of two independent variables (subsection 1.1.3).

1.1.1 SINGLE EQUATIONS AND SIMULTANEOUS SYSTEMS

An ordinary (partial) differential equation is a relation between one dependent variable y and one (several) independent variable(s) x $(x_1, ..., x_n)$, and the ordinary (partial) derivatives dy/dx, ..., $d^N y/dx^N$ $(\partial y/\partial x_n, \quad \partial^2 y/\partial x_n \partial x_m, ...)$ up to a certain order N. The extension to several dependent variables $y_1, ..., y_m$, leads to a simultaneous system of o.d.e.s (p.d.e.s), if there is one (are several) independent variable(s) $x(x_1, ..., x_n)$. This classification of differential equations is indicated in Table 1.1,

DOI: 10.1201/9781003186595-1

TABLE 1.1

Types of Differential and Integral Equations

Dependent Variable(s)	One y	Several $y_1,..., y_m$
ordinary derivatives with regard to one independent variable x	*ordinary differential equation: o.d.e.*	*simultaneous ordinary differential equations: s.o.d.e.*
partial derivatives with regard to several independent variables $x_1,..., x_n$	*partial differential equation: p.d.e.*	*simultaneous partial differential equations: s.p.d.e.*
only independent varaible and its integrals	*integral equation: i.e.*	*simultaneous integral equations: s.i.e.*
both derivatives and integrals of dependent variable	*integrodifferential equation: i.d.e.*	*simultaneous integrodifferential equations: s.i.d.e.*

Note: A single (simultaneous system of) differential equations has one (several) dependent variables y $(y_1,....,y_M)$ and the differential equations are ordinary (partial) if there is one (several) independent variables $(x_1,..., x_L)$

where are included also: (i) integral equations, which involve integrals of the dependent variable(s) y $(y_1, ..., y_m)$; (ii) integrodifferential equations which involve both integrals and derivatives. Considering ordinary and partial differential equations, there are both: (i) analogies, for example, certain types of o.d.e.s have analogue p.d.e.s which can be solved in a similar way, such as linear with constant coefficients; (ii) differences, that is some o.d.e.s have no analogue p.d.e.s, or the methods used to solve the former fail for the latter. An example of (ii) is the solution of a first-order linear o.d.e., which always reducible to quadratures, whereas the solution of a first-order linear p.d.e. depends on the solution of one (or a system) of o.d.e.s which may be non-linear. Since the solution of p.d.e.s is generally more difficult than that of o.d.e.s, a p.d.e. is considered as 'solved' if the problem is reduced to an o.d.e.. Simultaneous o.d.e.s (p.d.e.s) appear in the formulation of physical problems with one (several) independent variable(s), for example, in classical mechanics (physics of continua), where the variable(s) is time (are time and one or more spatial coordinates). Thus p.d.e.s have, if anything, even wider application than o.d.e.s, although the former should be studied after the latter. The classification in Table 1.1 is formalized next for a single partial differential equation of any order N with L variables (subsection 1.1.2).

1.1.2   PARTIAL DIFFERENTIAL EQUATION IN N VARIABLES

The solution or **integral** of a partial differential equation is a function (1.1b,c) of L **independent variables** (1.1a) and thus is the **dependent variable**:

$$\ell = 1,...,L: \qquad \Phi = \Phi\left(x_1,...,x_L\right) = \Phi\left(x_\ell\right); \tag{1.1a–c}$$

the function or dependent variable may have derivatives of the first (1.2b), second (1.2c) or higher orders (1.2d) up to N:

$$\ell_1,\ell_2,...,\ell_N = 1,...,L: \qquad \frac{\partial\Phi}{\partial x_{\ell_1}}, \ \frac{\partial^2\Phi}{\partial x_{\ell_1}\partial x_{\ell_2}},...,\frac{\partial^2\Phi}{\partial x_{\ell_1}...\partial x_{\ell_N}}, \tag{1.2a–d}$$

leading for L independent variables (1.1a) $\equiv$ (1.2a) to: (i) a vector (1.2b) of first-order partial derivatives with L components; (ii) an L $\times$ L matrix (1.2c) of second-order partial derivatives with L^2 components; (iii) a multiplicity (1.2d) of partial derivatives of order N with N indices with a range of L values each and hence L^N components. A **partial differential equation of order N** is a relation (1.3):

$$F\left(x_\ell; \Phi;, \frac{\partial^2\Phi}{\partial x_\ell};, \frac{\partial^2\Phi}{\partial x_{\ell_1}\partial x_{\ell_2}};,\ldots;, \frac{\partial^N\Phi}{\partial x_{\ell_1}\ldots\partial x_{\ell_N}}\right) = 0, \qquad (1.3)$$

among the dependent variable (1.1c), the independent variables (1.1a, b) and the partial derivatives of the former with regard to the latter (1.2a) up to (1.2b,c, d) order N.

The partial differential equation is **quasi-linear** iff it is linear in all partial derivatives:

$$0 = \sum_{n=0}^{N} \sum_{\ell_1,\ldots,\ell_n=1}^{L} A_{\ell_1\ldots\ell_n}\left(\Phi,x_\ell\right) \frac{\partial^n\Phi}{\partial x_{\ell_1}\ldots\partial x_{\ell_n}}, \qquad (1.4)$$

where the coefficients may involve the independent variable and the independent variables but not any partial derivatives. The partial differential equation is **linear** iff both the independent variable and all its derivatives appear linearly (1.5):

$$B(x_\ell) = \sum_{n=0}^{N} \sum_{\ell_1,\ldots,\ell_N=1}^{L} A_{\ell_1\ldots\ell_N}(x_\ell) \frac{\partial^N\Phi}{\partial x_{\ell_1}\ldots\partial x_{\ell_n}}, \qquad (1.5a)$$

and: (i) has **variable (constant) coefficients** iff at least one of the coefficients depends (1.5a) on the independent variables [are all constant (1.5b)]; (ii) is **forced (unforced)** if there is an independent term possibly involving the independent variable (1.5a) [there is no independent term (1.5c)]:

$$A_{\ell_1\ldots\ell_N} = \text{const}, \qquad B(x_\ell) = 0, \qquad (1.5b,c)$$

so that the p.d.e. (*1.5a*) does not have (has) a trivial solution $\Phi = 0$. The forcing term separates in the linear p.d.e. (*1.5a*) and is included in $A_0(x, y, \Phi)$ in the quasi-linear case (*1.4*).

The partial differential equation is of **fixed order** iff all derivatives are (1.6a) of the same order N:

$$0 = F\left(x_\ell; \Phi, \frac{\partial^N\Phi}{\partial x_{\ell_1}\ldots\partial x_{\ell_N}}\right);$$

$$A_0(x_\ell,\Phi) = \sum_{\ell_1,\ldots,\ell_N=1}^{L} A_{\ell_1\ldots\ell_N}(\Phi,x_\ell) \frac{\partial^N\Phi}{\partial x_{\ell_1}\ldots\partial x_{\ell_N}}, \qquad (1.6a, b)$$

$$B(x_\ell) = \sum_{\ell_1,\ldots,\ell_N=1}^{L} A_{\ell_1\ldots\ell_N}(x_\ell) \frac{\partial^N \Phi}{\partial x_{\ell_1} \ldots \partial x_{\ell_N}}, \qquad (1.6c)$$

for example, (1.6b) [(1.6c)] in the quasi-linear (linear) case. As examples (1.7a) is a non-linear third-order p.d.e., (1.7b) is a quasi-linear first-order p.d.e., (1.7c) [(1.7d)] is a linear fourth-order p.d.e. with variable (constant) coefficients:

$$\sum_{\ell_1,\ell_2,\ell_3=1}^{L} \left\{ A_{\ell_1\ell_2\ell_3}(x_\ell,\Phi) \frac{\partial^3 \Phi}{\partial x_{\ell_1} \partial x_{\ell_2} \partial x_{\ell_3}} + B_{\ell_1\ell_2\ell_3}(x_\ell,\Phi) \frac{\partial^2 \Phi}{\partial x_{\ell_1}} \frac{\partial^2 \Phi}{\partial x_{\ell_2} \partial x_{\ell_3}} \right\}$$
$$= B(x_\ell,\Phi), \qquad (1.7a)$$

$$\sum_{\ell_1=1}^{L} A_{\ell_1}(x_\ell,\Phi) \frac{\partial \Phi}{\partial x_{\ell_1}} + A_0(x_\ell,\Phi) = 0, \qquad (1.7b)$$

$$\sum_{\ell_1,\ell_2,\ell_3,\ell_4=1}^{L} A_{\ell_1\ell_2\ell_3\ell_4}(x_\ell) \frac{\partial^4 \Phi}{\partial x_{\ell_1} \partial x_{\ell_2} \partial x_{\ell_3} \partial x_{\ell_4}} + A(x_\ell)\Phi = B(x_\ell), \qquad (1.7c)$$

$$\sum_{\ell_1,\ell_2,\ell_3,\ell_4=1}^{L} A_{\ell_1\ell_2\ell_3\ell_4} \frac{\partial^4 \Phi}{\partial x_{\ell_1} \partial x_{\ell_2} \partial x_{\ell_3} \partial x_{\ell_4}} + A\Phi = B(x_\ell), \qquad (1.7d)$$

all with forcing. The formalization of a single partial differential equation with L variables is simplified and made more explicit for L = 2 variables (subsection 1.1.3).

1.1.3   ONE DEPENDENT AND TWO INDEPENDENT VARIABLES

In the case (1.8b) of one dependent and two independent variables with continuous derivatives up to order N, the partial differential equation (1.8c) of order N:

$$n = 1,\ldots,N; \qquad \Phi = \Phi(x,y) \in C^N(\mathbb{R}^2):$$
$$0 = F\left(x,y; \Phi, \frac{\partial\Phi}{\partial x}, \frac{\partial\Phi}{\partial y}, \frac{\partial^2\Phi}{\partial x^2}, \frac{\partial^2\Phi}{\partial x\partial y}, \frac{\partial^2\Phi}{\partial y^2}, \ldots, \right.$$
$$\left. \frac{\partial^N\Phi}{\partial x^N}, \ldots, \frac{\partial^N\Phi}{\partial x^n\partial y^{N-n}}, \ldots, \frac{\partial^N\Phi}{\partial y^N} \right), \qquad (1.8a\text{--}c)$$

involves: (i) the two independent variables x,y; (ii) the single dependent variable Φ; (iii) two first-order partial derivatives; (iv) three second-order partial derivatives; (v) N+1 partial derivatives (1.8a) of order N. In (iv) and (v) it was taken into account that continuous partial derivatives have interchangeable order, for example:

$$\Phi \in C^2(\mathbb{R}^2): \frac{\partial^2\Phi}{\partial x\partial y} = \frac{\partial^2\Phi}{\partial y\partial x}, \quad \Phi \in C^N(\mathbb{R}^2): \frac{\partial^N\Phi}{\partial x^n\partial y^{N-n}} = \frac{\partial^N\Phi}{\partial y^{N-n}\partial x^n}, \qquad (1.9a\text{--}d)$$

for the second (1.9a, b) and N-th (1.9c, d) order.

A quasi-linear partial differential equation involves the partial derivatives linearly (1.10):

$$0 = \sum_{n=0}^{N}\sum_{m=0}^{M} A_{n,m}(x,y,\Phi) \frac{\partial^{n+m}\Phi}{\partial x^n \partial y^m}, \tag{1.10}$$

with coefficients that may involve the dependent variable and two independent variables. If the partial differential equation is linear with variable (1.11b) [constant (1.11a)] coefficients:

$$A_{n,m} = \text{const}; \quad B(x,y) = \sum_{n=0}^{N}\sum_{m=0}^{M} A_{n,m}(x,y)\frac{\partial^{n+m}\Phi}{\partial x^n \partial y^m}, \tag{1.11a, b}$$

the forcing term may involve the independent variables. If the partial differential equation is of fixed order all partial derivatives (1.12a) are of the same order (1.12b):

$$n = 0,\dots,N: \quad F\left(x,y;\Phi, \frac{\partial^N \Phi}{\partial y^n\, \partial y^{N-n}}\right) = 0, \quad 0 = \sum_{n=0}^{N} A_n(x,y,\Phi)\frac{\partial^N \Phi}{\partial y^n\, \partial y^{N-n}}, \tag{1.12a--c}$$

$$A_n = \text{const}: \quad B(x,y) = \sum_{n=0}^{N} A(x,y)\frac{\partial^N}{\partial x^n\, \partial y^{N-n}}, \tag{1.12d,e}$$

for example, in the quasi-linear case (1.12c) or the linear forced case with variable (1.12e) [constant (1.12d)] coefficients, although the forcing term (1.12d) may still involve the independent variable. For example, (1.13a) is a non-linear p.d.e. of third-order, (1.13b) is quasi-linear p.d.e. of first-order, (1.13c) [(1.13d)] is linear p.d.e. of the fourth-order with variable (constant) coefficients:

$$A_1(x,y,\Phi)\frac{\partial^3 \Phi}{\partial x^2\, \partial y} + A_2(x,y,\Phi)\frac{\partial \Phi}{\partial x}\frac{\partial \Phi}{\partial y} = B(x,y,\Phi), \tag{1.13a}$$

$$A_1(x,y,\Phi)\frac{\partial \Phi}{\partial x} + A_2(x,y,\Phi)\frac{\partial \Phi}{\partial y} + A_0(x,y,\Phi) = 0, \tag{1.13b}$$

$$A_1(x,y)\frac{\partial^4 \Phi}{\partial x^4} + A_2(x,y)\frac{\partial^2 \Phi}{\partial x\, \partial y} = B(x,y), \tag{1.13c}$$

$$A_1\frac{\partial^4 \Phi}{\partial x^4} + A_2\frac{\partial^2 \Phi}{\partial x\, \partial y} = B(x,y), \tag{1.13d}$$

all forced. The classification of partial differential equations (Section 1.1) is a first step towards their solution, that is determining their integrals (Section 1.2).

1.2   GENERAL INTEGRAL AND ARBITRARY FUNCTIONS

The general integral of an ordinary differential equation of order N involves N arbitrary constants (sections IV.1.1 and IV.9.1), and likewise the general integral of a partial differential equation of order N involves N arbitrary functions (subsection 1.2.2), for example, one for the first order (subsection 1.2.1).

1.2.1   FIRST-ORDER P.D.E. AND ONE ARBITRARY FUNCTION

A **first-order p.d.e.** is a relation (1.14c) between one dependent variable Φ and L independent variables $x_1, \ldots, x_L$, and the partial derivatives $\partial\Phi/\partial x_1, \ldots, \partial\Phi/\partial x_L$ of the former with regard to the latter (1.14a, b):

$$\ell = 1,\ldots,L: \quad 0 = F\left(x_\ell; \Phi, \partial\Phi/\partial x_\ell\right) = F\left(x_1,\ldots,x_L; \Phi, \frac{\partial\Phi}{\partial x_1},\ldots,\frac{\partial\Phi}{\partial x_L}\right). \qquad (1.14\text{a--c})$$

The simplest first-order p.d.e. states that the dependent variable Φ does not depend on one independent variable, for example, x_1 in (1.15a):

$$\frac{\partial\Phi}{\partial x_1} = 0: \qquad \Phi = f\left(x_2,\ldots,x_n\right), \qquad f \in \mathcal{D}\left(|R^{L-1}\right), \qquad (1.15\text{a--c})$$

hence it can depend (1.15b) on all other variables $x_2, \ldots, x_L$, where f is an arbitrary differentiable function (1.15c). The example (1.15a–c) suggests that the **general integral** *of a first-order partial differential equation (1.14a–c) is a relation (1.16b) among the dependent Φ and independent $x_1, \ldots, x_M$ variables:*

$$f \in \mathcal{D}\left(|R^L\right): \qquad \Phi = f\left(x_1,\ldots,x_L\right), \qquad (1.16\text{a, b})$$

involving an arbitrary differentiable function (1.16a). The rigorous proof of this existence theorem is deferred (to section 9.1) and here given an explanation by comparison with o.d.e.s. If in p.d.e. (1.14a–c) x_1 is considered as the variable, and $x_2, \ldots, x_M$ as parameters, this leads an o.d.e. of first-order (1.17):

$$F\left(x_1; \Phi, \frac{d\Phi}{dx_1}; x_2,\ldots,x_L\right) = 0; \qquad (1.17)$$

The general integral involves (subsection IV.1.2) an arbitrary constant of integration C:

$$G\left(x_1, \Phi; C\left(x_2,\ldots,x_L\right)\right) = 0, \qquad (1.18)$$

which in this case (1.18) will generally depend on all the parameters $x_2, \ldots, x_L$, in (1.17). Considering again all $x_1, \ldots, x_L$ as variables, (1.18) specifies an arbitrary function $C(x_2,\ldots,x_L)$ in the general integral (1.16a, b) of the p.d.e. (1.14a–c). The form of the arbitrary function could be different; for example, if the variable x_L had been singled-out instead of x_1 the arbitrary function would have been $C(x_1,...,x_{L-1})$; the essential property is that the general integral of a first-order p.d.e. involves exactly one arbitrary function, no more and no less. These other possibilities can be excluded, by the following reasoning: (i) if there were more than one arbitrary function in the general integral of a first-order p.d.e., then the general integral of a first-order o.d.e. would involve more than one arbitrary constant of integration, which is false; (ii) if there were no arbitrary functions in the solution of the first-order p.d.e., then the solution of the first-order o.d.e. would involve no arbitrary constant of integration, that is the solution would be a particular or a special integral, but not the general integral.

1.2.2 P.D.E. OF ORDER N AND N ARBITRARY FUNCTIONS

In the case of a partial differential equation (1.3) of order N taking all $x_2, ..., x_L$ as parameters lead to (1.19) an ordinary differential equation of order N:

$$n = 1,...,N: \qquad 0 = G\left(x_1; \Phi, \frac{d\Phi}{dx_1},..., \frac{d^n\Phi}{dx_1^n},...., \frac{d^N\Phi}{dx_1^N}\right). \qquad (1.19)$$

The general integral involves (sections IV.1.1 and IV.9.1) N arbitrary constants (1.18a, b):

$$n = 1,...,N: \qquad \Phi = \Phi\left(x_1, C_n\left(x_2,...,x_L\right)\right), \qquad (1.20a, b)$$

that can depend on the order independent variables. Thus *the general integral of a partial differential equation (1.3) of order N in L variables (1.1a–c;1.2a–d) involves N arbitrary (1.21a,d) N-times differentiable functions (1.21c) of the L independent variables (1.21b):*

$$n = 1,...,N; \quad \ell = 1,...,L: \qquad g_n\left(x_\ell\right) \in \mathcal{D}^N\left(|R^L\right): \qquad \Phi\left(x_\ell\right) = f\left(g_n\left(x_\ell\right)\right). \ (1.21a\text{–}d)$$

A **particular integral** (section IV.1.1) is obtained by choosing one or more of the arbitrary functions. A **special integral** (sections IV.1.1 and IV.5.1-IV.5.4) involves less than N arbitrary functions and is distinct from all particular integrals, and thus is not contained in the general integral for any choice of arbitrary functions (sections V.1-V.3). Next is shown how the arbitrary function appears in the solution of the p.d.e. of first order (Section 1.3).

1.3 UNFORCED P.D.E. WITH FIRST-ORDER DERIVATIVES

Starting with p.d.e.s one of the simplest is linear without forcing term and only first-order derivatives and with variable coefficients (subsection 1.3.1). Its solution involves an arbitrary function of L variables (subsection 1.3.4) and represents the family of hypersurfaces (subsection 1.3.2) whose intersection is the characteristic curve (subsection 1.3.3) tangent to the vector of coefficients (Figure 1.1).

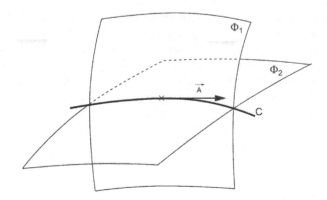

FIGURE 1.1   A linear unforced first-order partial differential equation with variable coefficients with three variables is specified by its three-dimensional vector of coefficients $\vec{A}$; its general integral is an arbitrary differentiable function of two characteristic variables representing two surfaces tangent to the vector field. Thus the vector field of coefficients is tangent to the intersection of the two characteristic surfaces.

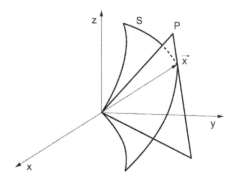

FIGURE 1.2   As an example of Figure 1.1, the three-dimensional position vector passing through the origin lies at the intersection of planes and curved surfaces passing through the origin.

The simplest particular cases are: (i) into two dimensions the plane curves tangent to a vector field (subsection 1.3.5); (ii) in three dimensions (subsection 1.3.6) the family of surfaces to which a vector field is tangent (subsection 1.3.6). The vector field chosen as an example is the two (three) dimensional position vector through the origin (Figure 1.2).

1.3.1   Classification of P.D.E.S of First Order

If in (1.1) the function F is linear in the partial derivatives $\partial/\partial x_\ell$, and the coefficients depend only on the independent variables $x_1, \ldots, x_\ell$, it is a **linear first-order p.d.e.:**

$$\sum_{\ell=1}^{n} A_\ell\left(x_1, \ldots, x_L\right) \frac{\partial \Phi}{\partial x_\ell} + A_0\left(x_1, \ldots, x_n\right)\Phi = B\left(x_1, \ldots, x_L\right). \tag{1.22}$$

If the coefficients depend also on the dependent variable Φ, it is a **quasi-linear first-order p.d.e.** (1.23):

$$\sum_{\ell=1}^{L} A_\ell (x_1,\ldots,x_L;\,\Phi)\frac{\partial\Phi}{\partial x_\ell} + A_0 (x_1,\ldots,x_L;\,\Phi) = 0. \tag{1.23}$$

If the coefficients are constant, it is a **linear first-order p.d.e. with constant coefficients** (1.24):

$$\sum_{i=1}^{L} A_\ell \frac{\partial\Phi}{\partial x_\ell} + \Phi = B(x_1,\ldots,x_L). \tag{1.24}$$

If the forcing term is present $B \neq 0$ (absent $B = 0$) the equation does not have (has) the trivial solution $\Phi = 0$. The forcing term separates in the linear p.d.e.s (1.22;1.24) and is included in A_0 in the quasi-linear case (1.23).

1.3.2 FAMILY OF HYPERSURFACES TANGENT TO A VECTOR FIELD

As a starting point consider a linear p.d.e. of first-order (1.22) with neither forcing term (1.25a) and nor (1.25b) the term proportional to the dependent variable (1.25c):

$$B = 0 = A_0: \qquad \sum_{\ell=1}^{L} A_\ell (x_1,\ldots,x_L)\frac{\partial\Phi}{\partial x_\ell} = 0. \tag{1.25a–c}$$

The general integral involves an arbitrary function, and represents a hypersurface in the L-dimensional space (1.26a) of coordinates $x_1, \ldots, x_L$:

$$\Phi (x_1,\ldots,x_L) = C; \qquad 0 = d\Phi = \sum_{\ell=1}^{L} \frac{\partial\Phi}{\partial x_\ell}\partial x_\ell = \nabla\Phi.\,d\bar{x}_\Phi\,, \qquad \bar{N} = \nabla\Phi, \tag{1.26a, b}$$

differentiation of (1.26b) shows that the gradient $\nabla\Phi$ is orthogonal to all displacements $d\bar{x}_\Phi$ on the surface, and thus lies (1.26c) in the normal direction; it becomes the unit normal vector if it is divided by its modulus. The p.d.e. (1.25c) states that:

$$0 = \sum_{\ell=1}^{L} A_\ell N_\ell = \bar{X}\cdot\bar{N} \Leftrightarrow \bar{X}\perp\bar{N}, \tag{1.27a, b}$$

the vector $\bar{A}_i$ of coefficients has zero inner product with the normal (1.27a) and thus is orthogonal to the normal $\bar{N}$ in (1.27b), and hence tangent to the hypersurface. Thus *the general integral (1.26a) of the linear (1.22) p.d.e. (1.25c) with only first-order derivatives (1.25b) without forcing term (1.25b), solves the following geometric problem: find the family of all hypersurfaces (1.26a) which are tangent to the vector $\bar{A}$ of coefficients.*

1.3.3   CHARACTERISTIC CURVE TANGENT TO A VECTOR FIELD

The geometric problem is illustrated in Figure 1.1, in the case of $L = 3$ dimensions. The intersection of all hypersurfaces in the general integral (1.24b) is the curve tangent to the vector of coefficients, which is called the **characteristic curve** of the first-order p.d.e. (1.25a–c), and given by the solution of the system of **characteristic equations** (1.28):

$$\frac{dx_1}{A_1} = \frac{dx_2}{A_2} = \cdots = \frac{dx_n}{A_L} \equiv \lambda. \tag{1.28}$$

This geometrical argument suggests that the solution of the system of L-1 independent o.d.e.s (1.28) specifying the characteristic curve, is connected with the general integral of the first-order p.d.e. (1.25a–c); analytically, if the general integral (1.26a) is differentiated it leads to (1.29a):

$$0 = d\vec{x} \cdot \nabla \Phi = \sum_{\ell=1}^{L} dx_\ell \frac{\partial \Phi}{\partial x_\ell}, \tag{1.29}$$

where the displacements $dx_1, \ldots, dx_L$ are arbitrary; it follows on comparison of (1.29) with (1.25c) that the vectors $d\vec{x}$ and $\vec{A}$ are parallel leading to the characteristic equations (1.28).

1.3.4   CHARACTERISTIC VARIABLES AS SOLUTIONS OF THE CHARACTERISTIC EQUATIONS

The system of characteristic equations (1.28) consists of L-1 independent o.d.e.s (1.30a) whose solutions are the **characteristic variables**:

$$s = 1, \ldots, L-1: \qquad U_s\left(x_1, \ldots, x_L\right) = C_s, \tag{1.30a, b}$$

where the $C_1, \ldots, C_{n-1}$ are constants. An arbitrary differentiable function (1.31a) of the characteristic variables (1.30b):

$$f \in \mathcal{D} : \left(|\mathbb{R}^{L-1}\right) : \qquad f\left(U_1\left(x_1, \ldots, x_L\right), \ldots, U_{L-1}\left(x_1, \ldots, x_L\right)\right) = C, \tag{1.31a, b}$$

satisfies the p.d.e. (1.25c):

$$0 = df = \sum_{s=1}^{L-1} \frac{\partial f}{\partial U_s} dU_s = \sum_{s=1}^{L-1} \frac{\partial f}{\partial U_s} \sum_{\ell=1}^{L} \frac{\partial U_s}{\partial x_\ell} dx_\ell = \lambda \sum_{s=1}^{L-1} \frac{\partial f}{\partial U_s} \sum_{\ell=1}^{L} A_\ell \frac{\partial U_s}{\partial x_\ell} = 0, \tag{1.32}$$

where were used (1.28) and (1.25c). Hence (1.31b) or (1.33a, b):

$$g \in \mathcal{D}\left(|\mathbb{R}^{L-1}\right) : \qquad \Phi\left(x_1, \ldots, x_L\right) = g\left(U_1\left(x_1, \ldots, x_L\right), \ldots, U_{L-1}\left(x_1, \ldots, x_L\right)\right). \tag{1.33a, b}$$

is a general integral of the p.d.e. (1.25a–c), since it is a solution involving one arbitrary function. In conclusion: *the general integral of the linear first-order p.d.e. without forcing term (1.25a–c), involves (1.33b) an arbitrary differentiable function (1.33a), and is specified by the L-1 characteristic variables $U_1, ..., U_{L-1}$ in (1.31a, b), that are independent solutions of the system (1.28) of characteristic o.d.e.s.*

1.3.5 PLANE CURVE TANGENT TO A VECTOR FIELD

The simplest example is a linear partial differential equation of first-order without forcing term (1.34a) that is equivalent to the first-order ordinary differential Equation (1.34b) where was used (1.34c) the implicit function theorem:

$$0 = A_1(x,y)\frac{\partial\Phi}{\partial x} + A_2(x,y)\frac{\partial\Phi}{\partial y} \Leftrightarrow \frac{A_2(x,y)}{A_1(x,y)} = -\frac{\partial\Phi/\partial x}{\partial\Phi/\partial y} = \left(\frac{dy}{dx}\right)_\Phi. \qquad (1.34a–c)$$

For example, the plane curves (1.35a) tangent to the position vector (1.35b) satisfy (1.35c):

$$\Phi(x,y) = C, \quad \vec{A} = (x,y): \qquad x\frac{d\Phi}{dx} + y\frac{d\Phi}{dy} = 0, \qquad (1.35a, b)$$

implying (1.36a, b):

$$\frac{dy}{dx} = -\frac{\partial\Phi/\partial x}{\partial\Phi/\partial y} = \frac{y}{x}, \qquad \frac{dy}{y} = \frac{dx}{x}, \qquad (1.36a, b)$$

and hence (1.32a):

$$\log y = \log C + \log x, \qquad y = Cx, \qquad (1.36c, d)$$

are straight lines through the origin (1.36d) with arbitrary slope. The general integral is (1.37c) involving a differentiable function (1.37a,d) of the **similarity variable** (1.37b) that is a linear combination of the two independent variables (x,y):

$$f \in \mathcal{D}(\mathbb{R}), \quad \xi = y - Cx: \qquad \Phi(x,y) = f(y - Cx) = f(\xi). \qquad (1.37a\ d)$$

If can be checked (1.38a–c) from (1.37b,d):

$$x\frac{\partial\Phi}{\partial x} + y\frac{\partial\Phi}{\partial y} = \frac{df}{d\xi}\left(x\frac{\partial\xi}{\partial x} + y\frac{\partial\xi}{\partial y}\right) = \frac{df}{d\xi}(-Cx + y) = 0, \qquad (1.38a–c)$$

that the partial differential equation (1.35b) ≡ (1.38c) is satisfied by (1.37d). The arbitrary function f is determined from a boundary condition, for example, (1.39a, b) leads to the integral (1.39c, d):

$$e^y = \Phi(0,y) = f(y): \qquad \Phi(x,y) = f(y - Cx) = \exp(y - Cx), \qquad (1.39a–d)$$

The plane curves (1.39c) reduce to straight lines (1.36d) if the function (1.37a) is a constant.

1.3.6 EXAMPLE OF THE FAMILY OF SURFACES TANGENT TO THE POSITION VECTOR

As an example in three-dimensional space of coordinates x, y, z, determine the family of surfaces (1.40a) tangent to the position vector (1.40b):

$$\Phi(x,y,z) = C: \qquad x\frac{\partial\Phi}{\partial x} + y\frac{\partial\Phi}{\partial y} + z\frac{\partial\Phi}{\partial z} = 0. \qquad (1.40\text{a, b})$$

The characteristic curves, that are tangent to the position vector (1.41a):

$$\vec{A} \equiv (x,y,z): \qquad \frac{dx}{x} = \frac{dy}{y} = \frac{dz}{z}, \qquad (1.41\text{a--c})$$

are specified by the characteristic equations $(1.28) \equiv (1.41\text{b,c})$; they are $(1.36\text{b};1.37\text{a, b})$ straight lines through the origin, respectively, in the (x,y) and (x,z) planes (1.42a, b):

$$\left\{\frac{y}{x}, \frac{z}{x}\right\} = (U,V); \quad f \in \mathcal{D}(\mathbb{R}^2) \quad \Phi(x,y) = f(U,V) = f\left(\frac{y}{x}, \frac{z}{x}\right), \qquad (1.42\text{a--d})$$

the general integral is (1.42d) is an arbitrary differentiable function (1.42c) of the two characteristic variables (1.42a, b). It can be checked that (1.42a–c) imply (1.43a, b):

$$\left\{\frac{\partial\Phi}{\partial x}, \frac{\partial\Phi}{\partial y}, \frac{\partial\Phi}{\partial z}\right\} = \frac{\partial f}{\partial U}\left\{\frac{\partial U}{\partial x}, \frac{\partial U}{\partial y}, \frac{\partial U}{\partial z}\right\} + \frac{\partial f}{\partial V}\left\{\frac{\partial V}{\partial x}, \frac{\partial V}{\partial y}, \frac{\partial V}{\partial z}\right\}$$

$$= \left\{-\frac{y}{x^2}, \frac{1}{x}, 0\right\}\frac{\partial f}{\partial U} + \left\{-\frac{z}{x^2}, 0, \frac{1}{x}\right\}\frac{\partial f}{\partial V}, \qquad (1.43\text{a, b})$$

and substitution in (1.43c):

$$x\frac{\partial\Phi}{\partial x} + y\frac{\partial\Phi}{\partial y} + z\frac{\partial\Phi}{\partial y} = \left(-\frac{y}{x} + \frac{y}{x} + 0\right)\frac{\partial f}{\partial U} + \left(-\frac{z}{x} + 0 + \frac{z}{x}\right)\frac{\partial f}{\partial V} = 0, \qquad (1.43\text{c, d})$$

proves that the p.d.e. $(1.40\text{b}) \equiv (1.43\text{c, d})$ is satisfied by the general integral (1.42a–d). The general integral (1.42c) can be rewritten (1.44b):

$$g \in \mathcal{D}(\mathbb{R}): \qquad \Phi(x,y,z) = \frac{z}{x} - g\left(\frac{y}{x}\right), \qquad (1.44\text{a, b})$$

where g(f) is an arbitrary differentiable functions of one (1.44a) [two (1.42a)] variable(s). For example, if g is a linear function (1.45a):

$$g(\xi) = a\xi + b: \qquad \Phi = 0 \Rightarrow z = xg\left(\frac{y}{x}\right) = x\left(a\frac{y}{x} + b\right) = ay + bx, \qquad (1.45\text{a--c})$$

the solution (1.45b) specifies planes through the origin (1.45c), which contain the position vector $\vec{A}$ in (1.41a); however (1.46c) is not the general integral of (1.40b), because there are non-plane surfaces included in (1.42a,d) that are tangent to (1.41a), as shown in Figure 1.2. The arbitrary function g is determined from a boundary condition, for example (1.46c):

$$\frac{y^2}{x^2} = \Phi(x, y, 0) = -g\left(\frac{y}{x}\right): \qquad \Phi(x, y, z) = \frac{z}{x} + \frac{y^2}{x^2}. \qquad (1.46a\text{--}c)$$

Two cases of linear unforced p.d.e.s with variable coefficients and all derivatives of the first order with two or three variables are considered in example 10.1. The solution of the linear first-order partial differential equations without forcing (Section 1.3) is generalized next to the of quasi-linear first-order partial differential equations (Section 1.4) that include as a particular case the forced linear first-order partial differential equation with variable coefficients.

1.4   QUASI-LINEAR AND FORCED FIRST-ORDER P.D.E.S

The quasi-linear p.d.e. (1.23) differs from the linear (1.22), both of first order, in that the coefficients may involve the dependent variable (subsection 1.4.1). The solution of a quasi-linear first-order p.d.e. (subsection 1.4.2) includes as a particular case a linear forced p.d.e. with all derivatives of first-order (subsection 1.4.3). In both cases, the solution can be obtained (subsection 1.4.1) from to that of a linear p.d.e. without forcing term and all derivatives of the first order and with an extra independent variable. An example is (subsection 1.4.4) is the family of plane curves with unit projection on the position vector through the origin.

1.4.1   QUASI-LINEAR P.D.E. OF THE FIRST-ORDER

Next the problem (1.25a–c) is generalized to the quasi-linear first-order p.d.e. (1.23), whose general integral is of the form (1.47b):

$$\Psi \in \mathcal{D}\left(|\mathbb{R}^{L+1}\right): \qquad \Psi(x_1, \ldots, x_L, \Phi) = C, \qquad (1.47a, b)$$

where Ψ is a differentiable function (1.47a), implying (1.48a, b):

$$\ell = 1, \ldots, L: \qquad 0 = \frac{d\Psi}{dx_\ell} = \frac{\partial \Psi}{\partial x_\ell} + \frac{\partial \Psi}{\partial \Phi} \frac{\partial \Phi}{\partial x_\ell}. \qquad (1.48a, b)$$

This result is equivalent to the rule of implicit differentiation:

$$\ell = 1, \ldots, L: \qquad \frac{\partial \Phi}{\partial x_\ell} = -\frac{\partial \Psi / \partial x_i}{\partial \Psi / \partial \Phi}. \qquad (1.49a, b)$$

Substituting (1.49b) into (1.23) leads to:

$$\sum_{\ell=1}^{L} A_\ell\left(x_1,\ldots,x_n,\Phi\right)\frac{\partial\Psi}{\partial x_\ell} - A_0\left(x_1,\ldots,x_n,\Phi\right)\frac{\partial\Psi}{\partial\Phi} = 0; \tag{1.50}$$

this is an equation of the type (1.25a–c) considered before, with: (i) L+1 independent variables and coefficients in (1.51a):

$$\left\{x_{L+1}, A_{L+1}\right\} = \left\{\Phi, -A_0\right\}; \qquad \sum_{r=1}^{L+1} A_r\frac{\partial\Psi}{\partial x_r} = 0, \tag{1.51a, b}$$

(ii) dependent variable Ψ in (1.51b).

1.4.2 SOLUTION AS IMPLICIT FUNCTION OF N VARIABLES

It has been shown that *the quasi-linear first-order p.d.e. (1.23) with one dependent Φ and L independent variables $x_1, \ldots, x_L$ and coefficients $A_0, \ldots, A_L$, can be transformed to a linear first-order p.d.e. without forcing term (1.50), by introducing (1.51a) one more independent variable $x_{L+1} \equiv \Phi$ and coefficient $A_{L+1} \equiv A_0$.* Since the solution of the latter type of p.d.e. is already known (Section 1.3), it can also be used to solve the former; the characteristic equation (1.28) associated with (1.51b) is specified by the characteristic system of o.d.e.s (1.52):

$$\frac{dx_1}{A_1\left(x_1,\ldots,x_L;\Phi\right)} = \cdots = \frac{dx_L}{A_L\left(x_1,\ldots,x_L;\Phi\right)} = -\frac{d\Phi}{A_0\left(x_1,\ldots,x_L;\Phi\right)}, \tag{1.52}$$

that adds one term or one more ordinary differential equation to (1.28). There are L independent (1.53a) solutions (1.53b) as characteristic variables of the system (1.52) of o.d.e.s:

$$\ell = 1,\ldots,L: \qquad\qquad U_\ell\left(x_1,\ldots,x_L,\Phi\right) = C_\ell; \tag{1.53a, b}$$

an arbitrary differentiable function (1.54a) of these specifies the general integral (1.54b):

$$f \in \mathcal{D}\left(\mathbb{R}^L\right): \qquad f\left(U_1\left(x_1,\ldots,x_L,\Phi\right),\ldots,U_L\left(x_1,\ldots,x_L,\Phi\right)\right) = 0, \tag{1.54a, b}$$

in implicit form. Thus *the general integral of the quasi-linear first-order p.d.e. (1.23), is an arbitrary differentiable function (1.54a, b) of L variables (1.53a, b) that are independent solutions of the characteristic system (1.52) of o.d.e.s.* Since the linear first-order forced p.d.e. with all derivatives of the first-order is a particular case of the quasi-linear first-order p.d.e. the preceding theorem also applies (subsection 1.4.3).

1.4.3 LINEAR FORCED P.D.E. WITH FIRST-ORDER DERIVATIVES

The linear forced partial differential equation (1.24) with (1.55a) all derivatives of first-order (1.55b):

$$A_0(x_\ell) = 0: \qquad \sum_{\ell=1}^{L} A_0(x_1,...,x_L) \frac{\partial \Phi}{\partial x_\ell} = B(x_\ell), \qquad (1.55a, b)$$

is a particular case of the quasi-linear first-order p.d.e. (1.23) with (1.56a–b):

$$\ell = 1,...,L: \quad A_\ell(x_1,...,x_L;\Phi) \to A_\ell(x_1,...,x_L),$$
$$A_0(x_1,...,x_L,\Phi) \to -B(x_1,...,x_L). \qquad (1.56a–c)$$

Thus the characteristic system (1.52) is replaced by (1.57):

$$\frac{dx_1}{A_1(x_1,...,x_L)} = \cdots = \frac{dx_L}{A_L(x_1,...,x_L)} = \frac{d\Phi}{B(x_1,...,x_L)}. \qquad (1.57)$$

The preceding theorem (subsection 1.4.2) can be thus be re-stated in this particular case: *a linear forced first-order partial differential equation (1.24) with (1.55a) only first-order derivatives has (1.57) as the characteristic system of L ordinary differential equations, whose L independent solutions (1.53a, b) specify the L characteristic variables in the general integral (1.54b) involving an arbitrary differentiable function (1.54a).*

1.4.4 PLANE CURVES WITH UNIT PROJECTION ON THE POSITION VECTOR

As an example in the plane (x,y) consider the family of curves (1.58a):

$$\Phi(x,y) = C; \qquad x\frac{\partial \Phi}{\partial x} + y\frac{\partial \Phi}{\partial y} = 1, \qquad (1.58a, b)$$

whose normal has unit projection on the position vector (1.58b). The characteristic curves in three dimensional space of coordinates (x,y,Φ) are specified by (1.24) ≡ (1.58a) leading (1.59a) to the characteristic equations (1.57) ≡ (1.59b,c):

$$\{A_1,A_2,B\} = \{x,y,1\}: \qquad \frac{dx}{x} = \frac{dy}{y} = d\Phi; \quad \left\{\frac{y}{x}, \Phi - \log x\right\} = \{U,V\}, \qquad (1.59a–e)$$

the latter (1.59b,c) have the independent pair of solutions (1.59d,e) specifying the characteristic variables. The general integral (1.54b) of (1.58b) is (1.60a):

$$0 = f(U,V) = f\left(\frac{y}{x}, \Phi - \log x\right), \qquad \Phi(x,y) = \log x + g\left(\frac{y}{x}\right), \qquad (1.60a, b)$$

that can be re-written (1.60b), where f,g are arbitrary differentiable functions. From (1.60b) follow (1.61a–c):

$$\xi = \frac{y}{x}: \quad \left\{ \frac{\partial\Phi}{\partial x}, \frac{\partial\Phi}{\partial y} \right\} = \left\{ \frac{1}{x}, 0 \right\} + \frac{dg}{d\xi} \left\{ \frac{\partial\xi}{\partial x}, \frac{\partial\xi}{\partial y} \right\}$$

$$= \left\{ \frac{1}{x}, 0 \right\} + \frac{dg}{d\xi} \left\{ -\frac{y}{x^2}, \frac{1}{x} \right\}, \tag{1.61a–c}$$

and substitution of (1.61b,c) in (1.62a):

$$x\frac{\partial\Phi}{\partial x} + y\frac{\partial\Phi}{\partial y} = 1 + \frac{dg}{d\xi}\left(-\frac{y}{x} + \frac{y}{x} \right) = 1, \tag{1.62a, b}$$

confirms (1.62b) $\equiv$ (1.58b) that (1.60b) is the general integral of the p.d.e. (1.58b). Two cases of linear forced partial differential equations with all derivatives of the first-order and two or three variables are considered in the Example 10.1. The relation between linear first-order partial differential equations and differentials is used next (Section 1.5) to give an alternate proof of the theorem (section IV.3.9) related to: (i) the first principle of thermodynamics (section IV.5.5 and chapter 2); (ii) the representation of three-dimensional vector fields by scalar or vector potentials (sections III.6.2 and IV.3.9).

1.5   DIFFERENTIALS OF FIRST-DEGREE IN THREE VARIABLES

The theory of solution of linear (quasi-linear) first-order p.d.e.s [Section 1.3 (1.4)] is related to first-order differentials (section IV.3.8-IV.3.9 and notes N.IV.3.1-N. IV.3.20). The particular case of first-order differentials in three variables, that involve a three-dimensional vector of coefficients, leads to the representation of a vector field in space (subsection 1.5.1) by: (i) one scalar potential if the vector of coefficients has zero curl (subsection 1.5.3); (ii) two scalar potentials if the vector of coefficients is orthogonal to its non-zero curl (subsection 1.5.4); (iii) three scalar Euler potentials if the latter condition is not met (subsection 1.5.5). A kinematic interpretation is given (subsection 1.5.2) for the velocity vector, whose curl is half the rotation, with their inner product specifying the helicity. To the proofs of (i, ii, iii) in the context of differentials (section IV.3.9) can be provided analogues or alternatives, some using first-order linear partial differential equations as an intermediate step (subsections 1.5.7–1.5.9). This provides a method to determine the representation of a spatial vector field by 3 scalar Euler potentials (subsection 1.5.10) in the most general case (iii), that includes (subsection 1.5.6) the particular cases (i) and (ii). The alternative representation of a three-dimensional vector by (iv) scalar and vector potentials (subsection 1.5.14) leads to an alternative general representation by (v) three scalar Clebsch potentials (subsection 1.5.15) distinct from the three scalar Euler potentials; (b) the particular cases i (vi) of an irrotational (solenoidal) vector field, that has zero curl (divergence), and is represented by a scalar (vector) potential alone [subsection 1.5.12 (1.5.11)]. The simplest case (vii) is a potential vector field (subsection 1.5.13) that is both irrotational and solenoidal. The seven cases (i) to (vii) of three-dimensional vector fields (subsection 1.5.16) are each illustrated by one further example (subsections 1.5.17–1.5.23) demonstrating the relations among differentials, vector fields and partial differential equations (subsections 1.5.24–1.5.26).

1.5.1 Exact, Inexact and Non-Integrable Differentials

There are some connections between p.d.e.s and differentials, that include second-order p.d.e.s and quadratic forms (Section 5.2), and also first-order p.d.e.s and differentials of first-degree that are considered next (Section 1.5) in the spatial case of three variables:

$$\vec{X} \cdot d\vec{x} = X(x,y,z)dx + Y(x,y,z)dy + Z(x,y,z)dz, \tag{1.63}$$

to prove the following theorem: *the differential of first-degree in three variables (1.63) falls under one of three classes: (I) an **exact differential** of a function (1.64a) iff its gradient coincides with the vector of coefficients (1.64b) or equivalently the latter has zero curl (1.64c):*

$$\vec{X} \cdot d\vec{x} = d\Phi \Leftrightarrow \vec{X} = \nabla\Phi \Leftrightarrow \nabla \wedge \vec{X} = 0, \tag{1.64a–c}$$

*corresponding to a **conservative or irrotational field**; (II) iff it has an integrating factor Ψ, that is when divided by Ψ becomes an exact differential (1.65a) so that the vector of coefficients is the product of a scalar **integrating factor** by a gradient (1.65b) or equivalently satisfies the **condition of integrability** stating that it is orthogonal to its curl (1.65c), that is the **helicity** is zero:*

$$\vec{X} \cdot d\vec{x} = \Psi d\Phi \Leftrightarrow \vec{X} = \Psi\nabla\Phi \Leftrightarrow H \equiv \vec{X} \cdot (\nabla \wedge \vec{X}) = 0, \tag{1.65a–c}$$

*corresponding to a **rotational non-helical vector field**; (III) an **inexact non-integrable differential** in the general case when the vector of coefficients is not orthogonal to its curl (1.66c) and three functions are needed to represent the differential (1.66a):*

$$\vec{X} \cdot d\vec{x} = \Psi d\Phi + d\Theta \Leftrightarrow X = \nabla\Theta + \Psi\nabla\Phi \Leftrightarrow H \equiv X \cdot (\nabla \wedge X) \neq 0, \tag{1.66a–c}$$

*and the vector of coefficients is a **helical vector field** that differs from a gradient by a variable multiple of another gradient (1.66b). The three cases concern the representation of a three-dimensional continuous vector field (1.67a), respectively, by one (1.67b), two (1.67c) or three (1.67d) **scalar potentials**:*

$$\vec{X}(\vec{x}) \in C^1(|R^3): \qquad \vec{X} = \nabla\Phi, \Psi\nabla\Phi, \nabla\Theta + \Psi\nabla\Phi. \tag{1.67a–d}$$

These 3 cases are considered (respectively, in subsections 1.5.3, 1.5.4 and 1.5.5) after giving a kinematic interpretation (subsection 1.5.2) to the conditions of (i) immediate

integrality (1.64c) and (ii) existence (1.65c) [(iii) non-existence (1.66c)] of an integrating factor.

1.5.2 LINEAR AND ANGULAR VELOCITIES AND HELICITY

Consider (Figure 1.3) a constant **angular velocity** $\vec{\Omega}$ along the z-axis (1.68a) and the
corresponding **linear velocity** $\vec{V}$ in the perpendicular (x,y) plane (1.68b) where $\vec{r}$ is
the position vector in the plane (1.68b).

$$\vec{\Omega} = \vec{e}_z\Omega, \qquad \vec{r} = \vec{e}_x x + \vec{e}_y y: \qquad \vec{v} = \vec{\Omega} \wedge \vec{r}; \qquad (1.68a\text{--}c)$$

The outer vector product $(1.68c) \equiv (1.69a, b)$:

$$\vec{v} = \Omega\vec{e}_z \wedge \left(\vec{e}_x x + \vec{e}_y y\right) = -\vec{e}_x\Omega y + \vec{e}_y\Omega x, \qquad (1.69a, b)$$

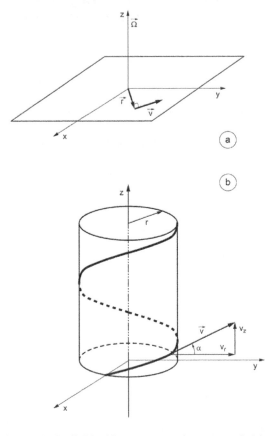

FIGURE 1.3 A plane velocity field with non-zero curl corresponds (a) to a rotation around
an axis perpendicular to the plane iff the helicity is zero. If the helicity is not zero there is a
translation along the axis in addition to a rotation around the same axis, resulting in a helical
trajectory.

leads to the components of the two-dimensional velocity vector in the plane (1.70a, b):

$$v_x = -\Omega y, v_y = \Omega x: \quad \nabla \wedge \vec{v} = \left(\partial_x v_y - \partial_y v_x \right) \vec{e}_z = 2\Omega \vec{e}_z = 2\vec{\Omega}, \qquad (1.70\text{a–c})$$

implying that the curl of the velocity is (1.70c) twice the angular velocity. Thus *the condition (1.64b) of irrotational velocity field is equivalent to: (i) absence of rotation; (ii) existence of orthogonal potential surfaces* (Figure 1.4).

Consider next (Figure 1.3) the **helical motion** on a cylinder of radius r with constant linear tangential velocity v making a constant angle α with the horizontal plane so that the velocity vector has cylindrical components (1.71a):

$$\vec{v} = v\left(\vec{e}_\phi \cos\alpha + \vec{e}_z \sin\alpha \right); \quad \nabla \wedge \vec{v} = \frac{\vec{e}_z}{r} \frac{\partial}{\partial r} \left(r v_\phi \right) = \frac{v_\phi}{r} \vec{e}_z = \frac{v}{r} \cos\alpha \, \vec{e}_z, \qquad (1.71\text{a, b})$$

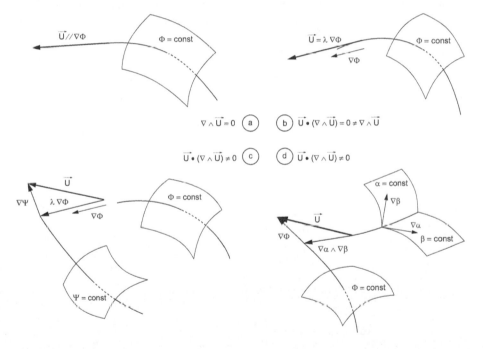

FIGURE 1.4 An irrotational vector (a) field is the gradient of a scalar potential; the latter is constant on a family of equipotential surfaces orthogonal to the vector field. A non-helical vector field (b) that is orthogonal to its curl, when divided by a second scalar potential, becomes an irrotational vector field that has a first scalar potential specifying a family of equipotential surfaces. A general helical vector field (c) that is not orthogonal to its curl becomes a non-helical vector field subtracting the gradient of a third vector potential. Another representation of a non-helical vector filed by three scalar potentials is possible (d): (i) two scalar potentials have gradients orthogonal to the equipotential surfaces; (ii) their outer product lies along the curves of intersection of the two families of equipotentials; (iii) adding the gradient of a third scalar potential species the vector.

the curl in cylindrical coordinates (III.6.41b) is (1.71b) showing that there is rotation $\nabla \wedge \vec{v} \neq 0$ for $\alpha \neq \pi/2$, that is except for axial motion. The inner vector product of the linear and angular velocities is one-half of the **helicity** (1.72a):

$$H \equiv 2(\vec{v}\cdot\vec{\Omega}) = \vec{v}\cdot(\nabla\wedge\vec{v}) = \frac{v^2}{r}\cos\alpha\,\sin\alpha = \frac{v^2\sin(2\alpha)}{2r}, \qquad (1.72\text{a–d})$$

that equals (1.72b) twice the inner product of the velocity by its curl, and is non-zero $H \neq 0$ in (1.72c) $\equiv$ (1.72d) in all cases of non-plane motion $\alpha \neq 0$. Thus *the case (ii) of rotational motion with zero helicity: (i) allows for rotation in a plane; (ii) implies that (1.65b) the vector divided by a suitable scalar has orthogonal surfaces (Figure 1.4). In the most general case (iii) of rotational motion with non-zero helicity: (i) the motion is not plane and there is rotation; (ii) it is (1.66b) necessary to subtract a gradient from the velocity, and then divide by a scalar to find orthogonal surfaces* (Figure 1.4). Next are given simple examples of the three cases of: (i) exact differential and immediate integrability (subsection 1.5.3); (ii) inexact differential with integrating factor (subsection 1.5.4); (iii) inexact differential without integrating factor (subsection 1.5.5).

1.5.3 EXACT DIFFERENTIAL AND IMMEDIATE INTEGRABILITY

The first-order differential in three variables (1.73a) has the position vector (1.73b) as the vector of coefficients:

$$x\,dx + y\,dy + z\,dz = \vec{X}\cdot d\vec{x}: \qquad \vec{X} = \{x,y,z\}. \qquad (1.73\text{a, b})$$

The zero curl (1.74):

$$\vec{e}_z \nabla \wedge \vec{X} = \begin{vmatrix} \vec{e}_x & \vec{e}_y & \vec{e}_z \\ \partial_x & \partial_y & \partial_z \\ x & y & z \end{vmatrix} = 0, \qquad (1.74)$$

confirms immediate integrability (1.75a) of the exact differential (1.75b):

$$x\,dx + y\,dy + z\,dz = d\left(\frac{x^2+y^2+z^2}{2}\right) = d\Phi, \qquad (1.75\text{a, b})$$

with a single scalar potential (1.75b) $\equiv$ (1.76a):

$$\Phi(x,y,z) = \frac{x^2+y^2+z^2}{2}, \qquad \nabla\Phi = \{\partial_x\Phi, \partial_y\Phi, \partial_z\Phi\} = \{x,y,z\} = \vec{X}, \qquad (1.76\text{a–d})$$

whose gradient (1.76b) coincides (1.76c) with the vector of coefficients (1.76d) $\equiv$ (1.73b). Thus *the three-dimensional position vector (1.73b) is irrotational (1.74) and equals (1.76b–d) the gradient of the scalar potential (1.76a) specified by the exact differential (1.75a, b).* Next is considered an inexact integrable first-order differential in 3 variables, whose vector of coefficients is rotational and non-helical, and involves two scalar potentials (subsection 1.5.4).

1.5.4   INEXACT DIFFERENTIAL WITH AN INTEGRATING FACTOR

It suffices to divide the exact differential (1.73a) by x to obtain the inexact first-order differential with three variables (1.77a) with vector of coefficients (1.77b):

$$dx + \frac{y}{x} dy + \frac{z}{x} dz = \vec{X} \cdot d\vec{x}: \qquad \vec{X} = \left\{ 1, \frac{y}{x}, \frac{z}{x} \right\}. \qquad (1.77a, b)$$

The curl is non-zero (1.78a):

$$\nabla \wedge \vec{X} = \begin{vmatrix} \vec{e}_x & \vec{e}_y & \vec{e}_z \\ \partial_x & \partial_y & \partial_z \\ 1 & \dfrac{y}{x} & \dfrac{z}{x} \end{vmatrix} = \frac{z}{x^2} \vec{e}_y - \frac{y}{x^2} \vec{e}_z, \qquad H = \vec{X} \cdot \left(\nabla \wedge \vec{X} \right) = 0, \qquad (1.78a, b)$$

and the helicity is zero (1.78b), confirming that (1.77a) is an inexact differential with (1.79a, b) an integrating factor 1/x:

$$dx + \frac{y}{x} dy + \frac{z}{x} dx = \frac{xdx + ydy + zdz}{x} = \frac{1}{x} d\left(\frac{x^2 + y^2 + z^2}{2} \right), \qquad (1.79a, b)$$

Thus the potential (1.76a) and integrating factor (1.80a) specify through (1.80b.c):

$$\Psi = \frac{1}{x}: \qquad \Psi \nabla \Phi = \frac{1}{x} \left\{ \partial_x, \partial_y, \partial_z \right\} \frac{x^2 + y^2 + z^2}{2} = \left\{ 1, \frac{y}{x}, \frac{z}{x} \right\} = \vec{X}, \qquad (1.80a\text{--}d)$$

the vector of coefficients (1.80d) ≡ (1.77b). Thus *the three-dimensional vector field (1.77b) is rotational (1.78a) and non-helical (1.78b), and equals the product (1.80b–d) of the second (1.80a) [gradient of the first (1.76a)] scalar potential, that is the integrating factor (solution) of (1.79a, b) the first-order inexact differential (1.77a).* Next is considered a first-order inexact differential with three variables without integrating factor whose vector of coefficients is helical, and involves 3 scalar potentials (subsection 1.5.5).

1.5.5   INEXACT DIFFERENTIAL WITHOUT INTEGRATING FACTOR

The first-order differential in three variables (1.81a) has vector of coefficients (1.81b):

$$dx + \frac{yz}{x} \left(dy + dz \right) = \vec{X} \cdot d\vec{x}: \qquad \vec{X} = \left\{ 1, \frac{yz}{x}, \frac{yz}{x} \right\}. \qquad (1.81a, b)$$

whose curl is not zero (1.82):

$$\nabla \wedge \vec{X} = \begin{vmatrix} \vec{e}_x & \vec{e}_y & \vec{e}_z \\ \partial_x & \partial_y & \partial_z \\ 1 & \dfrac{zy}{x} & \dfrac{zy}{x} \end{vmatrix} = \dfrac{z-y}{x}\vec{e}_x + \dfrac{zy}{x^2}(\vec{e}_y - \vec{e}_z),$$ (1.82)

and also the helicity is not zero (1.83a):

$$H \equiv \vec{X} \cdot (\nabla \wedge \vec{X}) = \dfrac{z-y}{x}; \qquad dx + \dfrac{yz}{x}(dy+dz) = d\Theta + \Psi d\Phi,$$ (1.83a–b)

the differential (1.81a) is of the form (1.83b) involving 3 potentials (1.84a–c):

$$\Theta(x) = x, \qquad \Psi(x,y,z) = \dfrac{yz}{x}, \qquad \Phi(y,z) = y+z,$$ (1.84a–c)

and (1.85a) leads (1.85b,c):

$$\nabla\Theta + \Psi\nabla\Phi = \{\partial_x, \partial_y, \partial_z\}x + \dfrac{yz}{x}\{\partial_x, \partial_y, \partial_z\}(y+z) = \left\{1, \dfrac{yz}{x}, \dfrac{yz}{x}\right\} = \vec{X},$$ (1.85a–d)

to the vector of coefficients (1.85d) ≡ (1.81b). Thus *the three-dimensional helical (1.82;1.83a) vector field (1.81a) equals (1.85a–d) the sum of the gradient of the third (1.84a) [with the second (1.84b) multiplied by the gradient of the first (1.84c)] scalar potential that appear in the general form (1.83b) of the inexact non-integrable first-order differential in three variables (1.81a).* The three cases of first-order differential in three variables (subsections 1.5.1–1.5.2), considered in the preceding partial examples (subsections 1.5.3–1.5.5) are reconsidered (subsection 1.5.6) as a preliminary step to the general proofs (subsections 1.5.7–1.5.9).

1.5.6 IRROTATIONAL/ROTATIONAL AND NON-HELICAL/HELICAL VECTOR FIELDS

Concerning the first-order differential in three variables (1.63), the theory of ordinary differential equations (section IV.3.9) provides proofs of: (i) immediate integrability (1.64a–c) for exact differentials (subsections IV.3.8.1 and IV3.9.1); (ii/iii) existence (1.65a–c) [non-existence (1.66a–c)] integrating factor [subsections IV.3.9.2-IV.3.9.5 (IV.3.9.6-IV.3.9.9)] for inexact differentials. Next is given an alternative proof of the more general case (iii) using linear first-order partial differential equations (subsections 1.5.6–1.5.7) that also provides a method to determine (subsection 1.5.8) the 3 scalar vector potentials. Thus the three cases in the theorem on first-order differentials with three variables (subsection 1.5.1) are illustrated by examples (proved in general) for: (i) an irrotational or conservative vector field that is the gradient of a scalar potential corresponding to an exact differential [subsection 1.5.3 (1.5.7)]; (ii) rotational non-helical velocity field that is with non-zero curl and zero helicity, that corresponds to an inexact differential with an integrating factor, and equals the integrating factor as a second potential multiplied by the gradient of a first potential [subsection 1.5.5 (1.5.8)]; (iii) helical velocity field, that is with non-zero helicity and hence rotational, corresponding to an inexact differential without an integrating factor, and equals the gradient of a third scalar potential added to the second scalar potential multiplied by the gradient of first scalar potential [subsection 1.5.5 (1.5.9)].

1.5.7 IRROTATIONAL OR CONSERVATIVE VECTOR FIELD AS THE GRADIENT OF A SCALAR POTENTIAL

In case I the equivalence of (1.64a) ≡ (1.64b) ≡ (1.64c) must be proved in two opposite directions: (a) that (1.64a) implies (1.64b) that implies (1.64c); (b) that (1.64c) implies (1.64b) that implies (1.64a). Starting with the sequence (a) the exact differential (1.64a) ≡ (1.86a) coincides with (1.86b) [(1.86c)] in the explicit (vector) notation:

$$\vec{X} \cdot d\vec{x} = d\Phi = \frac{\partial \Phi}{\partial x} dx + \frac{\partial \Phi}{\partial y} dy + \frac{\partial \Phi}{\partial x} dz = \nabla\Phi \cdot d\vec{x}, \quad \vec{X} = \nabla\Phi, \qquad (1.86a\text{–}d)$$

that for arbitrary $d\vec{x}$ implies (1.86d) ≡ (1.64b). The following property: *curl of the gradient of twice continuously differentiable function (1.87a) is zero (1.87b)*:

$$\Phi \in C^2\left(|R^3\right): \qquad 0 = \nabla \wedge \left(\nabla\Phi\right) = \nabla \wedge \vec{X}, \qquad (1.87a\text{–}c)$$

proves (1.87c) ≡ (1.64c) and completes the sequence (a). QED. The proof of the property (1.87b) ≡ (1.88a, b):

$$\nabla \wedge \left(\nabla\Phi\right) = \frac{\partial}{\partial x_i}\left(\frac{\partial \Phi}{\partial x_j}\right) - \frac{\partial}{\partial x_j}\left(\frac{\partial \Phi}{\partial x_i}\right) = \frac{\partial^2 \Phi}{\partial x_i \partial x_j} - \frac{\partial^2 \Phi}{\partial x_j \partial x_i} = 0, \qquad (1.88a\text{–}c)$$

is immediate from (1.9a, b) ≡ (1.87a;1.88c).

For the proof of (1.64a–c) in the opposite sequence (b) from (1.64c) to (1.64a) the starting point (subsection III.5.7.7) is the **Stokes or curl loop theorem**: *the circulation of the of a continuously differentiable vector field (1.89a) along a closed regular curve or loop L equals (1.89b) the flux of the curl across a regular surface D_2 supported on the loop*:

$$\vec{X} \in C^1\left(|R^3\right): \qquad \int_L \vec{X} \cdot d\vec{x} = \int_{D_2} \left(\nabla \wedge \vec{X}\right) \cdot d\vec{S}. \qquad (1.89a, b)$$

In the case of an irrotational vector field (1.90a) ≡ (1.64c) the flux of its curl is zero across any regular surface (1.90b) and thus the circulation is also zero along any closed regular curve (1.90c), implying that the integrand is an exact differential (1.90d):

$$\nabla \wedge \vec{X} = 0: \qquad\qquad 0 = \int_{D_2} \left(\nabla \wedge \vec{X} \right) \cdot d\vec{S} = \int_L \vec{X} \cdot d\vec{x} = d\Phi. \qquad (1.90a\text{--}d)$$

From (1.90c) $\equiv$ (1.90d) follows (1.64a), that by (1.86a–c) is equivalent to (1.86d) $\equiv$ (1.64b). This completes the proof of (1.64a–c) in the second reverse direction. QED. Case II is a rotational non-helical vector field (subsection 1.5.8).

1.5.8   ROTATIONAL NON-HELICAL VECTOR FIELD AND TWO SCALAR POTENTIALS

The case II of a first-order differential with three variables with an integrating factor (1.65a) = (1.91a) is equivalent to (1.91b) [(1.91c)] in the explicit (vector) notation:

$$\vec{X} \cdot d\vec{x} = \Psi d\Phi = \Psi \left(\frac{\partial \Phi}{\partial x} dx + \frac{\partial \Phi}{\partial y} dy + \frac{\partial \Phi}{\partial z} dz \right) = \Psi \nabla \Phi \cdot d\vec{x}, \quad \vec{X} = \Psi \nabla \Phi, \quad (1.91a\text{--}d)$$

and implies (1.91d) $\equiv$ (1.65b). To prove (1.65c) is needed the following property: *the curl of the product of a scalar by a vector is given by (1.92).*

$$\nabla \wedge \left(\Psi \vec{A} \right) = \Psi \nabla \wedge \vec{A} + \nabla \Psi \wedge \vec{A}, \qquad (1.92)$$

that is equivalent in index notation to (1.93) $\equiv$ (1.92):

$$\frac{\partial}{\partial x_i} \left(\Psi \vec{A}_j \right) - \frac{\partial}{\partial x_j} \left(\Psi \vec{A}_i \right) = \Psi \left(\frac{\partial A_j}{\partial x_i} - \frac{\partial A_i}{\partial x_j} \right) + A_j \frac{\partial \Psi}{\partial x_i} - A_i \frac{\partial \Psi}{\partial x_j}. \qquad (1.93)$$

In the particular case when the vector is the gradient of a scalar (1.94a) follows (1.94b):

$$\vec{A} = \nabla \Phi: \qquad \nabla \wedge \left(\Psi \nabla \Phi \right) = \Psi \nabla \wedge \left(\nabla \Phi \right) + \nabla \Psi \wedge \nabla \Phi = \nabla \Psi \wedge \nabla \Phi, \qquad (1.94a\text{--}c)$$

with (1.87b) implying (1.94c). Thus the vector field (1.91d) has: (i) non-zero curl (1.94b,c); (ii) zero helicity (1.95a) since the mixed vector product (1.95b,c):

$$H = \vec{X} \cdot \left(\nabla \wedge \vec{X} \right) = \Psi \nabla \Phi \cdot \left(\nabla \Psi \wedge \nabla \Phi \right) = 0, \qquad (1.95a\text{--}c)$$

involves two coincident vectors. The proof of (1.95c) $\equiv$ (1.65c) completes the sequence from (1.65a) to (1.65c). QED.

For the opposite sequence the condition of zero helicity $(1.65c) \equiv (1.96a)$ leads to the characteristic system $(1.96b)$:

$$0 = \vec{X} \cdot \left(\nabla \wedge \vec{X} \right): \qquad \frac{dx}{\left(\nabla \wedge \vec{X} \right)_x} = \frac{dy}{\left(\nabla \wedge \vec{X} \right)_y} = \frac{dz}{\left(\nabla \wedge \vec{X} \right)_z}, \qquad \text{(1.96a, b)}$$

that (1.63) is equivalent to:

$$\frac{dx}{\partial Z/\partial y - \partial Y/\partial z} = \frac{dy}{\partial X/\partial z - \partial Z/\partial x} = \frac{dz}{\partial Y/\partial x - \partial X/\partial y}. \qquad \text{(1.96c)}$$

If (U, Φ) are two solutions of $(1.96c)$ the general integral is $(1.97d)$ $[(1.97b)]$ involving an arbitrary differentiable function $(1.97c)$ $[(1.97a)]$ of one (two) variable(s):

$$f \in \mathcal{D}\left(\mathbb{R}^2 \right): \qquad f\left(U, \Phi \right) = 0 \Leftrightarrow g \in D\left(\mathbb{R} \right): \qquad U = g\left(\Phi \right). \qquad \text{(1.97a–d)}$$

From $(1.97d)$ follows $(1.98a, b)$ and using $(1.98d)$ leads to $(1.98c)$:

$$\vec{X} \cdot d\vec{x} = dU = \frac{dg}{d\Phi} d\Phi = \Psi d\Phi \qquad \Psi \equiv \frac{dg}{d\Phi}. \qquad \text{(1.98a–d)}$$

The coincidence $(1.98c) \equiv (1.65a)$ also proves $(1.91d)$, completing the sequence from $(1.65c)$ to $(1.65a)$. The remaining case III is that of a helical vector field (subsection 1.5.9).

1.5.9   HELICAL VECTOR FIELD AND THREE SCALAR POTENTIALS

The theorem stated in the subsection 1.5.1 has important applications in vector field theory (subsections IV.3.9.8-IV.3.9.9 and 1.5.8) and in thermodynamics (subsections IV.3.9.10, IV.5.5.4-VI.5.5.25 and Section 2.1). The proof of the most general case (iii) requires showing that three scalar potentials Φ, Ψ, Θ exist, satisfying $(1.66a)$ so that $(1.66b) \equiv (1.66c) \equiv (1.66a)$ follow as equivalent re-statements. Starting with the vector $\vec{X}$, its curl is given by $(1.99a)$:

$$\nabla \wedge \vec{X} = \nabla \wedge \left(\nabla \Theta + \Psi \nabla \Phi \right) = \nabla \Psi \wedge \nabla \Phi \equiv \vec{Y}, \qquad \text{(1.99a, b)}$$

where were used the properties $(1.87b; 1.94c)$ leading to $(1.99b)$. From $(1.99b)$ follow $(1.100a, b)$:

$$\vec{Y} \cdot \nabla \Phi = \left(\nabla \Psi \wedge \nabla \Phi \right) \cdot \nabla \Phi = 0 = \nabla \Psi \cdot \left(\nabla \Psi \wedge \nabla \Phi \right) = \nabla \Psi \cdot \vec{Y}. \qquad \text{(1.100a, b)}$$

Thus two scalar potentials Φ, Ψ satisfy the same $(1.101a, b)$ linear first-order p.d.e. without forcing term $(1.101c)$:

$$F \equiv \Psi, \Phi: \qquad 0 = \vec{Y} \cdot \nabla F = \left(\nabla \wedge \vec{X} \right) \cdot \nabla F. \qquad \text{(1.101a–c)}$$

Because the integrability condition is not met (1.66c):

$$0 \neq \vec{X} \cdot \vec{Y} = \vec{Y} \cdot \nabla\Theta + \Psi\vec{Y} \cdot \nabla\Phi = \left(\nabla \wedge \vec{X}\right) \cdot \nabla\Theta, \qquad (1.102)$$

the third scalar potential Θ satisfies an Equation (1.102), that is distinct from that (1.101b,c) satisfied by Φ, Ψ.

Both (1.101a–c) and (1.102) are linear p.d.e.s with the same vector $\vec{Y}$ of coefficients (1.103):

$$\vec{Y} \equiv \nabla \wedge \vec{X} = \nabla\Psi \wedge \nabla\Phi = \vec{e}_x A + \vec{e}_y B + \vec{e}_z C, \qquad (1.103)$$

where $\vec{X}$ is given by (1.66b) $\equiv$ (1.104a, b):

$$\vec{X} = \vec{e}_x a + \vec{e}_y b + \vec{e}_z c: \qquad \{a, b, c\} = \{\partial_x, \partial_y, \partial_z\}\Theta + \Psi\{\partial_x, \partial_y, \partial_z\}\Phi. \qquad (1.104a, b)$$

From (1.103) and (1.104b) follow the components of the vector of coefficients $\vec{Y}$ in terms of the three scalar potentials in (1.105a–c):

$$A = \left(\nabla \wedge \vec{X}\right)_x = \partial_y c - \partial_z b = \partial_y \partial_z \Theta + \partial_y\left(\Psi\partial_z\Phi\right) - \partial_z\partial_y\Theta - \partial_z\left(\Psi\partial_y\Phi\right)$$
$$= \left(\partial_y\Psi\right)\left(\partial_z\Phi\right) - \left(\partial_z\Psi\right)\left(\partial_y\Phi\right), \qquad (1.105a)$$

$$B = \left(\nabla \wedge \vec{X}\right)_y = \partial_z a - \partial_x c = \left(\partial_z\Psi\right)\left(\partial_x\Phi\right) - \left(\partial_x\Psi\right)\left(\partial_z\Phi\right), \qquad (1.105b)$$

$$C = \left(\nabla \wedge \vec{X}\right)_z = \partial_x b - \partial_y a = \left(\partial_x\Psi\right)\left(\partial_y\Phi\right) - \left(\partial_y\Psi\right)\left(\partial_x\Phi\right). \qquad (1.105c)$$

Since (1.101b,c) [(1.102)] is a linear first-order p.d.e. without (with) forcing term, its characteristic equations are (1.28) [(1.57)] leading to the first two (all three) equalities in (1.106):

$$\frac{dx}{A} = \frac{dy}{B} = \frac{dz}{C} = \frac{d\chi}{Aa + Bb + Zc}; \qquad (1.106)$$

thus the first two scalar potentials Φ, Ψ are two independent solutions of the first two equalities in (1.106), and the third scalar potential Θ is a third independent solution, obtained when the third equality is included. This shows that the three scalar potentials Φ, Ψ, Θ exist, proving the statement III in (1.66a–c). The preceding proof of existence also shows the method by which the three functions can be determined: *let (1.66a) be an inexact non integrable first-order differential in three variables with vector of coefficients (1.104a, b). The curl of the vector of coefficients (1.103; 1.105a–c) leads to the system of Equations (1.106); the third Θ (first two Ψ and Φ) scalar potentials is (are) a solution (two independent solutions) of the set of three (two) equalities (1.106). This specifies the decomposition of the vector field into three* *scalar potentials (1.66b)*. Two (three) cases of exact and inexact differentials are given in the example 10.2, with the inexact differentials always having (having or not) an integrating factor. Next is given an example of a first-order differential with three variables without integrating factor, illustrating the determination of the three scalar potentials (subsection 1.5.10).

1.5.10   EXISTENCE AND DETERMINATION OF THE THREE SCALAR POTENTIALS

As an example consider the differential of first degree in three variables (1.107a) with vector of coefficients (1.107b):

$$ydx + zdy + xdz = \vec{X} \cdot d\vec{x}: \qquad\qquad \vec{X} = \{y, z, x\}. \qquad (1.107a, b)$$

The curl (11.49a) is not zero:

$$\vec{Y} = \nabla \wedge \vec{X} = \begin{vmatrix} \vec{e}_x & \vec{e}_y & \vec{e}_z \\ \partial_x & \partial_y & \partial_z \\ y & z & x \end{vmatrix} = -\vec{e}_x - \vec{e}_y - \vec{e}_z, \qquad (1.108)$$

and coincides with minus the unit radial vector $(1.108) \equiv (1.109a)$ implying that the helicity is also non-zero (1.109b):

$$\vec{Y} = \{-1, -1, -1\}, \qquad\qquad H \equiv \vec{X} \cdot \vec{Y} = -y - z - x; \qquad (1.109a, b)$$

it follows that the integrability condition (1.66c) is not met and three scalar potentials are needed.

The three components (1.109a) of the curl $(1.108) \equiv (1.109a)$ are all equal (1.110a) and lead $(1.28) \equiv (1.108)$ to the characteristic system (1.110a–b):

$$A = B = C = -1: \qquad dx = dy = dz = \frac{d\chi}{x + y + z}, \qquad (1.110a\text{--}d)$$

where: (i) the first two equalities (1.110b,c) have independent solutions (1.111a, b):

$$x - y = \Phi, \qquad\qquad y - z = \Psi, \qquad (1.111a, b)$$

that determine two of the potentials; (ii) the third potential follows from the integration of the exact differential (1.112a) that results from (1.66a; 1.107a):

$$d\Theta = \vec{X} \cdot d\vec{x} - \Psi d\Phi = ydx + zdy + xdz - (y - z)(dx - dy)$$
$$= zdx + ydy + xdz = d\left(xz + \frac{y^2}{2}\right), \qquad \Theta = xz + \frac{y^2}{2}, \qquad (1.112a\text{--}e)$$

leading (1.112b–d) to the third potential (1.112e). This *specifies the decomposition (1.113a) of the inexact non-integrable first-order differential in three variables (1.107a) in terms of the three scalar potentials (1.111a, b;1.112e):*

$$ y\,dx + z\,dy + x\,dz = d\left(xz + \frac{y^2}{2} \right) + (y - z)\,d(x - y), \qquad (1.113a) $$

$$ \{\partial_x, \partial_y, \partial_z\}\left(xz + \frac{y^2}{2} \right) + (y - z)\{\partial_x, \partial_y, \partial_z\}(x - y) = \{y, z, x\}, \qquad (1.113b) $$

and equivalently the representation (1.113b) of the helical (1.109b) vector field (1.107b) as the sum of the gradient of the third scalar potential (1.112e) with the second potential (1.111b) multiplied by the gradient of the first potential (1.111a).

Two alternatives for the general representation of a three-dimensional vector field are the use of: (i) three scalar potentials (subsections 1.5.1–1.5.10); (ii) one scalar and one vector potential (subsections 15.11–1.5.15). There are particular cases (subsection 1.5.16) of both (i) and (ii), and the simplest is: an irrotational (solenoidal) vector field, that is with zero curl (divergence), equals the gradient (curl) of a scalar (vector) potential, that satisfies a scalar (vector) Poisson equation forced by the divergence (curl) of the vector field [subsection 1.5.11 (1.5.12)].

1.5.11   SCALAR POISSON EQUATION FORCED BY THE DIVERGENCE OF A VECTOR FIELD

*In case I of a conservative or irrotational vector field (1.64c) ≡ (1.114a), that equals the gradient of a scalar potential (1.64b) ≡ (1.114b) the potential satisfies a **scalar Poisson equation** (III.8.12a) ≡ (1.114c) forced by the divergence of the vector field:*

$$ \nabla \wedge \vec{X} = 0 \Leftrightarrow \vec{X} = \nabla\Phi: \qquad \nabla \cdot \vec{X} = \nabla \cdot (\nabla\Phi) = \nabla^2\Phi, \qquad (1.114a\text{–}c) $$

whose solution (III.8.13a)≡(1.115) determines the potential:

$$ \Phi(\vec{x}) = \frac{1}{4\pi} \int_{D_3} \frac{\nabla \cdot \vec{X}(\vec{y})}{|\vec{x} - \vec{y}|}\, d^3\vec{y}. \qquad (1.115) $$

In (1.114c) was used the property that *the divergence of the gradient of a twice differentiable function (1.116a) equals the **Laplace operator*** (1.116b):

$$ \Phi \in \mathcal{D}^2(\mathbb{R}^3): \qquad \nabla \cdot (\nabla\Phi) = \sum_{n=1}^{N} \frac{\partial}{\partial x_n}\left(\frac{\partial\Phi}{\partial x_n} \right) = \sum_{n=1}^{N} \frac{\partial\Phi}{\partial x_n^2} = \nabla^2\Phi. \qquad (1.116a, b) $$

The reverse case is a solenoidal vector field, that is with zero divergence, specified by the curl of a vector potential, that satisfies a vector Poisson equation forced by minus the curl of the vector field (subsection 1.5.12).

1.5.12   Vector Poisson Equation Forced by the Curl of a Vector Field

A **solenoidal vector field,** that is with zero divergence (1.117a) equals the curl
(1.117b) of a **vector potential** leading to (1.117c) by taking the curl:

$$\nabla \cdot \vec{X} = 0: \qquad \vec{X} = \nabla \wedge \vec{A}, \qquad \nabla \wedge \vec{X} = \nabla \wedge \left(\nabla \wedge \vec{A} \right). \qquad (1.117a\text{–}c)$$

Using the identity (III.6.48a) $\equiv$ (1.118) for the Laplacian of a vector:

$$\nabla^2 \vec{A} = \nabla \left(\nabla \cdot \vec{A} \right) - \nabla \wedge \left(\nabla \wedge \vec{A} \right), \qquad (1.118)$$

leads from (1.117c) to (1.119):

$$\nabla^2 \vec{A} - \nabla \left(\nabla \cdot \vec{A} \right) = -\nabla \wedge \vec{X}. \qquad (1.119)$$

The vector potential is specified by its curl (1.117b) to within its divergence, which
may be set to zero by the **gauge condition** (1.120a):

$$\nabla \cdot \vec{A} = 0: \qquad \nabla^2 \vec{A} = -\nabla \wedge \vec{X}, \qquad (1.120a, b)$$

leading to a **vector Poisson equation** (III.8.91a) $\equiv$ (1.120b) forced by minus the curl
of the vector field whose solution is (II.8.92) $\equiv$ (1.121):

$$\vec{A}(x) = -\frac{1}{4\pi} \int_{D_3} \frac{\nabla \wedge \vec{X}(\vec{y})}{|\vec{x} - \vec{y}|} d^3\vec{y}, \qquad (1.121)$$

Thus *a solenoidal vector field, that is with zero divergence (1.117a) equals the curl
of a vector potential (1.117b); imposing a gauge condition that the divergence of the
vector potential is zero (1.120a) implies that the vector potential satisfies a vector
Poisson equation (1.120b) forced by minus the curl of the vector field, whose solution
is (1.121).*

1.5.13   Potential Vector Field and Laplace Equation

Combining the results of subsections 1.5.11 and 1.5.12 it follows that a **potential
vector field,** *that is whose curl (divergence) are both zero (1.122A) [(1.122b)] equals
both the gradient (curl) of a scalar (vector) potential (1.122c) [(1.122d)], that satisfies a scalar (vector) Laplace equation (1.122e) [(1.122f)]:*

$$\nabla \wedge \vec{X} = 0 = \nabla \cdot \vec{X}: \qquad \nabla \Phi = \vec{X} = \nabla \wedge \vec{A}, \qquad \nabla^2 \Phi = 0 = \nabla^2 \vec{A}. \qquad (1.122a\text{–}f)$$

The opposite general case of a vector field that is neither irrotational nor solenoidal
can be decomposed in the sum of an irrotational and a solenoidal component (subsection 1.5.14).

1.5.14 SCALAR AND VECTOR POTENTIALS FOR A GENERAL VECTOR FIELD

It is shown next that a general continuously differentiable vector field (1.123a) can be decomposed into the sum of an irrotational (and solenoidal) part, that is equals the sum of the gradient (curl) of a scalar (vector) potential (1.123b):

$$\vec{X} \in C^1\left(|R^3\right): \qquad\qquad \vec{X} = \nabla\Phi + \nabla \wedge \vec{A}, \qquad\qquad (1.123a, b)$$

Taking the divergence of (1.123b) and bearing in mind that the divergence of the curl is zero (1.124a) follows (1.124b) $\equiv$ (1.114c):

$$\nabla \cdot \left(\nabla \wedge \vec{A}\right) = 0: \qquad\qquad \nabla \cdot \vec{X} = \nabla \cdot \left(\nabla\Phi\right) = \nabla^2\Phi, \qquad (1.124a, b)$$

that the scalar potential is the solution of a Poisson equation forced by the divergence of the vector field, whose solution is (1.115). The property that *the divergence of the curl of a twice differentiable vector field (1.125a) is zero (1.124a)* follows from (1.125b,c):

$$\vec{A} \in \mathcal{D}^2\left(|R^3\right): \qquad \nabla \cdot \left(\nabla \wedge \vec{A}\right) = \begin{vmatrix} \partial_1 & \partial_2 & \partial_3 \\ \partial_1 & \partial_2 & \partial_3 \\ A_1 & A_2 & A_3 \end{vmatrix} = 0, \qquad (1.125a\text{–}c)$$

Taking the curl of (1.123b) and bearing in mind that the curl of the gradient is zero (1.87b) follows (1.126a):

$$\nabla \wedge \vec{X} = \nabla \wedge \left(\nabla \wedge \vec{A}\right) = -\nabla^2\vec{A} + \nabla\left(\nabla \cdot \vec{A}\right), \qquad (1.126a, b)$$

using also the identity (III.648a) $\equiv$ (1.118) $\equiv$ (1.126b). Imposing the gauge condition (1.120a) $\equiv$ (1.127a) of zero divergence of the vector potential it follows that the vector potential satisfies (1.127b) $\equiv$ (1.120b) a vector Poisson equation forced by minus curl of the vector field, whose solution is (1.129):

$$\nabla \cdot \vec{A} = 0: \qquad\qquad \nabla^2\vec{A} = -\nabla \wedge \vec{X}. \qquad\qquad (1.127a, b)$$

Thus a continuously differentiable three-dimensional vector field (1.123a) can be represented (1.123b) $\equiv$ (1.128a–c):

$$\vec{X} = \vec{X}_1 + \vec{X}_2: \qquad\qquad \vec{X}_1 = \nabla\Phi, \qquad \vec{X}_2 = \nabla \wedge \vec{A}, \qquad (1.128a\text{–}c)$$

as the sum of an irrotational (1.129a, b) [and a solenoidal (1.129c, d)] vector field components:

$$\nabla \cdot \vec{X} = \nabla \cdot \vec{X}_1, \qquad \nabla \wedge \vec{X}_1 = 0 = \nabla \cdot \vec{X}_2, \qquad \nabla \wedge \vec{X} = \nabla \wedge \vec{X}_2, \qquad (1.129a\text{–}d)$$

that is as the sum of the gradient (curl) of a scalar Φ (vector $\vec{A}$) potential, that satisfies a scalar (1.124b) [vector (1.127b)] Poisson equation forced by the divergence (minus the curl) of the vector field [assuming the gauge condition (1.127a) of zero divergence of the vector potential] whose solution is (1.115) [(1.121)]. If the gauge condition is not met the vector potential satisfies a double **curl vector equation** *(1.126a, b) forced by the curl of the vector field.* There are two general representations (subsections IV.3.9.8-iv.3.9.9) of a continuously differentiable three-dimensional vector field by three scalar potentials: (i) the **Euler potentials** (1.66b) already considered (subsections 1.5.1,1.5.5,1.5.9); (ii) the Clebsch potentials considered next (subsection 1.5.15) via scalar and vector potentials (subsections 1.5.11–1.5.14).

1.5.15 THREE ALTERNATIVE SCALAR EULER OR CLEBSCH POTENTIALS

If the vector potential is continuously differentiable (1.130a) then its curl, equals (1.99b) the outer vector product of two scalar potentials (1.130b):

$$\vec{A} \in \mathcal{D}\left(|\mathbb{R}^3\right): \qquad\qquad \nabla \wedge \vec{A} = \nabla\alpha \wedge \nabla\beta. \qquad (1.130a, b)$$

Substituting (1.130b) in (1.123b) requires (1.130a) that the vector field be continuously differentiable (1.131a) for the representation (1.131b) by **three scalar Clebsch potentials**:

$$\vec{X} \in \mathcal{D}\left(|\mathbb{R}^3\right): \qquad\qquad \vec{X} = \nabla\Xi + \nabla\alpha \wedge \nabla\beta. \qquad (1.131a, b)$$

Two of the Clebch potentials satisfy the linear unforced partial differential equations with only first-order partial derivatives (1.132a, b):

$$\left(\vec{X} - \nabla\Xi\right)\cdot\nabla\alpha = 0 = \nabla\beta\cdot\left(\vec{X} - \nabla\Xi\right), \qquad (1.132a, b)$$

and thus are solutions (1.28) of the characteristic system (1.133):

$$\frac{dx}{X - \partial\Xi/\partial x} = \frac{dy}{Y - \partial\Xi/\partial y} = \frac{dz}{Z - \partial\Xi/\partial y}, \qquad (1.133)$$

that involves the other potential. It has been shown that *a continuously differentiable three-dimensional vector field (1.131a) has a representation (1.131b) in terms of three Clebsch potentials, as the sum of the gradient of the first potential Ξ with the outer vector product of the gradients of the second α and third β potentials, that satisfy the characteristic equations (1.133).* The eight cases of three-dimensional vector fields are summarized next (subsections 1.5.16).

1.5.16   EIGHT CASES OF THREE-DIMENSIONAL VECTOR FIELDS

The preceding results (subsections (1.5.1–1.5.15) can be collected in *eight cases of three-dimensional vector fields (Table 1.2), with one example of each in the sequel. There are four representations (Figure 1.4) of a three-dimensional continuously differentiable vector field in terms of scalar potentials: (II) an irrotational (subsections 1.5.3,1.5.7 and 1.5.18) vector field (1.64c), corresponds to the exact differential (1.64a) of a scalar potential (1.64b) that is constant on surfaces orthogonal to the vector field (Figure 1.4a); (VII) a rotational non-helical (subsections 1.5.4,1.5.8 and 1.5.20) vector field with zero helicity (1.65c) corresponds to an inexact differential with an integrating factor (1.65a), and dividing the vector field by the integrating factor (1.65b) leads to a vector orthogonal to the equipotential surfaces (Figure 1.4b); (VI) a helical (subsections 1.5.5,1.5.9 and 1.5.21) vector field with non-zero helicity (1.66c), corresponds to an inexact differential without integrating factor (1.66a), and the three Euler potentials and to obtain equipotential surfaces (Figure 1.4c) it is necessary (1.66b) to subtract the gradient of a scalar and then divide by another scalar; (V) an alternative representation (subsections I.5.15 and 1.5.23) in terms of three scalar Clebsch potentials (1.131b) adds (Figure 1.4d) the normal to one surface to the tangent vector to the intersection of two other surfaces.*

The alternative (IV) general representation (subsections 1.5.14 and 1.5.22) in terms of scalar and vector potentials (1.123b) adds to the curl of a vector potential (Figure 1.5a) the normal to a surface (Figure 1.5b). Taking as reference the representation of a continuously differentiable vector field (1.123a) by scalar and vector potentials (1.123b) two particular cases are [II (III)] the vector field is irrotational (solenoidal) that is [subsections 1.5.3, 1.5.7 and 1.5.18 (1.5.12 and 1.5.19)] has zero curl (1.64c) [divergence (1.117a)] then it is represented only by the gradient (1.64b) [curl (1.117b)] of a scalar (1.64a) $\equiv$ (1.134a) [vector (1.134b)] potential:

$$\vec{X} \cdot d\vec{x} = \nabla\Phi \cdot d\vec{x} = d\Phi, \qquad \vec{X} \cdot d\vec{x} = \left(\nabla \wedge \vec{A}\right) \cdot d\vec{x}. \qquad (1.134a, b)$$

The solenoidal vector field (1.117a) can be represented (VIII) by (1.117b;1.130b) the vector outer product of the gradients of two Clebsch potentials. The combination (I) of irrotational (1.64c) $\equiv$ (1.122a) [an solenoidal (1.117a) $\equiv$ (1.122b)] cases [II(III)] is the simplest case (I) of a potential field (subsections 1.5.13 and 1.5.17) that has both scalar (1.122c) [and vector 1.122d)] potentials.

1.5.17   POTENTIAL VECTOR FIELD WITH SCALAR OR VECTOR POTENTIALS

The simplest three-dimensional vector field is potential (subsections 1.5.13) for example, (1.135a) that has zero curl (1.135b) and divergence (1.135c):

$$\vec{X}_1 = \{x, y, -2z\}: \qquad \nabla \wedge \vec{X}_1 = 0 = \nabla \cdot \vec{X}_1. \qquad (1.135a\text{–}c)$$

TABLE 1.2
Representations of Three-Dimensional Vector Field

Case	Designation	Vector field Properties	Representation Potentials	Names	Section
I	Potential	$\nabla_\wedge \vec{X} = 0 = \nabla \cdot \vec{X}$	$\nabla\Phi = \vec{X} = \nabla_\wedge \vec{A}$	scalar or vector	1.5.13,1.5.17
II	Irrotational	$\nabla_\wedge \vec{X} = 0 \neq \nabla \cdot \vec{X}$	$\vec{X} = \nabla\Phi$	Scalar	1.5.3,1.5.7,1.5.11,1.5.18
III	Solenoidal	$\nabla_\wedge \vec{X} \neq 0 = \nabla \cdot \vec{X}$	$\vec{X} = \nabla_\wedge \vec{A}$	Vector	1.5.12,1.5.19
IV	General	$\nabla_\wedge \vec{X} \neq 0 \neq \nabla \cdot \vec{X}$	$\vec{X} = \nabla\Phi + \nabla_\wedge \vec{A}$	scalar and vector	1.5.14,1.5.20
V	General	$\nabla_\wedge \vec{X} \neq 0 \neq \nabla \cdot \vec{X}$	$\vec{X} = \nabla\Xi + \nabla\alpha_\wedge \nabla\beta$	3 Clebsch scalar	1.5.15,1.5.23
VI	Helical	$H \equiv \left(\nabla_\wedge \vec{X}\right)\cdot \vec{X} \neq 0$	$\vec{X} = \nabla\Theta + \Psi\nabla\Phi$	2 Euler scalar	1.5.5,1.5.9,1.5.10,1.5.22
VII	Rotational non-helical	$\nabla_\wedge \vec{X} \neq 0 = H$	$\vec{X} = \Psi\nabla\Phi$	2 Euler scalar	1.5.4,1.5.5,1.5.8,1.5.21
VIII	Solenoidal	$\nabla \cdot \vec{X} = 0 \neq \nabla \cdot \vec{X}$	$\nabla\alpha_\wedge \nabla\beta$	2 Clebsch scalar	E10.3

Note: Eight cases of the representation of a three-dimensional vector field by scalar and/or vector potentials.

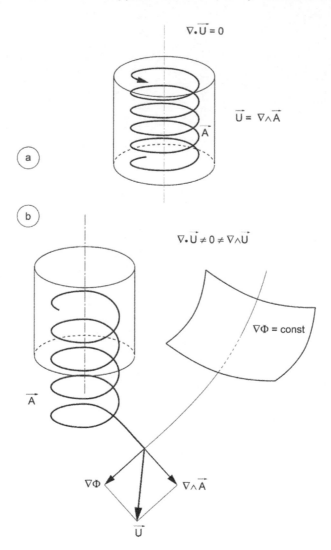

FIGURE 1.5 A solenoidal, divergence-free or equivoluminar vector field is (a) the curl of a vector potential. A general vector field (b) which is neither solenoidal nor irrotational can be decomposed into the sum of an solenoidal (irrotational) vector field, and thus represented by the sum of curl (gradient) of a vector (scalar) potential.

Since the vector field is irrotational (1.122a) ≡ (1.135b) or conservative it must equal (1.122e) ≡ (1.136b) the gradient of a scalar potential specified by an exact differential (1.64b) ≡ (1.136b) with vector of coefficients (1.135a):

$$\vec{X}_1 = \nabla\Phi_1; \quad \vec{X}_1 \cdot d\vec{x} = xdx + ydy - 2zdz$$
$$= d\left(\frac{x^2 + y^2}{2} - z^2\right); \quad \Phi_1(x,y,z) = \frac{x^2 + y^2}{2} - z^2: \qquad (1.136a\text{–}d)$$

from (1.136c) ≡ (1.136d) follows the scalar potential (1.137a) whose gradient (1.137a) ≡ (1.137b):

$$\nabla\Phi_1 = \left\{\frac{\partial}{\partial x}, \frac{\partial}{\partial y}, \frac{\partial}{\partial z}\right\}\left(\frac{x^2+y^2}{2} - z^2\right) = \{x, y, -2z\} = \vec{X}_1, \qquad (1.137a\text{--}c)$$

specifies the vector field (1.137d) ≡ (1.135a) ≡ (1.136a).

The potential vector field (1.135a) is also solenoidal (1.122b) ≡ (1.135c) and thus must equal the curl of a vector potential (1.122d) ≡ (1.138a, b) satisfying the gauge condition (1.120a) ≡ (1.138c) of zero divergence:

$$\nabla \wedge \vec{A}_1 = \vec{X}_1 = \{x, y, -2z\}, \qquad \nabla \cdot \vec{A}_1 = 0. \qquad (1.138a\text{--}c)$$

Since the vector field (1.135a) is a linear function of the Cartesian coordinates (x,y,z), the vector potential must be a quadratic function of (x,y,z). The squares of the coordinates (x^2, y^2, z^2) vanish when the curl operator is applied, and hence each component of the vector potential must be a linear combination of the cross-products (xy,xz,yz) of two Cartesian coordinates (1.139):

$$\begin{bmatrix} A_{1x} \\ A_{1y} \\ A_{1z} \end{bmatrix} = \begin{bmatrix} a_{11} & a_{12} & a_{13} \\ a_{21} & a_{22} & a_{23} \\ a_{31} & a_{32} & a_{33} \end{bmatrix}\begin{bmatrix} xy \\ xz \\ yz \end{bmatrix}. \qquad (1.139)$$

The components of the constant matrix a_{ij} in (1.139) must satisfy (1.138a, b) leading to (1.140a–c):

$$x = X_{1x} = \left(\nabla \wedge \vec{A}_1\right)_x = \partial_y A_{1z} - \partial_z A_{1y} = a_{31}x + a_{33}z - a_{22}x - a_{23}y, \qquad (1.140a)$$

$$y = X_{1y} = \left(\nabla \wedge \vec{A}_1\right)_y = \partial_z A_{1x} - \partial_x A_{1z} = a_{12}x + a_{13}y - a_{31}y - a_{32}z, \qquad (1.140b)$$

$$-2z = X_{1z} = \left(\nabla \wedge \vec{A}_1\right)_z = \partial_x A_{1y} - \partial_y A_{1x} = a_{21}y + a_{22}z - a_{11}x - a_{13}z. \qquad (1.140c)$$

The identities (1.140a/b/c) show that six components of the matrix a_{ij} are zero (1.141a, b/c, d/e,f) and the other three satisfy (1.141g/h/i):

$$a_{23} = 0 = a_{33}, \qquad a_{12} = 0 = a_{32}, \qquad a_{11} = 0 = a_{21},$$
$$1 = a_{31} - a_{22}, \qquad 1 = a_{13} - a_{31}, \qquad 2 = a_{13} - a_{22}. \qquad (1.141a\text{--}i)$$

From (1.139) with use of (1.141a–f) follows (1.142a–c):

$$\nabla \cdot \vec{A}_1 = \partial_x A_{1x} + \partial_y A_{1y} + \partial_z A_{1z}$$
$$= a_{11}y + a_{12}z + a_{21}x + a_{23}z + a_{32}x + a_{33}y = 0, \qquad (1.142a\text{--}c)$$

confirming that the vector potential satisfies the gauge condition of zero divergence (1.142c) $\equiv$ (1.138c). The three equations (1.141g,h,i) are redundant since the sum of the first two (1.141g,h) equals the third (1.141i). Thus one of the three values a_{13}, a_{22}, a_{31} can be chosen arbitrarily, implying that the vector potential is not unique (subsection 1.5.25). A unique solution is obtained choosing (1.143a) that implies (1.143b,d) by (1.141g,h,c) leading by (1.139) to the vector potential (1.143d):

$$ a_{31} = 1, \quad a_{22} = 0, \quad a_{13} = 2: \qquad \vec{A}_1 = \{2yz, 0, xy\}. \qquad (1.143a\text{–}d) $$

It can be confirmed that the curl (1.144a) of the vector potential (1.143d):

$$ \nabla \wedge \vec{A}_1 = \begin{bmatrix} \vec{e}_x & \vec{e}_y & \vec{e}_z \\ \partial_x & \partial_y & \partial_z \\ 2yz & 0 & xy \end{bmatrix} = \vec{e}_x x + \vec{e}_y y - \vec{e}_z 2z = \vec{X}_1, \qquad (1.144a\text{–}c) $$

equals the vector field (1.144b) $\equiv$ (1.144c) $\equiv$ (1.135a).

The scalar (1.136d) [vector (1.143d)] potentials both satisfy the Laplace equation (1.145a, b) [(1.145c, d)] since the vector field (1.135a) is potential (1.122e) [(1.122f)]:

$$ \nabla^2 \Phi_1 = \left\{ \frac{\partial^2}{\partial x^2} + \frac{\partial^2}{\partial y^2} + \frac{\partial^2}{\partial z^2} \right\} \left(\frac{x^2 + y^2}{2} - z^2 \right) = 0, \qquad (1.145a, b) $$

$$ \nabla^2 \vec{A}_1 = \left\{ \frac{\partial^2}{\partial x^2} + \frac{\partial^2}{\partial y^2} + \frac{\partial^2}{\partial z^2} \right\} [2yz, 0, xy] = 0. \qquad (1.145c, d) $$

It has been shown that *the three-dimensional potential vector field (1.135a), that is both irrotational (1.135b) [and solenoidal (1.135c)], equals the gradient (1.137a–c) [curl (1.144a–c)] of the scalar (1.136b) [vector (1.143d)] potential that satisfies a scalar (1.145a, b) [vector (1.145c, d)] Laplace equation.* In the case of a potential vector field (1.122a, b) $\equiv$ (1.135b,c) when the representations by scalar (1.122c) $\equiv$ (1.137a–d) [vector (1.122d) $\equiv$ (1.143d;1.144a–c)] potentials both exist, the former is simpler (the latter may be considered for completeness). The next case (subsection 1.5.18) is an irrotational non-solenoidal vector field that equals the gradient of a scalar potential and has no vector potential.

1.5.18   IRROTATIONAL VECTOR FIELD AND ONE SCALAR POTENTIAL

The three-dimensional vector field (1.146a) is (subsection 1.5.11) irrotational (1.64c) $\equiv$ (1.114a) $\equiv$ (1.146b) and non-solenoidal (1.146c):

$$ \vec{X}_2 = \{x^a, y^b, z^a\}: \qquad \nabla \wedge \vec{X}_2 = 0, \qquad \nabla \cdot \vec{X}_2 = ax^{a-1} + by^{b-1} + cz^{c-1}, \qquad (1.146a\text{–}c) $$

and thus a scalar (vector) potential does (does not) exist. The position vector (1.73b) is the particular case $a = b = c = 1$ of (1.146a). The first-order differential in three variables (1.64a) $\equiv$ (1.147a) with vector of coefficients (1.146a) is exact (1.64b)$\equiv$(1.147c):

$$\vec{X}_2 \cdot d\vec{x} = x^a dx + y^b dy + z^c dz = d\left(\frac{x^{a+1}}{a+1} + \frac{y^{b+1}}{b+1} + \frac{z^{c+1}}{c+1}\right) = d\Phi_2, \qquad (1.147a\text{–}c)$$

and determines the scalar potential (1.147c)$\equiv$(1.148a):

$$\Phi_2(x,y,z) = \frac{x^{a+1}}{a+1} + \frac{y^{b+1}}{b+1} + \frac{z^{c+1}}{c+1}:$$

$$\nabla\Phi_2 = \left\{\frac{\partial}{\partial x}, \frac{\partial}{\partial y}, \frac{\partial}{\partial z}\right\}\left(\frac{x^{a+1}}{a+1} + \frac{y^{b+1}}{b+1} + \frac{z^{c+1}}{c+1}\right) = \left\{x^a, y^b, z^c\right\}, \qquad (1.148a\text{–}c)$$

whose gradient (1.148b) $\equiv$ (1.148c) = (1.146a) specifies the vector field. The scalar potential (1.148a) satisfies (1.114c) $\equiv$ (1.149a) a scalar Poisson equation (1.149b) forced by (1.149c) $\equiv$ (1.146c) the divergence of the vector field:

$$\nabla^2\Phi_2 = \left\{\frac{\partial^2}{\partial x^2} + \frac{\partial^2}{\partial y^2} + \frac{\partial^2}{\partial z^2}\right\}\left(\frac{x^{a+1}}{a+1} + \frac{x^{b+1}}{b+1} + \frac{x^{c+1}}{c+1}\right) \qquad (1.149a\text{–}c)$$

$$= a x^{a-1} + b x^{b-1} + c x^{c-1} = \nabla\cdot\vec{X}_2.$$

Thus *the irrotational (1.146b), non-solenoidal (1.146c) three-dimensional vector field (1.146a) equals the gradient (1.148b,c) of the scalar potential (1.148a) that satisfies a scalar Poisson equation (1.149a–c) forced by the divergence of the vector field (1.146c).* Next is considered a solenoidal rotational vector field that equals the curl of a vector potential (subsection 1.5.19) and has no scalar potential.

1.5.19 SOLENOIDAL VECTOR FIELD AND VECTOR POTENTIAL

The 3 dimensional vector field (1.150a) is solenoidal (subsection 1.5.12) since its divergence is zero (1.117a) $\equiv$ (1.150b):

$$\vec{X}_3 = \left\{x + 2y, x - 2y, z\right\}: \qquad \nabla\cdot\vec{X}_3 = 1 - 2 + 1 = 0, \qquad (1.150a, b)$$

and is rotational since the curl is non-zero (1.151a):

$$\nabla\wedge\vec{X}_3 = \begin{vmatrix} \vec{e}_x & \vec{e}_y & \vec{e}_z \\ \partial_x & \partial_y & \partial_z \\ x+2y & x-2y & z \end{vmatrix} = -\vec{e}_z, \qquad \vec{X}_3 = \nabla\wedge\vec{A}_3, \qquad (1.151a, b)$$

and hence has a vector potential (1.117b) $\equiv$ (1.151b) but no scalar potential. The vector potential is of the form (1.139) and substitution of (1.150a) in (1.140a–c) leads to (1.152a–c):

$$x + 2y = \vec{X}_{3x} = \partial_y A_{3z} - \partial_z A_{3y} = a_{31}x + a_{33}z - a_{21}x - a_{23}y, \qquad (1.152a)$$

$$x - 2y = \vec{X}_{3y} = \partial_z A_{3x} - \partial_x A_{3z} = a_{12}x + a_{13}y - a_{31}y - a_{32}z, \qquad (1.152b)$$

$$z = \vec{X}_{3z} = \partial_x A_{3y} - \partial_y A_{3x} = a_{21}y + a_{22}z - a_{11}x - a_{13}z. \qquad (1.152c)$$

From (1.152a–c) follows that the matrix a_{ij} has 4 zero components (1.153a–d), 2 non-zero components (1.153e,f):

$$0 = a_{33} = a_{32} = a_{11} = a_{21}, \qquad a_{23} = -2, \qquad a_{12} = 1, \qquad (1.153a\text{–}f)$$

$$a_{31} - a_{21} = 1, \qquad a_{31} - a_{13} = 2, \qquad a_{22} - a_{13} = 1, \qquad (1.153g\text{–}i)$$

and the remaining three components satisfy the system of equations (1.153g,h,i). Using (1.153d) in (1.153g) gives (1.154a), and (1.153h) implies (1.154b), and then (1.153i) implies (1.154c):

$$a_{31} = 1, \qquad a_{13} = -1, \qquad a_{22} = 0; \qquad \nabla \cdot \vec{A}_3 = -z, \qquad (1.154a\text{–}d)$$

using (1.153a–f;1.154a–c) in (1.142b) it follows that the vector potential has non-zero divergence (1.154d) and thus does not meet the gauge condition (1.120a). Substituting (1.153a–f;1.154a–c) in (1.139) specifies the vector potential (1.155a):

$$\vec{A}_3 = \{xz - yz, -2yz, xy\}: \qquad \nabla \cdot \vec{A}_3 = -z, \qquad (1.155a, b)$$

confirming (1.155b) ≡ (1.154d). The curl (1.156a) of the vector potential (1.155a):

$$\nabla \wedge \vec{A}_3 = \begin{vmatrix} \vec{e}_x & \vec{e}_y & \vec{e}_z \\ \partial_x & \partial_y & \partial_z \\ z(x-y) & -2yz & xy \end{vmatrix} \begin{aligned} &= (x+2y)\vec{e}_x \\ &+ (x-2y)\vec{e}_y \\ &+ z\vec{e}_z = \vec{X}_3, \end{aligned} \qquad (1.156a, b)$$

coincides with the vector field (1.156b) ≡ (1.150a). The Laplacian (1.157a) of the vector potential (1.155a) is zero (1.157b):

$$\nabla^2 \vec{A}_3 = \left\{ \frac{\partial^2}{\partial x^2} + \frac{\partial^2}{\partial y^2} + \frac{\partial^2}{\partial z^2} \right\} \left[(x-y)z, -2yz, xy \right] = 0, \qquad (1.157a, b)$$

and thus the Poisson equation (1.120b) is not met, because the gauge condition (1.120a) is also not met by (1.154d) ≡ (1.155b). When the gauge condition (1.120a) is not met (1.158a) by the vector potential the Poisson equation (1.120b) is replaced by the double curl equation (1.117c) ≡ (1.158b):

$$\nabla \cdot \vec{A} \neq 0: \qquad \nabla \wedge (\nabla \wedge \vec{A}) = \nabla \wedge \vec{X}. \qquad (1.158a, b)$$

It can be confirmed from (1.156b) ≡ (1.159a) that (1.159b,c) ≡ (1.151a):

$$\nabla \wedge \left(\nabla \wedge \vec{A}_3 \right) = \begin{vmatrix} \vec{e}_x & \vec{e}_y & \vec{e}_z \\ \partial_x & \partial_y & \partial_z \\ x+2y & x-2y & z \end{vmatrix} = -\vec{e}_z = \nabla \wedge \vec{X}_3, \qquad (1.159a\text{--}c)$$

satisfies (1.158b). Thus *the solenoidal (1.150b) rotational (1.151a) three-dimensional vector field (1.150a) equals the curl (1.156a, b) of the vector potential (1.155a) that does not satisfy the gauge condition (1.154d) = (1.155b) ≠ (1.120a) and thus the Poisson equation (1.120b) ≠ (1.157a, b) is replaced by the double curl equation (1.117a, b) = (1.158b) = (1.159a–c) forced by the curl (1.151a) of the vector field (1.150a).* Next is considered a rotational non-solenoidal vector field whose representation requires both scalar and vector potentials (subsection 1.5.20).

1.5.20 NON-SOLENOIDAL, ROTATIONAL VECTOR FIELD AND SCALAR AND VECTOR POTENTIALS

The 3-dimensional vector field (1.160a) is non-solenoidal (1.160b):

$$\vec{X}_4 = \{x+y, -2y, 2z\}: \qquad\qquad \nabla \cdot \vec{X}_4 = 1-2+2 = 1, \qquad (1.160a, b)$$

and rotational (1.161a):

$$\nabla \wedge \vec{X}_4 = \begin{vmatrix} \vec{e}_x & \vec{e}_y & \vec{e}_z \\ \partial_x & \partial_y & \partial_z \\ x+y & -2y & 2z \end{vmatrix} = -\vec{e}_z, \quad X_4 = \nabla \Phi_4 + \nabla \wedge \vec{A}_4, \qquad (1.161a, b)$$

and thus (subsection 1.5.14) equals (1.123b) ≡ (1.161b) the sum of the gradient (curl) of a scalar (vector) potential. The two distinct vector fields (1.150a)≠(1.160a) have the same curl (1.151a) = (1.161a) and hence must have the same vector potential (1.155a) ≡ (1.162a, b) and their difference (1.162c) must be the gradient of a scalar potential (1.162d,e):

$$\vec{A}_3 = \{z(x-y), 2yz, xy\} \equiv \vec{A}_4: \qquad (1.162a, b)$$

$$\{-y, -x, z\} = \vec{X}_4 - \vec{X}_3 = \nabla \wedge \vec{A}_4 + \nabla \Phi_4 - \nabla \wedge \vec{A}_3 = \nabla \Phi_4. \qquad (1.162c\text{--}e)$$

The first-order differential in three variables (1.163a) with vector of coefficients (1.162c) is exact (1.163b) specifying (1.163c) the scalar potential (1.163d):

$$\left(\vec{X}_4 - \vec{X}_3 \right) \cdot d\vec{x} = -ydx - xdy + zdz = d\left(\frac{z^2}{2} - xy \right) = d\Phi_4:$$

$$\Phi_4 = \frac{z^2}{2} - xy. \qquad (1.163a\text{--}d)$$

It can be confirmed that the curl (1.156b) of the vector potential (1.162a, b) added to the gradient of the scalar potential (1.163d) leads to (1.164a–d):

$$\nabla \wedge \vec{A}_4 + \nabla\Phi_4 = \nabla \wedge \vec{A}_3 + \left\{\partial_x, \partial_y, \partial_z\right\}\left(\frac{z^2}{2} - xy\right)$$
$$= \left\{x+2y, x-2y, z\right\} + \left\{-y, -x, z\right\} = \left\{x+y, -2y, 2z\right\} \equiv \vec{X}_4,$$

(1.164a–e)

that coincides with the vector (1.164e) =(1.160a). The vector potential (1.162a, b) as before satisfies neither (1.154d) ≡ (1.155b)≠(1.127a) the gauge condition nor the Poisson equation (1.157a, b) ≠ (1.127b) and satisfies instead the double curl equation (1.158b) ≡ (1.159a–c). The scalar potential (1.163d) satisfies the scalar Poisson equation (1.165a) forced (1.165b) by the divergence (1.165c) ≡ (1.160b) of the vector field (1.160a):

$$\nabla^2\Phi_4 = \left\{\frac{\partial^2}{\partial x^2} + \frac{\partial^2}{\partial y^2} + \frac{\partial^2}{\partial z^2}\right\}\left(\frac{z^2}{2} - xy\right) = 1 = \nabla \cdot \vec{X}_4.$$

(1.165a–c)

Thus *the three-dimensional non-solenoidal (1.160b) rotational (1.161a) vector field (1.160a) equals (1.164a–e) the sum of the gradient (curl) of the scalar (1.163d) [vector (1.162b)] potential, that satisfies the Poisson (1.124b) ≡ (1.165a–c) [double curl (1.126a, b) ≡ (1.159a–c)] equation forced by the divergence (1.160b) [curl (1.161a)] of the vector field (1.160a).* Next is considered a rotational non-helical vector field that has two Euler scalar potentials (subsection 1.5.21).

1.5.21   ROTATIONAL, NON-HELICAL VECTOR FIELD AND TWO EULER SCALAR POTENTIALS

The three-dimensional vector field (1.166a) is rotational, that is has non-zero curl (1.166b)

$$\vec{X}_5 = \{xz, yz, 1\}: \quad \nabla \wedge \vec{X}_5 = \begin{vmatrix} \vec{e}_x & \vec{e}_y & \vec{e}_z \\ \partial_x & \partial_y & \partial_z \\ xz & yz & 1 \end{vmatrix} = -\vec{e}_x y + \vec{e}_y x,$$

(1.166a–c)

to which it is orthogonal (1.166d) so the helicity (1.166c) is zero:

$$H_5 = \vec{X}_5 \cdot \left(\nabla \wedge \vec{X}_5\right) = -xyz + xyz = 0.$$

(1.166c, d)

It follows that the rotational non-helical vector field (subsections 1.5.4 and 1.5.8) vector field must equal (1.167a) the product of second Euler potential as integrating factor by the gradient the first Euler potential as solution. The vector field (1.166a) leads to the differential (1.167c):

$$\vec{X}_5 = \Psi_5 \nabla\Phi_5: \quad \vec{X}_5 \cdot d\vec{x} = zxdx + yzdy + dz = \Psi_5 d\Phi_5,$$

(1.167a–c)

The differential (1.167b) in the form (1.168a) specifies (1.168b) the first (1.168c) [second (1.168d)] Euler scalar potential:

$$zxdx + yzdy + dz = z\left(xdx + ydy + \frac{dz}{z} \right) = zd\left(\frac{x^2 + y^2}{2} + \log z \right), \qquad (1.168a, b)$$

$$\Phi_5 = \frac{x^2 + y^2}{2} + \log z, \qquad \Psi_5 = z. \qquad (1.168c, d)$$

It can be confirmed that (1.169a) the gradient of (1.168c) multiplied by (1.168d) specifies (1.169b):

$$\Psi_5 \nabla \Phi_5 = z\{\partial_x, \partial_y, \partial_z\}\left(\frac{x^2 + y^2}{2} + \log z \right) = \{xz, yz, 1\} = \vec{X}_5, \qquad (1.169a-c)$$

the vector field (1.169c) ≡ (1.166a). Thus *the three-dimensional vector field (1.166a) that is rotational (1.166b) and non-helical (1.166c) equals (1.169a–c) the product of the second scalar Euler potential (1.168d) by the gradient of the first scalar Euler potential (1.168c).* Next is considered a helical vector field whose representation requires 3 scalar Euler potentials (subsection 1.5.22).

1.5.22   HELICAL VECTOR FIELD AND THREE SCALAR EULER POTENTIALS

The three-dimensional vector field (1.170a) is rotational (1.170b):

$$\vec{X}_6 = \{z, x, y\}: \qquad \nabla \wedge \vec{X}_6 = \begin{vmatrix} \vec{e}_x & \vec{e}_y & \vec{e}_z \\ \partial_x & \partial_y & \partial_z \\ z & x & y \end{vmatrix} = \vec{e}_x + \vec{e}_y + \vec{e}_z, \qquad (1.170a-c)$$

and helical (1.170):

$$H_6 = \vec{X}_6 \cdot \left(\nabla \wedge \vec{X}_6 \right) = x + y + z, \qquad (1.170c, d)$$

and thus (subsections 1.5.5, 1.5.9 and 1.5.10) has a representation (1.171a) as the sum of the gradient of the third Euler potential with the second Euler potential multiplied by the gradient of first Euler potential:

$$\vec{X}_6 = \nabla \Theta_6 + \Psi_6 \nabla \Phi_6; \qquad \vec{X}_6 \cdot d\vec{x} = zdx + xdy + ydz = d\Theta_6 + \Psi_6 d\Phi_6, \qquad (1.171a-c)$$

equivalently the first-order differential (1.171b) with vector of coefficients (1.170a) must be of the form (1.171c) without integrating factor. The curl of the vector field

(1.170b,c) leads (1.96b) to the characteristic equations (1.172a) whose solutions are first (1.172b) and second (1.172c) Euler scalar potentials:

$$dx = dy = dz: \qquad \Phi_6 = x - y, \qquad \Psi_6 = z - y. \qquad (1.172a\text{-}c)$$

Substituting (1.172b,c) in (1.171c) leads to an exact differential (1.173a):

$$zdx + xdy + ydz - (z - y)(dx - dy) = ydx + (x + z - y)dy + ydz$$
$$= d\left(xy + yz - \frac{y^2}{2} \right) = d\Theta_6, \qquad \Theta_6 = y\left(x + z - \frac{y}{2} \right), \qquad (1.173a\text{-}d)$$

that specifies (1.173b,c) the third Euler scalar potential (1.173d). It can be confirmed that the three Euler scalar potentials (1.172b,c;1.173d) specify through (1.174a–c):

$$\nabla\Theta_6 + \Psi_6\nabla\Phi_6 = \{\partial_x, \partial_y, \partial_z\}\left[y\left(x - \frac{y}{2} + z \right) \right] + (z - y)\{\partial_x, \partial_y, \partial_z\}(x - y) \qquad (1.174a\text{-}d)$$
$$= \{y, x - y + z, y\} + \{z - y, y - z, 0\} = \{z, x, y\} \equiv \vec{X}_6,$$

the vector field (1.174d) ≡ (1.170a). Thus *the three-dimensional vector field (1.170a) that is helical (1.170b–d) equals (1.174a–d) the sum of the gradient of the third Euler scalar potential (1.173d) with the second Euler scalar potential (1.172b) multiplied by the gradient of the first Euler scalar potential (1.172c). There are two representations in terms of 3 scalar potentials, namely the Euler (Clebsch) potentials for a helical (rotational non-solenoidal) vector field [subsection 1.5.22 (1.5.23)].*

1.5.23   ROTATIONAL, NON-SOLENOIDAL VECTOR FIELD AND THREE SCALAR CLEBSCH POTENTIALS

The three-dimensional vector field (1.175a) is non-solenoidal (1.175b):

$$\vec{X}_7 = \{x + y - z, x + y - z, x - y + z\}: \qquad \nabla \cdot \vec{X}_7 = 3, \qquad (1.175a, b)$$

and rotational (1.176a):

$$\nabla \wedge \vec{X}_7 = \begin{vmatrix} \vec{e}_x & \vec{e}_y & \vec{e}_z \\ \partial_x & \partial_y & \partial_z \\ x + y - z & x + y - z & x - y + z \end{vmatrix} = -2\vec{e}_y, \vec{X}_7 = \nabla\Phi_7 + \nabla \wedge \vec{A}_7, \qquad (1.176a, b)$$

and hence (subsection 1.5.15) has representation (1.131b) ≡ (1.176b) in terms of scalar and vector potentials. The scalar potential satisfies (1.114c) a Poisson equation (1.177b) forced by the divergence (1.175b) = (1.177a) vector field (1.177c):

$$3 = \nabla \cdot \vec{X}_7 = \nabla^2\Phi_7 = \frac{\partial^2\Phi_7}{\partial x^2} + \frac{\partial^2\Phi_7}{\partial y^2} + \frac{\partial^2\Phi_7}{\partial z^2}: \qquad \Phi_7 = \frac{x^2 + y^2 + z^2}{2}, \qquad (1.177a\text{-}d)$$

and a solution is the first scalar Clebsch potential (1.177d). The curl of the vector potential (1.123b) ≡ (1.178a) equals (1.178b) the vector field (1.175a) minus the gradient of the first Clebsch scalar potential:

$$\nabla \wedge \vec{A}_7 = \vec{X}_7 - \nabla \Phi_7 = \{y - z, x - z, x - y\}, \qquad (1.178a, b)$$

and is given by (1.178b).

The second and third scalar Clebsch potentials satisfy (1.133) the characteristic equations (1.179) specified by the curl of the vector potential (1.178b):

$$\frac{dx}{y-z} = \frac{dy}{x-z} = \frac{dz}{x-y} \equiv \lambda. \qquad (1.179)$$

A linear combination (1.180a) of the denominators of (1.179) with constant coefficients (a, b) is zero (1.180b):

$$0 = y - z + a(x - z) + b(x - y) = x(a + b) + y(1 - b) - z(1 + a) \equiv P_2(a, b), \quad (1.180a, b)$$

if (1.182a, b) hold implying (1.182c, d) from (1.179):

$$a = -1, b = 1: \qquad 0 = \lambda\left[(y - z) - (x - z) + (x - y)\right] = dx - dy + dz. \qquad (1.181a–d)$$

Thus one solution is the second Clebsch potential (1.183a) whose gradient is (1.183b):

$$\alpha_7 = x - y + z, \qquad \nabla \alpha_7 = \{1, -1, 1\}. \qquad (1.182a, b)$$

The curl of the vector potential (1.178a) ≡ (1.183b) equals (1.130b) ≡ (1.183c) the outer vector product (1.183d) of the gradients of the second (1.182b) and third (1.183a) Clebsch scalar potentials:

$$\nabla \beta_7 = \{A, B, C\}: \qquad (y - z)\vec{e}_x + (x - z)\vec{e}_y + (x - y)\vec{e}_z$$

$$= \nabla \wedge \vec{A}_7 = \nabla \alpha_7 \wedge \nabla \beta_7 = \begin{vmatrix} \vec{e}_x & \vec{e}_y & \vec{e}_x \\ 1 & -1 & 1 \\ A & B & C \end{vmatrix}. \qquad (1.183a–d)$$

From (1.183d) follow the equalities (1.184a–c):

$$y - z = -B - C, \qquad x - z = A - C, \qquad x - y = B + A, \qquad (1.184a–c)$$

whose solutions are (1.185a):

$$\{x, -y, z\} = \{A, B, C\} = \nabla \beta_7: \qquad \beta_7 = \frac{x^2 - y^2 + z^2}{2}, \qquad (1.185a–c)$$

specifying (1.185b) ≡ (1.183a) the gradient of the third Clebsch potential (1.185c).

It can be checked that (1.186a):

$$\lambda\left[x(y-z)-y(x-z)+z(x-y)\right]=0: \qquad xdx-ydy+zdz=0, \qquad (1.186a, b)$$

implies (1.186b) from (1.179); from (1.186b) follows the third Clebsch scalar potential (1.185b). It follows *that the non-solenoidal (1.175b) rotational (1.176a) three-dimensional vector field (1.175a) is specified (1.176b) ≡ (1.187a) by the sum of the gradient of the first Clebsch scalar potential (1.177d) ≡ (1.187b) with the outer vector product of the gradient of the second (1.182a) ≡ (1.187c) and third (1.185c) ≡ (1.187d) Clebsch scalar potentials:*

$$\vec{X}_7 = \nabla\Phi_7 + \nabla\alpha_7 \wedge \nabla\beta_7:$$

$$\{\Phi_7, \alpha_7, \beta_7\} = \left\{ \frac{x^2+y^2+z^2}{2}, x-y+z, \frac{x^2-y^2+z^2}{2} \right\}. \qquad (1.187a\text{–}d)$$

The general theory and particular examples (Sections 1.3–1.5 and examples 10.1–10.3) show the relations (subsections 1.5.24–1.5.26) between; (i) linear first-order partial differential equations; (ii) first-order differentials; (iii) representation of vector fields by potentials.

1.5.24 PARTIAL DIFFERENTIAL EQUATIONS AND CHARACTERISTIC SYSTEMS

Table 1.3 compares the solutions of first-order partial differential equations that: (i) in the linear unforced case (1.25c) lead to the characteristic system (1.28); (ii) in the linear forced case (1.55b) lead to the characteristic system (1.57); (iii) in the quasi-linear case (1.23) lead to the characteristic system (1.52). *The characteristic system (1.28) ≡ (1.188a) implies (1.188b) for any choice of the L functions E_ℓ:*

$$\lambda = \frac{dx_1}{A_1(x_\ell)} = \frac{dx_1}{A_1(x_\ell)} = \cdots = \frac{dx_L}{A_1(x_\ell)} = \frac{\displaystyle\sum_{s=1}^{L} E_s(x_\ell)dx_s}{\displaystyle\sum_{s=1}^{L} A_s(x_\ell)E_s(x_c)}. \qquad (1.188a, b)$$

The proof follows immediately from (1.189a):

$$\lambda \sum_{s=1}^{L} A_s(x_\ell)E_s(x_c) = \sum_{s=1}^{L} E_s(x_\ell)dx_s \iff \lambda\vec{A}\cdot\vec{E} = \vec{E}\cdot d\vec{x}, \qquad (1.189a, b)$$

that is equivalent to (1.189b) in vector notation. From (1.189b) follows that: *con-sider the characteristic system (1.28) ≡ (1.188a) with vector of coefficients $\vec{A}$; an*

TABLE 1.3
First-Order Partial Differential Equations

Type	I	II	III
Section	1.3	1.4.1 – 1.4.2	1.4.3 – 1.4.4
Class p. d. e.	*linear unforced*	*Quasi-linear*	*Linear forced*
	$\displaystyle\sum_{\ell=1}^{L} A_\ell\left(x_1,\ldots,x_L\right)\frac{\partial \Phi}{\partial x_\ell} = 0$	$\displaystyle\sum_{\ell=1}^{L} A_\ell\left(x_1,\ldots,x_L,\Phi\right)\frac{\partial \Phi}{\partial x_\ell} + A_0\left(x_1,\ldots,x_L,\Phi\right) = 0$	$\displaystyle\sum_{\ell=1}^{L} A_\ell\left(x_1,\ldots,x_L,\Phi\right)\frac{\partial \Phi}{\partial x_\ell} = B\left(x_1,\ldots,x_L\right)$
Characteristic System	$\dfrac{dx_1}{A_1} = \cdots = \dfrac{dx_L}{A_L}$	$\dfrac{dx_1}{A_1} = \cdots = \dfrac{dx_L}{A_L} = -\dfrac{d\Phi}{A_0}$	$\dfrac{dx_1}{A_1} = \cdots = \dfrac{dx_L}{A_L} = \dfrac{d\Phi}{B}$
Characteristic variable	$U_1(x_1,\ldots,x_L),\ldots, U_{L-1}(x_1,\ldots,x_L)$	$U_1(x_1,\ldots,x_L:\Phi),\ldots,U_L(x_1,\ldots,x_L:\Phi)$	$U_1(x_1,\ldots,x_L),\ldots U_{L-1}(x_1,\ldots,x_L),\ U_L(x_1,\ldots,x_L:\Phi)$
General integral	$f(U_1,\ldots,U_{L-1}) = 0$	$f(U_1,\ldots,U_L) = 0$	$U_L(x_1,\ldots,x_L:\Phi) = g(U_1,\ldots,U_{L-1})$

Note: Three cases of the solution of a first-order partial differential equations via characteristic equations and variables first-order.

orthogonal (1.190a) vector of coefficients $\vec{E}$ *corresponds to a zero differential (1.190b):*

$$\vec{A} \cdot \vec{E} = \sum_{s=1}^{L} A_s(x_\ell) E_s(x_\ell) = 0 \Leftrightarrow 0 = \vec{E} \cdot d\vec{x} = \sum_{s=1}^{L} B_s(x_\ell) dx_s. \qquad (1.190a, b)$$

With a suitable selection of the auxiliary vector $\vec{E}$ *the property (1.190a, b) provides a method of solution of characteristic systems; for example, if the orthogonal vector (1.190a) is constant (1.191a) a solution (1.190b) of the characteristic system (1.188a, b) is (1.191b,c):*

$$\vec{E} = \text{cosnst:} \qquad\qquad U(\vec{x}) = \vec{E} \cdot \vec{x} \equiv \sum_{s=1}^{L} E_s x_s. \qquad (1.191a\text{--}c)$$

The property (1.191a–c) [(1.190a, b)] was used before, namely for the characteristic system (1.179) associated with the vector of coefficients (1.178b), namely in (1.180a, b;1.181a–d;1.182a) [(1.186a, b;1.185c)] with constant (variable) orthogonal vector (1.192a) [(1.19b)]:

$$\vec{B} = \{1, -1, 1\}, \{x, -y, z\}. \qquad (1.192a, b)$$

Some concluding remarks are made next concerning: (i) the unicity of scalar and vector potentials (subsection 1.5.25); (ii) the comparison of the three scalar Euler and Clebsch potentials for a general helical non-solenoidal vector field (subsection 1.5.26).

1.5.25   UNICITY OF SCALAR AND VECTOR POTENTIALS

The example of the redundant equations (1.141g,h,i) shows, in a particular instance, that the vector potential is not unique. In general, *the scalar potential of a vector field is unique to within an added constant (1.193a) that does not change its gradient (1.193b):*

$$\Phi_1(\vec{x}) = \Phi(\vec{x}) + \text{const:} \qquad\qquad \nabla \Phi_1 = \nabla \Phi. \qquad (1.193a, b)$$

The vector potential of a vector field is unique to within the gradient of a scalar potential (1.194a) that does not change its curl (1.194b):

$$\vec{A}_1(\vec{x}) = \vec{A}(\vec{x}) + \nabla \Phi: \qquad \nabla \wedge \vec{A}_1 = \nabla \wedge \vec{A} + \nabla \wedge (\nabla \Phi) = \nabla \wedge \vec{A}; \qquad (1.194a, b)$$

the imposition of the gauge condition (1.120a) $\equiv$ *(1.194c,f) still allows (1.194e) the scalar potential to be (1.194d) a solution of the Laplace equation (1.194g):*

$$0 = \nabla \cdot \vec{A}_1 = \nabla \cdot \vec{A} + \nabla \cdot (\nabla \Phi) = \nabla \cdot \vec{A} + \nabla^2 \Phi = \nabla \cdot \vec{A}: \qquad \nabla^2 \Phi = 0. \qquad (1.194c\text{--}g)$$

The scalar potentials are of Euler or Clebsch type that are compared next (subsection 1.5.26).

1.5.26 CHARACTERISTIC SYSTEMS FOR EULER AND CLEBSCH POTENTIALS

The characteristic system (Table 1.4) for the first two Clebsch (1.133) ≡ (1.195a)
[Euler (1.106, 1.105a–c) ≡ (1.196a)] potentials involves (1.195b) [(1.196b)]:

$$\alpha, \beta: \qquad \frac{dx}{X - \partial\Xi / \partial x} = \frac{dy}{Y - \partial\Xi / \partial y} = \frac{dz}{Z - \partial\Xi / \partial z}, \qquad (1.195a, b)$$

$$\Phi, \Psi: \qquad \frac{dx}{\partial Z/\partial y - \partial Y/\partial z} = \frac{dy}{\partial Z/\partial x - \partial X/\partial x} = \frac{dz}{\partial Y/\partial x - \partial X/\partial y}, \qquad (1.196a, b)$$

the vector field (1.197a) [its curl (1.197b)]:

$$\vec{X} = \{X, Y, Z\}, \qquad \nabla \wedge \vec{X} = \left\{ \frac{\partial Z}{\partial y} - \frac{\partial Y}{\partial z}, \frac{\partial Z}{\partial x} - \frac{\partial X}{\partial z}, \frac{\partial Y}{\partial x} - \frac{\partial X}{\partial y} \right\}. \qquad (1.197a, b)$$

The third Euler (Clebsch) potential is determined by the exact differential (1.66a) ≡
(1.198a) [(1.131b) ≡ (1.198b)]:

$$d\Theta = \vec{X} \cdot d\vec{x} - \Psi d\Phi, \qquad d\Xi = \vec{X} \cdot d\vec{x} - \nabla\alpha \wedge \nabla\beta. \qquad (1.198a, b)$$

that specify the helical (1.199a) [rotational (1.199c) non-solenoidal (1.199d)] vector
field by (1.199b) [(1.199e)]:

$$\vec{X} \cdot \left(\nabla \wedge \vec{X} \right) \neq 0: \quad \vec{X} = \nabla\Theta + \Psi\nabla\Phi, \qquad (1.199a\text{–}b)$$

$$\nabla \wedge \vec{X} \neq 0 \neq \nabla \cdot \vec{X}: \quad \vec{X} = \nabla\Xi + \nabla\alpha \wedge \nabla\beta = \nabla\Xi + \nabla \wedge \vec{A}, \qquad (1.199e\text{–}f)$$

TABLE 1.4
Comparison of Euler and Clebsch Potentials

Case	Euler	Clebsch
	1.5.1–1.5.10	1.5.11–1.5.20
Section	1.5.16–1.5.18	1.5.23
Three scalar potentials	Φ, Ψ, Θ	Ξ, α, β
Condition	$\vec{X} \cdot \left(\nabla \wedge \vec{X} \right) \neq 0$	$\nabla \wedge \vec{X} \neq 0 \neq \nabla . \vec{X}$
	Helical	Rotational Non-solenoidal
Vector field: $\vec{X} = \{X, Y, Z\}$	$\vec{X} = \nabla\Theta + \Psi\nabla\Phi$	$\vec{X} = \nabla\Xi + \nabla\alpha \wedge \nabla\beta$
Characteristic equations	(1.196a,b) for Φ, Ψ	(1.195a,b) for α, β

Note: Comparison of two general representations of a three-dimensional vector field by three Euler or Clebsch potentials.

The representation (1.199f) of the helical (1.199a) non-solenoidal (1.199c) vector field by scalar (1.199e) and vector (1.199f) potentials relates (1.200b,c) to the first and second Clebch potentials to the vector potential by (1.200a):

$$\vec{A} = \alpha \nabla \beta: \qquad \nabla \wedge \vec{A} = \nabla \alpha \wedge \nabla \beta = \vec{X} - \nabla \Xi. \qquad (1.200a\text{–}c)$$

The sequence (1.196a, b; 1.197b; 1.198b) [(1.195a, b; 1.197a; 1.198a)] is explicit (implicit) in the sense that it doses (does not) provide a method of direct determination of the three Euler (Clebsch) potentials. These methods and properties (subsections 1.5.24–1.5.26) are illustrated by additional cases of scalar and vector potential representations of continuously differentiable three-dimensional vector fields are given in Example 10.3. The consideration of linear and quasi-linear partial differential equations (Sections 1.1–1.2) of the first-order with variable coefficients (Sections 1.3–1.5) is followed by equations of fixed order N in the linear case with constant coefficients and derivatives of the same order (Sections 1.6–1.9).

1.6   P.D.E.S WITH CONSTANT COEFFICIENTS AND ALL DERIVATIVES OF SAME ORDER

The solution of linear (quasi-linear) partial differential equations of the first-order [Section 1.3 (1.4)] has been applied to first-order differentials in three variables (Section 1.5). The simplest partial differential equation of order N is linear with constant coefficients and has all derivatives of the same order (subsection 1.6.1) and is associated with a characteristic polynomial, that may be factorized. The district roots (subsection 1.6.3) lead to linear partial differential equations with constant coefficients and all derivatives of the first-order that have similarity solutions (subsection 1.6.2). The solutions for multiple roots (subsections 1.6.4) of the characteristic polynomial can be obtained by the method of variation of parameters (subsection 1.6.5) or by parametric differentiation (subsection 1.6.6). The general integral (subsection 1.6.7) is illustrated for the second order that is the lowest for which characteristic polynomial can have single or non-single, that is, double roots (subsection 1.6.8).

1.6.1   P.D.E. OF CONSTANT ORDER AND CHARACTERISTIC POLYNOMIAL

The simplest partial differential equation of order N in two variables (1.8c) is linear (1.11b) with constant coefficients (1.11a) and with all derivatives of the same order (1.12d,e) including forcing in (1.201a):

$$B(x,y) = \sum_{n=0}^{N} A_n \frac{\partial^N \Phi}{\partial x^n \partial y^{N-m}} \equiv \left\{ P_N\left(\frac{\partial}{\partial x}, \frac{\partial}{\partial y} \right) \right\} \Phi(x,y), \qquad (1.201a, b)$$

and corresponds (1.201b) to a polynomial (1.120a) of partial derivatives with regard to the independent variables applied to the dependent variable and equated to the forcing term:

$$P_N(\xi, \eta) = \sum_{n=0}^{N} A_n \xi^n \eta^{N-n} = A_N \prod_{m=1}^{N} (\xi - \alpha_m \eta). \qquad (1.202a, b)$$

The characteristic polynomial (1.202a) can be factorized (1.202b) and so can the p.d.e.. (1.201a, b), that is considered first (1.203b) in the unforced case (1.203a):

$$B(x,y) = 0: \qquad 0 = A_N \left\{ \prod_{m=1}^{N} \left(\frac{\partial}{\partial x} - \alpha_m \frac{\partial}{\partial y} \right) \right\} \Phi(x,y), \qquad (1.203a, b)$$

to obtain the general integral.

1.6.2 SIMILARITY SOLUTIONS FOR A LINEAR COMBINATION OF VARIABLES

In order to satisfy (1.203b) it is sufficient that one factor vanishes (1.204a) leading to (1.204b):

$$0 = \left(\frac{\partial}{\partial x} - \alpha_m \frac{\partial}{\partial y} \right) \Phi = \frac{\partial \Phi}{\partial x} - \alpha_m \frac{\partial \Phi}{\partial y}: \qquad \frac{dy}{dx} = -\frac{\partial \Phi / \partial x}{\partial x / \partial y} = -\alpha_m; \qquad (1.204a, b)$$

the first-order ordinary differential equation (1.204b) has solution (1.205a) that would also follow (1.205b):

$$\xi_m \equiv y + \alpha_m x = const, \qquad dx = -\frac{dy}{\alpha_n}, \qquad (1.205a, b)$$

from the characteristic equation (1.28) applied to (1.204a).

It follows that a solution of the partial differential equation (1.204a) is (1.206b,c) where f is a differentiable function (1.206a):

$$f \in D(\mathbb{R}): \qquad \Phi_m(x,y) = f_m(\xi_m) = f(y + \alpha_m x). \qquad (1.206a\text{--}c)$$

It can be checked that (1.206b) ≡ (1.206c) is a solution (1.207a) ≡ (1.207b) of the partial differential equation (1.207c) ≡ (1.204a):

$$\left(\frac{\partial}{\partial x} - \alpha_m \frac{\partial}{\partial y} \right) \Phi_m(x,y) = \frac{df_m}{d\xi_m} \left(\frac{\partial \xi_m}{\partial x} - \alpha_m \frac{\partial \xi_m}{\partial y} \right) = \frac{df_m}{d\xi_m} (\alpha_m - \alpha_m) = 0. \qquad (1.207a\text{--}c)$$

The similary solutions (1.206c) are differentiable functions (1.206a) of a single variable (1.205a) that is a linear combination of the two independent variables.

1.6.3 GENERAL INTEGRAL FOR DISTINCT ROOTS

If the roots of the characteristic polynomial are distinct (1.202b) there are N distinct solutions (1.206c) of the p.d.e.; since the unforced (1.203a) p.d.e. (1.201a) is linear, the sum (1.208c) of (1.208a) solutions (1.206c) is also a solution provided that the N arbitrary functions (1.208b) are N times differentiable:

$$m = 1,\ldots; N: \qquad f_m \in D^N(\mathbb{R}), \qquad \Phi(x,y) = \sum_{m=1}^{N} \Phi_m(x,y). \qquad (1.208a\text{--}c)$$

It has been shown that *the general integral of an unforced (1.203a) linear partial differential equation with constant coefficients (1.201a) and all derivatives of the same order (1.209):*

$$0 = \sum_{n=0}^{N} A_n \frac{\partial^N \Phi}{\partial x^n \partial y^{N-n}},$$

(1.209)

*is the sum (1.208c) of N **similarity solutions** (1.206c):*

$$f_m \in \mathcal{D}^N(\mathbb{R}): \qquad \Phi(x,y) = \sum_{m=1}^{N} f_m(y + \alpha_m x),$$

(1.210a, b)

*and involves N times differentiable arbitrary functions (1.208b) ≡ (1.210a) of the **similarity variables** (1.205a), where the α_m are distinct roots (1.211a, b) of the **characteristic polynomial** (1.211c, d):*

$$P_N(\alpha_m) = 0 \neq P_N'(\alpha_m):$$

$$0 = P_N(a) = \sum_{m=1}^{N} A_m a^n = A_N \prod_{m=1}^{N} (a - \alpha_m).$$

(1.211a–d)

The case of multiple roots of the characteristic polynomial is considered next (subsection 1.6.4).

1.6.4   SINGLE OR MULTIPLE ROOTS OF THE CHARACTERISTIC POLYNOMIAL

If in (1.209) the characteristic polynomial has M roots α_m with multiplicity β_m in (1.212a) its factorization is (1.212b):

$$\sum_{m=1}^{M} \beta_m = N: \qquad P_N(a) = A_N \prod_{m=1}^{M} (a - \alpha_m)^{\beta_m},$$

(1.212a, b)

and correspondingly the p.d.e. (1.209) is factorized:

$$\left\{ \prod_{m=1}^{M} \left(\frac{\partial}{\partial x} - \alpha_m \frac{\partial}{\partial y} \right)^{\beta_m} \right\} \Phi(x,y) = 0.$$

(1.213)

Note that if α_m is a root of multiplicity $\beta_m = 2, 3, \ldots$ then β_m arbitrary functions in (1.210b) coincide, and (1.210b) cannot be the general integral of the p.d.e. (1.209) of order N because it involves less than N arbitrary functions. To recover the general integral β_m independent solutions of each factor of (1.213) must be obtained.

1.6.5 Method of Variation of Parameters

Since (1.206c) is a solution of (1.213) the method of variation of parameters is used seeking a solution of the form (1.214a):

$$\Phi_m(x,y) = g(x)f(y+\alpha_m x): \quad \left(\frac{\partial}{\partial x} - \alpha_m \frac{\partial}{\partial y}\right)\Phi_m(x,y) = \frac{dg}{dx}f(y+\alpha_m x), \quad (1.214a, b)$$

implying (1.214b) because (1.206c) satisfies (1.207a–c). Applying derivatives (1.215a) up to the order $\beta_m - 1$ leads to (1.215b).

$$s_m = 0,1,\ldots,\beta_m - 1: \quad \left(\frac{\partial}{\partial x} - \alpha_m \frac{\partial}{\partial y}\right)^{s_m} \Phi_m(x,y)$$

$$= f(y+\alpha_m x)\left(\frac{d}{dx}\right)^{s_m} g(x), \quad (1.215a, b)$$

since all differentiations pass to the function g. The vanishing (1.213) of (1.215b) implies that the function g has zero derivative of order β_m, and thus is (1.216a) a polynomial of y of degree $\beta_m - 1$:

$$g(x) = \sum_{r=1}^{\beta_m-1} a_r x^r; \quad \Phi_{m,s_m}(x,y) = x^{s_m} f_{s_m}(y+\alpha_m x), \quad (1.216a, b)$$

hence (1.216b) are β_m independent solutions of (1.213), explicitly (1.217a–c):

$$\Phi_{m,0}(x,y) = f_0(y+\alpha_m x), \quad \Phi_{m,1}(x,y) = xf_1(y+\alpha_m x), \quad (1.217a, b)$$

$$\ldots,\Phi_{m,\beta_m-1}(x,y) = x^{\beta_m-1} f_{\beta_m-1}(y+\alpha_m x). \quad (1.217c)$$

The function (1.216b) with larger integer values of $s_m = \beta_m, \beta_m + 1\ldots$ would no longer be a solution of the p.d.e. (1.213) since (1.215b) is no longer would hold.

1.6.6 Method of Parametric Differentiation

The same conclusion could be obtained by parametric differentiation of each factor in (1.213) with regard to α_m:

$$0 = \frac{\partial}{\partial \alpha_m}\left[\left(\frac{\partial}{\partial x} - \alpha_m \frac{\partial}{\partial y}\right)^{\beta_m} \Phi_m(x,y)\right]$$

$$(1.218)$$

$$= \left(\frac{\partial}{\partial x} - \alpha_m \frac{\partial}{\partial y}\right)^{\beta_m} \frac{\partial \Phi_m}{\partial \alpha_m} - \beta_m \frac{\partial}{\partial y}\left[\left(\frac{\partial}{\partial x} - \alpha_m \frac{\partial}{\partial y}\right)^{\beta_m-1} \Phi_m(x,y)\right].$$

The second term on the r.h.s. of (1.218) is zero and implies (1.219a):

$$0 = \left(\frac{\partial}{\partial x} - \alpha_m \frac{\partial}{\partial y}\right)^{\beta_m} \frac{\partial \Phi_m}{\partial \alpha_m}, \qquad \Phi_m(x,y) = f_0(y + \alpha_m x) = \Phi_{m,0}(x,y), \quad (1.219\text{a–c})$$

that in addition to the preceding solution $(1.217\text{a}) \equiv (1.219\text{b,c})$ there is another solution $(1.220\text{a}) \equiv (1.220\text{b})$:

$$\Phi_{m,1}(x,y) = \frac{\partial}{\partial \alpha_m} f(y + \alpha_m x) = x \frac{df_{m,0}}{d\xi_m}. \qquad (1.220\text{a, b})$$

that coincides with (1.217b). The differentiation with regard to α_m can continue as long as the second term on the r.h.s. of (1.218) vanishes, that is up to order $\beta_m - 1$ leading to the solutions (1.221a–c):

$$s_m = 0, 1, \ldots, \beta_m - 1: \quad \Phi_{m,s_m}(x,y) = \left(\frac{\partial}{\partial \alpha_m}\right)^{s_m} f(y + \alpha_m x)$$
$$= x^{s_m} f_{s_m}(y + \alpha_m x), \qquad (1.221\text{a–c})$$

similar to (1.217a–c).

1.6.7 GENERAL INTEGRAL FOR MULTIPLE ROOTS

The solutions of the p.d.e. (1.213) for multiple roots of the characteristic polynomial (1.212a, b) obtained (1.217a–c) [(1.221a–c)] by the method [subsection 1.6.5 (1.6.6)] of variation of parameters (parametric differentiation) are identical. If in the method of variation of parameters (subsection 1.6.5) a function j(y) had been used instead of g(x) in (1.214a) the result would have been similar exchanging x for y. Also x or y in the power factor could be replaced by a linear combination $y + b_m x$ for any $b_m \neq a_m$; the case $b_m = a_m$ is excluded because then all of (1.217a–c) or (1.221a–c) would be the same similarity solution. It has been shown that *if the linear unforced partial differential equation $(1.209) \equiv (1.213)$ with two independent variables, constant coefficients and all derivatives of order N has characteristic polynomial (1.202a, b) with M roots α_m of multiplicity β_m in $(1.212\text{a, b}) \equiv (1.222\text{c})$ the general integral is (1.222d)*:

$$b_m \neq a_m, f_{m,s_m} \in \mathcal{D}^N(\mathbb{R}); \quad P_N(\alpha_m) = \cdots = P_N^{(\beta_m - 1)}(\alpha_m) = 0 \neq P_N^{(\beta_m)}(\alpha_m):$$
$$\Phi(x,y) = \sum_{m=1}^{M} \sum_{s_m=0}^{\beta_m - 1} (y + b_m x)^{s_m} f_{m,s_m}(y + \alpha_m x), \qquad (1.222\text{a–d})$$

involving N arbitrary functions that are N times differentiable (1.222b) and M arbitrary constants (1.222a) distinct from the roots. The simplest example of p.d.e. with possible multiple roots of the characteristic polynomial is of the second-order (subsection 1.6.8) and can have two distinct single roots or a double root.

1.6.8 Linear P.D.E. with Constant Coefficients and Second-Order Derivatives

A general linear p.d.e. with constant coefficients, second-order derivatives only, and no forcing term is (1.223a):

$$0 = A \frac{\partial^2 \Phi}{\partial x^2} + 2B \frac{\partial^2 \Phi}{\partial x \partial y} + C \frac{\partial^2 \Phi}{\partial y^2}$$

$$= A \left(\frac{\partial^2}{\partial x} - a_+ \frac{\partial}{\partial y} \right) \left(\frac{\partial}{\partial x} - a_- \frac{\partial}{\partial y} \right) \Phi(x, y), \qquad \text{(1.223a, b)}$$

and can be factorized (1.224c) as the quadratic polynomial (1.224b) of variable (1.224a):

$$a \equiv \frac{\partial/\partial x}{\partial/\partial y}: \quad 0 = Aa^2 + 2Ba + C = A(a - a_+)(a - a_-) \equiv P_2(a); \qquad \text{(1.224a–d)}$$

the roots (1.225a, b) specify the similarity variables (1.225c):

$$a_\pm = -\frac{B}{A} \pm \frac{\sqrt{D}}{A}, \quad D \equiv B^2 - AC: \qquad \xi_\pm = y + a_\pm x. \qquad \text{(1.225a–c)}$$

If the roots are distinct (1.226a):

$$B^2 \neq AC: f, g \in \mathcal{D}(|R^2): \quad \Phi(x, y) = f(\xi_+) + g(\xi_-)$$

$$= f(y + a_+ x) + g(y + a_- x), \qquad \text{(1.226a–d)}$$

the general integral (1.226c, d) involves two arbitrary twice differentiable functions (1.226b); their arguments, (1.225a–c) for real coefficients A, b,C in (1.223a), are real (complex) for the hyperbolic D > 0 (elliptic D<0) type as classified subsequently (Section 5.2).

The case D=0 of a double root (1.227a):

$$B^2 = AC: \quad 0 = A \frac{\partial^2 \Phi}{\partial x^2} + 2B \frac{\partial^2 \Phi}{\partial x \partial y} + \frac{B^2}{A} \frac{\partial^2 \Phi}{\partial y^2} = A \left(\frac{\partial}{\partial x} - a \frac{\partial}{\partial y} \right)^2 \Phi(x, y), \qquad \text{(1.227a–c)}$$

corresponds (1.227b) to a parabolic type the (1.227c) with general integral (1.228a–b):

$$a = \frac{-B}{A} = -\sqrt{C/A} = a_\pm:$$

$$\Phi(x, y) = f(Ay - Bx) + (Ay + Bx) g(Ay - Bx), \qquad \text{(1.228a–b)}$$

and is an instance of (1.222d) with $\beta = 2$, with: (i) variable (1.229a–c); (ii) coefficient (1.229a–c):

$$y + ax = y - \frac{B}{A} x \Rightarrow Ay - Bx = \xi, \qquad (1.229\text{a--c})$$

$$Ay + Bx = A\left(y + \frac{B}{A} x \right) = A(y - ax). \qquad (1.229\text{d,e})$$

The general linear partial diffential eqauation with constant coefficients and all derivatives of second-order (1.223a, b) has characteristic polynomial (1.224a–d) with roots (1.225a, b) and characteristic variables (1.225c). In the unforced case, the general integral is given by (1.226a–d) [(1.228a–d)] in the case of distinct or single (1.226a) [coincident or doube (1.278a–c)] roots of the characteristic polynomial.

The scalar Laplace operator appears without (with) forcing in the Laplace (Poisson) equation that applies, among others, to: (i) the gravity field outside (inside) masses (chapter I.18); (ii) electrostatic field outside (inside) electric charges (chapters I.24 and III.8); (iii) irrotational flow outside (inside) flow sources (chapters I.12,14,16,28,34,36,38; II.2 and II.8; III.6); (iv) steady heat conduction outside (inside) heat sources (chapter I.32); (v) deflection of elastic membranes (sections II.6.1–II.6.2); (vi) torsion of elastic rods (sections II.6.5-II.6.8); (vii) flow in a rotating vessel (section II.6.9). The vector Laplace (Poisson) equation applies, among others, to: (vii) magnetostatic field outside (inside) electric currents (chapters I.26 and III.8); (ix) rotational flow outside (inside) vortices (chapters I.12,16,18,28,34,36,38; II.2 and II.8; III.6). The double Laplace or biharmonic operator appears in the biharmonic equation, that occurs, among others: (x) in plane elasticity (chapter II.4); (xi) steady plane viscous flow (notes II.4.1-II.4.11). The Laplace and biharmonic operators [Section 1.7 (1.9)] are taken as examples of [unforced (forced)] linear partial differential equations with constant coefficients [Section 1.6 (1.8)] with all derivatives of the same order, respectively, two and four.

1.7   HARMONIC AND BIHARMONIC FUNCTIONS ON THE PLANE

Two examples of linear unforced p.d.e.s with constant coefficients and all derivatives of order two (four) are the Laplace (biharmonic) equation [subsections 1.7.1 (1.7.3)] in Cartesian coordinates; they apply to potential fields (e.g., gravity field, electro-magnetostatics, potential flow, etc…) [plane elasticity and viscous creeping flow] in volume I (II). Their solutions are obtained for two Cartesian coordinates in the plane and the similarity variables are complex variables that may be used to obtain real or complex solutions [subsections 1.7.2 (1.7.4)].

1.7.1   LAPLACE EQUATION IN CARTESIAN COORDINATES

A particular case of (1.223a) with coefficients (1.230a–c), and hence of elliptic type (1.230d) is the Laplace equation in the Cartesian plane (1.230e,f):

$$B = 0, \; A = 1 = C, \; D = -1 < 0: \qquad 0 = \nabla^2 \Phi \equiv \Delta \Phi \equiv \frac{\partial^2 \Phi}{\partial x^2} + \frac{\partial^2 \Phi}{\partial y^2}. \qquad (1.230\text{a–f})$$

The characteristic polynomial has roots $\pm i$:

$$0 = \left(\frac{\partial^2}{\partial x^2} + \frac{\partial^2}{\partial y^2} \right) \Phi(x,y) = \left(\frac{\partial}{\partial y} - i \frac{\partial}{\partial x} \right) \left(\frac{\partial}{\partial y} + i \frac{\partial}{\partial x} \right) \Phi(x,y). \qquad (1.231\text{a, b})$$

Hence (1.210b) the general integral is (1.232c):

$$f, g \in \mathcal{D}^2\left(\mathbb{R}^2 \right): \qquad \Phi(x,y) = f\left(x + iy \right) + g\left(x - iy \right), \qquad (1.232\text{a–c})$$

where (1.232a, b) are arbitrary twice differentiable functions. The result (1.232b) shows that the characteristic variables in this case are:

$$z = x + iy, \qquad z^* = x - iy, \qquad (1.233\text{a, b})$$

the complex number (1.233a) and its conjugate (1.233b):
   Using z, z^* as variables instead of x, y, leads to (1.234a, b):

$$\frac{\partial}{\partial x} = \frac{\partial z}{\partial x} \frac{\partial}{\partial z} + \frac{\partial z^*}{\partial x} \frac{\partial}{\partial z^*} = \frac{\partial}{\partial z} + \frac{\partial}{\partial z^*},$$
$$\frac{\partial}{\partial y} = \frac{\partial z}{\partial y} \frac{\partial}{\partial z} + \frac{\partial z^*}{\partial y} \frac{\partial}{\partial z^*} = i \left(\frac{\partial}{\partial z} - \frac{\partial}{\partial z^*} \right); \qquad (1.234\text{a, b})$$

conversely $x(y)$ is the real (1.235a) [imaginary (1.235b)] part of the complex number (1.151a):

$$x = \frac{z + z^*}{2} = \operatorname{Re}(z) = \operatorname{Re}(z^*), \qquad y = \frac{z - z^*}{2i} = \operatorname{Im}(z) = -\operatorname{Im}(z^*), \qquad (1.235\text{a, b})$$

lead to (1.231a, b):

$$\frac{\partial}{\partial z} = \frac{\partial x}{\partial z} \frac{\partial}{\partial x} + \frac{\partial y}{\partial z} \frac{\partial}{\partial y} = \frac{1}{2} \left(\frac{\partial}{\partial x} - i \frac{\partial}{\partial y} \right), \qquad (1.236\text{a})$$

$$\frac{\partial}{\partial z^*} = \frac{\partial x}{\partial z^*} \frac{\partial}{\partial x} + \frac{\partial y}{\partial z^*} \frac{\partial}{\partial y} = \frac{1}{2} \left(\frac{\partial}{\partial x} + i \frac{\partial}{\partial y} \right). \qquad (1.236\text{b})$$

Using (1.234a, b) leads to (1.237a, b):

$$\frac{\partial}{\partial x} - i\frac{\partial}{\partial y} = 2\frac{\partial}{\partial z}, \qquad \frac{\partial}{\partial x} + i\frac{\partial}{\partial y} = 2\frac{\partial}{\partial z*}. \tag{1.237a, b}$$

that coincides with (1.237a, b) ≡ (1.236a, b).

Substituting (1.237a, b) in (1.231b) leads to (1.238a) that coincides with (1.238b) arising from the substitution of (1.234a, b) in (1.231a):

$$\left(\frac{\partial}{\partial x} - i\frac{\partial}{\partial y}\right)\left(\frac{\partial}{\partial x} + i\frac{\partial}{\partial y}\right) = 4\frac{\partial^2}{\partial z \partial z*} = \left(\frac{\partial}{\partial z} + \frac{\partial}{\partial z*}\right)^2 + \left[i\left(\frac{\partial}{\partial z} - \frac{\partial^2}{\partial z*}\right)\right]^2. \tag{1.238a, b}$$

Thus *the solution of the Laplace equation:*

$$\Phi(x,y) = \Phi\left(\frac{z+z*}{2}, \frac{z-z*}{2i}\right) \equiv \Psi(z,z*) = \Psi(x+iy, x-iy):$$

$$0 = \frac{\partial^2 \Phi}{\partial x^2} + \frac{\partial \Phi}{\partial y^2} = 4\frac{\partial^2 \Psi}{\partial z \partial z*}, \tag{1.239a, b}$$

is specified by:

$$f, g \in \mathcal{D}(|C): \quad \Psi(z,z*) = f(z) + g(z*)$$
$$= f(x+iy) + g(x-iy) = \Phi(x,y), \tag{1.240a–c}$$

in agreement with (1.232a–c;1.233a, b) ≡ (1.239a, b;1.240a–c), noting that complex holomorphic functions (1.240a–c) are infinitely differentiable (Section I.15.4) and thus have second-order derivatives (1.232a, b).

1.7.2   REAL/COMPLEX HARMONIC FUNCTIONS AND BOUNDARY CONDITIONS

The general integral of Laplace equation in the Cartesian plane consists (1.232a, b) ≡ (1.240b) of a function f(z) of a complex variable (1.233a), plus a function g(z*) of its conjugate (1.233b). If g (z) is an analytic function, then g(z*) is not analytic (Section I.33.2), and an analytic solution of the Laplace equation corresponds to g = 0, that is the first term of (1.240a–c) alone; this is the solution of Laplace's equation which follows from the Cauchy-Rieman conditions for an holomorphic function (chapter I.11). If the function f(z) is analytic, and real for real z, then it satisfies (Section I.31.2) the reflection principle (1.241a):

$$f(z*) = f*(z); \quad g(z) = f(z): \quad \Phi(x,y) = f(z) + f*(z)$$
$$= 2\,\mathrm{Re}\{f(z)\}, \tag{1.241a–c}$$

and choosing (1.241b) in (1.240a–c) leads to (1.241c), that is a real solution of the Laplace equation.

In conclusion: *the Laplace equation in the Cartesian plane (1.230e,f) has: (i) general integral (1.232c), consisting of the sum of two arbitrary twice differentiable functions (1.232a, b) of the complex variable (1.233a) and its conjugate (1.233b); (ii) a complex analytic (1.242a) solution is obtained taking only the first function (1.242b):*

$$f \in \mathcal{D}(|C): \qquad f(x+iy) = \Phi(x,y) + i\Psi(x,y), \quad \nabla^2\Phi = 0 = \nabla^2\Psi; \quad (1.242\text{a--d})$$

(iii) a real solution is obtained by taking the real (1.242c) ≡ (1.243c) [or imaginary (1.242d)] part of an analytic function (1.243b), that is (1.243a) real for real z:

$$f(x) \in R \wedge f(z) \in \mathcal{D}(|C): \qquad \Phi(x,y) = \text{Re}\{f(x+iy)\}. \qquad (1.243\text{a--c})$$

The arbitrary functions that appear in the solution of p.d.e.s are determined uniquely by suitable boundary (or initial) conditions.

To illustrate the preceding results, consider a real solution of the Laplace equation (1.244a) satisfying the boundary condition (1.244b):

$$\frac{\partial^2\Phi}{\partial x^2} + \frac{\partial^2\Phi}{\partial y^2} = 0, \qquad \Phi(x,0) = x^3. \qquad (1.244\text{a, b})$$

Since $f(z) = z^3$ is analytic, and $f(z) = x^3$ real for real x, the solution should be:

$$\Phi(x,y) = \text{Re}(z^3) = \text{Re}\{(x+iy)^3\} = x^3 - 3xy^2; \qquad (1.245\text{a--c})$$

it can be checked that (1.245c) satisfies (1.244a, b). This brief account (articles 1.7.1–1.7.2) shows how the main results of the theory of the potential (volume I and II) can be obtained in the context of the solution of p.d.e.s. It is extended next from harmonic to biharmonic functions (articles 1.7.3–1.7.4).

1.7.3 BIHARMONIC EQUATION IN THE CARTESIAN PLANE

The solution of the Laplace equation are called **harmonic functions**, and appear in the study (volume I and II) of the potential flow, electro- and magnetostatic field, gravity field and steady heat conduction. The **biharmonic equation:**

$$0 = \Delta^2\Phi = \nabla^4\Phi = \left(\frac{\partial^2}{\partial x^2} + \frac{\partial^2}{\partial y^2}\right)^2 \Phi = \frac{\partial^4\Phi}{\partial x^4} + 2\frac{\partial^4\Phi}{\partial x^2\partial y^2} + \frac{\partial^4\Phi}{\partial y^4}, \qquad (1.246\text{a--d})$$

appears in plane elasticity (chapter II.6) and also for plane steady incompressible viscous creeping flow (note N.II.6.6). It is a p.d.e. with constant coefficients, no forcing term and all derivatives of fourth-order. The factorization of the polynomial operator is the same as for Laplace equation (1.231a, b), except that the roots $\pm i$ are double (1.165a, b):

$$0 = \left(\frac{\partial^2}{\partial x^2} + \frac{\partial^2}{\partial y^2}\right)^2 \Phi = \left(\frac{\partial}{\partial y} - i\frac{\partial}{\partial x}\right)^2 \left(\frac{\partial}{\partial y} + i\frac{\partial}{\partial x}\right)^2 \Phi; \qquad (1.247a, b)$$

the characteristic variables are again the complex number (1.233a) and its conjugate (1.233b), and the general integral (1.222a–d) is:

$$f,g,h,j \in \mathcal{D}(|C):$$
$$\Phi(x,y) = f(z) + z * h(z) + g(z*) + zj(z*) \equiv \Psi(z,z*), \qquad (1.248a\text{–}e)$$

involving (1.248e) four arbitrary complex differentiable functions (1.248a–d) corresponding to (1.248e) ≡ (1.249e):

$$f,g,h,j \in \mathcal{D}^4\left(IR^2\right):$$
$$\Phi(x,y) = f(x+iy) + (x-iy)h(x+iy) + g(x-iy) + (x+iy)j(x-iy), \qquad (1.249a\text{–}e)$$

where the functions have fourth-order derivatives (1.249a–d) with regard to the real variables x, y.

This result could also be obtained by taking the biharmonic operator in the complex form (1.250) analogous to (1.238a, b):

$$\frac{\partial^4}{\partial x^4} + 2\frac{\partial^4}{\partial x^2 \partial y^2} + \frac{\partial^4}{\partial y^4} = 16\frac{\partial^4}{\partial z^2 \partial z *^2}. \qquad (1.250)$$

Integrating (1.251a): (i) twice with regard to z^* leads to (1.251b) a polynomial of first degree in z^*, whose coefficients may depend on z:

$$0 = \frac{\partial^4 \Psi}{\partial z *^2 \partial z^2} : \qquad \frac{\partial^2 \Psi}{\partial z^2} = f''(z) + z *^2 h''(z); \qquad (1.251a, b)$$

(ii) the coefficients are written as second-order derivatives of the functions f, g with regard to z^*, so that a double integration with regard to z regains the functions f, h in (1.248d); (iii) the double integration of (1.251b) with regard to z leads to a polynomial of first degree in z added in the r.h.s. of (1.248d) whose coefficients g, j may be functions of z^*; (iv) the four functions are analytic (1.248a–d) [four times differentiable (1.249a–d)] with regard to one complex (two real) variable(s).

1.7.4 REAL/COMPLEX BIHARMONIC FUNCTIONS AND FORCING

It has been shown that *the general integral of the biharmonic (1.246a–d) equation in the Cartesian plane is (1.248e) [(1.249e)], where f,g,h,j are arbitray functions that are complex analytic (1.248a–d) [have fourth-order derivatives with regard to two real variables (1.249a–d)]. A real solution is given by (1.252e):*

$$f,g \in \mathcal{A}(|C); \qquad f,g(x) \in |R: \qquad \Phi(x,y) = \operatorname{Re}\{f(z) + z^* h(z)\}, \qquad (1.252a\text{–}e)$$

involving two analytic functions of complex variable (1.252a, b) that are real (1.252c, d) for real z. The last result is obtained from (1.248e) choosing (1.253a, b) and using (1.241a) the reflection principle (1.252c, d):

$$\{g(z), j(z)\} = \{f(z), h(z)\}, \qquad \{f(z^*), h(z^*)\} = \{f^*(z), h^*(z)\}, \quad (1.253a\text{–}d)$$

leading to (1.254a–d):

$$\Psi(z,z^*) = f(z) + f(z^*) + z^* h(z) + z h^*(z)$$
$$= f(z) + f^*(z) + z^* h(z) + \{z^* h(z)\}^* \qquad (1.254a\text{–}d)$$
$$= 2\operatorname{Re}\{f(z) + z^* h(z)\} = 2\Phi(x,y).$$

The two arbitrary functions in (1.254d) are determined from two independent and compatible boundary conditions.

As an example consider the solution of the biharmonic equation (1.255a) with the boundary conditions (1.255b,c):

$$0 = \frac{\partial^4 \Phi}{\partial x^4} + 2\frac{\partial^4 \Phi}{\partial x^2 \partial y^2} + \frac{\partial^4 \Phi}{\partial y^4}: \qquad \Phi(x,0) = x - x^4, \qquad \Phi(0,y) = y^4. \quad (1.255a\text{–}c)$$

Using the complex variable (1.233a) and its conjugate (1.233b) the solution of the biharmonic equation (1.255a) must be (1.256):

$$\Phi(x,y) = \operatorname{Re}\{f(x+iy) + (x-iy)h(x+iy)\}, \qquad (1.256)$$

with the functions f, h determined by the boundary conditions (1.255b,c). Assuming that the functions are real variable (1.257a, b) the boundary condition (1.255b) substituted in (1.256) leads to (1.257c).

$$f(x) = \operatorname{Re}\{f(x)\}, \quad h(x) = \operatorname{Re}\{h(x)\}:$$
$$x - x^4 = \Phi(x,0) = \operatorname{Re}\{f(x) + x h(x)\} = f(x) + x h(x). \qquad (1.257a\text{–}c)$$

Assuming that the functions are imaginary for imaginary variable (1.258a, b) the boundary condition (1.255c) substituted in (1.256) leads to (1.285c):

$$f(iy) = if(y), \quad h(iy) = -ih(y):$$
$$y^4 = \Phi(0,y) = \operatorname{Re}\{f(iy) - iyh(iy)\} = \operatorname{Re}\{if(y) - y h(y)\} = -y h(y). \qquad (1.258a\text{–}c)$$

From (1.258c) follows the second function (1.259a) confirming that is satisfies (1.257b) and (1.258b); substituting (1.259a) in (1.257c) specifies the first function (1.259b) confirming that it satisfies (1.257a) and (1.258a):

$$h(z) = -z^3; \quad f(z) = z; \quad \Phi(x,y) = \text{Re}\left\{x + iy - (x - iy)(x + iy)^3\right\}. \quad (1.259a\text{--}e)$$

Substituting (1.259a, b) in (1.256) leads to (1.259c) that is simplified (1.260a–c):

$$\Phi(x,y) = x - (x^2 + y^2)\text{Re}\left\{(x + iy)^2\right\} = x - (x^2 + y^2)(x^2 - y^2)$$
$$= x - x^4 + y^4. \quad (1.260a\text{--}c)$$

It can be confirmed that *(1.260c) is a biharmonic function satisfying the partial differential equation (1.255a) and boundary conditions (1.255b,c).* The similarity solutions apply to linear partial differential equations with constant coefficients and derivatives all of the same order in the unforced case (Section 1.6), like the Laplace and biharmonic equations (Section 1.7) and extend to forcing by similarity type functions (Sections 1.8–1.9).

1.8   FORCED LINEAR P.D.E. WITH OF DERIVATIVES CONSTANT ORDER

The solution of the linear partial differential equation with constant coefficients and all derivatives of the same order is extended from the unforced case (Section 1.6) to forcing (subsection 1.8) by a similarity function (subsection 1.8.2) in the: (i) non-resonant case (subsection 1.8.1).; (ii) case of single or multiple resonances (subsection 1.8.5) using alternative methods [subsection 1.8.3 (1.8.4)] of parametric differentiation (L`Hôspital rule). The second-order p.d.e. is used as example to obtain the particular integrals for exponential (sinusoidal) forcing [subsection 1.8.6 (1.8.7)] and also the complete integral in non-resonant (resonant) cases [subsection 1.8.8 (1.8.9).

1.8.1   FORCING OF P.D.E. BY A SIMILARITY FUNCTION

The linear partial differential equation with constant coefficients and derivatives all of the same order (1.201a) is considered for forcing by a similarity function (1.261a) whose argument is a similarity variable that is a linear combination of the two impendent variables leading to (1.261b):

$$B(x,y) = g(y + cx): \quad \sum_{n=0}^{N} A_n \frac{\partial^N \Phi}{\partial x^n \partial y^{N-n}} = g(y + cx). \quad (1.261a, b)$$

A particular integral of the forced p.d.e. (1.261b) is sought as a similarity function (1.262b) similar to (1.261a) of the same similarity variable (1.262a) leading to the derivatives (1.262c, d):

$$\xi \equiv y + cx, \quad \Phi(x,y) = f(x,y) = f(\xi):$$
$$\left\{ \frac{\partial \Phi}{\partial x}, \frac{\partial \Phi}{\partial y} \right\} = \frac{df}{d\xi} \left\{ \frac{\partial \xi}{\partial x}, \frac{\partial \xi}{\partial y} \right\} = \{c, 1\} \frac{df}{d\xi}.. \qquad (1.262\text{a–d})$$

Substituting the similarity solution (1.262b) in the p.d.e. (1.261b) with simility forcing (1.261a) leads to (1.263a):

$$g(y + cx) = \frac{d^N f}{d\xi^N} \sum_{n=0}^{N} A_n c^n = P_N(c) \frac{d^N f}{d\xi^N}, \qquad (1.263\text{a, b})$$

were appears (1.263b) the characteristic polynomial (1.211c, d).

1.8.2 SIMILARITY FORCING IN THE NON-RESONANT CASE

In the **non-resonant case** when c is not a root of the characteristic polynomial (1.264a) it is legitimate to perform the division in (1.263b) leading to (1.264b):

$$P_N(c) \neq 0: \quad \frac{d^N f}{d\xi^N} = \frac{g(\xi)}{P_N(c)}, \quad f(\xi) = \frac{1}{P_N(c)} \left(\frac{d}{d\xi} \right)^{-N} g(\xi), \qquad (1.264\text{a–c})$$

so that the solution is (1.264c), where the **primitive of order N** is the inverse (1.265a) of the derivative of order N and corresponds to N indefinite integrations (1.265b):

$$\frac{d^N}{d\xi^N} \frac{d^{-N}}{d\xi^{-N}} = 1;$$

$$g^{(-N)}(\xi) = \frac{d^{-N} g}{d\xi^{-N}} \equiv \left(\frac{d}{d\xi} \right)^{-N} g(\xi) \equiv \int^{\xi} dx_1 \int^{x_1} dx_2 \dots \int^{x_{N-1}} g(x_N) dx_N. \qquad (1.265\text{a, b})$$

It has been shown that *a linear partial differential equation whith constant coefficients and derivatives all of the same order N forced (1.261b) by a similarity function (1.261a) in the non-resonant case when c is not (1.264a) a root of the characteristic polynomial (1.211c, d) has particular integral (1.264c) as a similarity function (1.262a, b) involving the N-th primitive (1.264a) of forcing function (1.265b).*In the case when c is a root of the characteristic polynomial the passage from (1.263b) to (1.264b) is not legitimate since it is a division by zero; in this resonant case may be used either the method of parametric differentiation (subsection 1.8.3) or L`Hôspital rule (subsection 1.8.4).

1.8.3 PARAMETRIC DIFFERENTIATION FOR MULTIPLE RESONANCE

Consider next the particular integral (1.263b) of the partial differential equation (1.261b) in case of **multiple resonance of order s** when c is a root (1.266a) of multiplicity s of the characteristic polynomial (1.211c, d)

$$P_N(c) = P_N'(c) = \cdots = P_N^{(s-1)}(c) = 0 \neq P_N^{(s)}(c): \quad \frac{\partial^s g}{\partial c^s} = \frac{\partial^s}{\partial c^s}\left[P_N(c)\frac{d^N f}{d\xi^N}\right]. \quad (1.266a, b)$$

and differentiate (1.263b) s times with regard to c leading to (1.266b). Using (1.262a) on the l.h.s. and the Leibniz rule (I.13.31) on the r.h.s. of (1.266b) leads to (1.267).

$$\frac{d^s g}{d\xi^s}\left(\frac{d\xi}{dc}\right)^s = \sum_{r=0}^{s} \frac{s!}{p!(s-r)!}P_N^{(r)}(c)\left(\frac{d}{dc}\right)^{s-r}\left[\frac{d^N f}{d\xi^N}\right]. \quad (1.267)$$

Since c is a root of multiplicity s of the characteristic polynomial (1.266a) all terms on the r.h.s. of (1.267) vanish except r=s leading to (1.268a) where (1.262a) is used on the l.h.s.:

$$x^s \frac{d^s g}{d\xi^s} = P_N^{(s)}(c)\frac{d^N f}{d\xi^N}; \quad f(\xi) = \frac{x^s}{P_N^{(s)}(c)}\left(\frac{d}{d\xi}\right)^{s-N} g(\xi). \quad (1.268a, b)$$

The result (1.268a) also follows from (1.266b) noting that all derivatives must be applied to the characteristic polynomial otherwise a zero arises from (1.266a). From (1.268a) follows the particular integral (1.268b), that can be obtained alternatively (subsection 1.8.4) from L`Hôpital´s rule (subsection I.19.8).

1.8.4 MULTIPLE RESONANCE VIA L'HÔSPITAL RULE

The particular integral (1.264c) of the partial differential equation (1.261b) forced by a similarity function (1.261a), in the case of multiple resonance of order s, when c is a root of multiplicity s of the characteristic polynomial (1.266a) and then (1.264c) has a zero of order s in the denominator. In the numerator of (1.264c) a solution of the unforced partial differential equation (1.203b) can be subtracted giving also a zero of order s. The indetermination of type 0:0 can be lifted by L`Hôspital rule (I.19.35) differentiating separately both the numerator and denominator s times with regard to c. Starting with (1.264c) modified by differentiation of order s with regard to c leads to (1.269a) where the denominator is not zero (1.269b) for a root of multiplicity s.

$$f(\xi) = \frac{(\partial/\partial c)^s g^{(-N)}(\xi)}{(d/dc)^s P_N(c)}; \qquad \frac{d^s}{dc^s}\left[P_N(c)\right] \equiv P_N^{(s)}(c) \neq 0; \quad (1.269a, b)$$

Differentiating s times the numerator of (1.269a) leads to (1.270b) using (1.270a) that follows from (1.262a):

$$\frac{d\xi}{dc} = x: \qquad \frac{d^s}{dc^s}\left[g^{(-N)}(\xi)\right] = g^{(s-N)}(\xi)\frac{d^s\xi}{dc^s} = x^s g^{(s-N)}(\xi). \qquad (1.270a\text{–}c)$$

The ratio of (1.270c) to (1.269b) in (1.269a) provides an alternative proof of (1.268b).

1.8.5 COMPARISON OF RESONANT AND NON-RESONANT CASES

It has been shown that *the linear partial differential equation with constant coefficients and derivatives all of order N forced (1.261b) by a similarity (1.261a) function, in the case of multiple resonance of order s, when c is a root (1.266a) ≡ (1.271a) of multiplicity s of the characteristic polynomial (1.211c, d) has particular integral (1.271b):*

$$P_N(c) = P_N'(c) = \cdots P_N^{(s-1)}(c) = 0 = P_N^{(s-1)}(c):$$

$$\Phi(x,y) = \frac{x^s}{P_N^{(s)}(c)}g^{(s-N)}(y+cx). \qquad (1.271a, b)$$

In the particular non-resonant case (1.272a), when c is not (1.272b) a root of the characteristic polynomial the partial differential equation (1.261b) has the particular integral (1.272c):

$$s = 0: \qquad P_N(c) \neq 0, \qquad \Phi(x,y) = \frac{g^{(-N)}(y+cx)}{P_N(c)}, \qquad (1.272a\text{–}c)$$

in agreement with (1.264c) ≡ (1.272c). As examples are considered [subsection 1.8.6 (1.8.7)] the forcing by an exponential (1.273a) [sinusoidal (1.273b)] similarity function:

$$g(y+cx) = e^{y+cx}, \qquad g(y+cx) = \cos(y+cx) = \text{Re}\left\{e^{i(y+cx)}\right\}, \qquad (1.273a, b)$$

of the second-order p.d.e (1.223a).

1.8.6 FORCING BY AN EXPONENTIAL SIMILARITY FUNCTION

A particular integral of the second-order partial differential equation (1.223a) forced by (1.273a) an exponential similarity function (1.274):

$$A\frac{\partial^2\Phi}{\partial x^2} + 2B\frac{\partial^2\Phi}{\partial x\partial y} + C\frac{\partial^2\Phi}{\partial y^2} = e^{y+cx}, \qquad (1.274)$$

is sought in all three possible cases; non-resonant and single or double resonance. The characteristic polynomial (1.224b) for (1.274) is (1.275a, b) with roots (1.225a, b):

$$P_2(c) = Ac^2 + 2Bc + D = (c - a_+)(c - a_-):$$
$$P_2'(c) = 2(Ac + B), \quad P_2''(c) = 2A, \tag{1.275a-d}$$

leading to non-zero derivatives of the first (1.275c) [second (1.275d)] order.

In the non-resonant case (1.276a) when c is not (1.276b) a root (1.224b) of the characteristic polynomial (1.211c, d) the particular integral (1.272c) is (1.276c, d):

$$s = 0, \ c \neq a_\pm: \quad \Phi(x,y) = \frac{e^{y+cx}}{Ac^2 + 2Bc + C} = \frac{e^{y+cx}}{A(c - a_+)(c - a_-)}. \tag{1.276a-d}$$

In the singly resonant (1.277a) case when c coincides with one of the roots (1.277b) the particular integral (1.271b) is (1.277c–e):

$$s = 1; \quad c = a_\pm: \quad \Phi_\pm(x,y) = \frac{x}{2} \frac{e^{y+a_\pm x}}{Aa_\pm + B} = \pm \frac{xe^{y+a_\pm x}}{2\sqrt{D}} = \frac{x}{A} \frac{e^{y+a_\pm x}}{a_\pm - a_\mp}, \tag{1.277a-e}$$

corresponding to the forced partial differential equation (1.274, 1.277a)≡(1.277f)

$$A\frac{\partial^2 \Phi}{\partial x^2} + 2B\frac{\partial^2 \Phi}{\partial x \partial y} + C\frac{\partial^2 \Phi}{\partial y^2} = \exp(y + a_\pm x) \tag{1.277f}$$

The doubly resonant case (1.278a) is possible only for a double root (1.278b,c) of the characteristic polynomial, implying (1.278d):

$$s = 2: \qquad c = a_+ = a_- \equiv a = -\frac{B}{A}: \qquad B^2 = AC, \tag{1.278a-d}$$

so that: (i) the partial differential equation (1.274) becomes (1.278e); (ii) the particular integral (1.183b) is (1.278f):

$$A\frac{\partial^2 \Phi_*}{\partial x^2} - 2B\frac{\partial^2 \Phi_*}{\partial x \partial y} + \frac{B^2}{A}\frac{\partial^2 \Phi_*}{\partial y^2} = \exp\left(y - \frac{B}{A}x\right):$$
$$\Phi_*(x,y) = \frac{x^2}{2A}\exp\left(y - \frac{B}{A}x\right). \tag{1.278e,f}$$

It has been shown that *the linear partial differential equation with constant coefficients and all derivatives of the second order (1.223a, b) forced by an exponential similarity function (1.274) has particular integral specified: (i) by (1.276c, d) in the non-resonant case (1.276a) when c is not (1.276b) a root (1.225a, b) of the characteristic polynomial (1.224c, d) ≡ (1.275a, b); (ii) by (1.277c–e) in the singly resonant case (1.277a) when c coincides (1.277b) with one of the roots of the characteristic polynomial; (iii) by (1.278f) in the doubly resonant case (1.278a) that is possible only (1.278c) for a double root (1.278b) of the characteristic polynomial coincident with c leading to the forced p.d.e. (1.278e).* In the derivation of the preceding results it has been taken into account that the exponential coincides with its derivatives and primitives of all orders (1.279a) and in fact is the only function having this property (sections II.3.1-II.3.3):

$$\left(\frac{d}{d\xi}\right)^{\pm N} e^{\xi} = e^{\xi}; \quad \left(\frac{d}{d\xi}\right)^{\pm 2N} \cos\xi = (-)^{N} \cos\xi,$$

$$\left(\frac{d}{d\xi}\right)^{\pm 2N+1} \cos\xi = (-)^{N+1} \sin\xi, \tag{1.279a–c}$$

thus other analytic functions do not have this property, for example, for the circular cosine the derivatives and primitives of even (1.279b) [odd (1.279c)] order are cosines (sines). This property is used next (subsection 1.8.7) when considering the forcing of a partial differential equation by a sinusoidal similarity forcing function (subsection 1.8.7).

1.8.7 FORCING BY A SINUSOIDAL SIMILARITY FUNCTION

The forcing of the partial differential equation (1.261b) by a circular cosine similarity function (1.261a) leading to (1.280a, b):

$$A\frac{\partial^2\Phi}{\partial x^2} + 2B\frac{\partial^2\Phi}{\partial x\partial y} + C\frac{\partial^2\Phi}{\partial y^2} = \cos(y + cx) = \text{Re}\left\{e^{i(y+cx)}\right\}, \tag{1.280a, b}$$

for which the characteristic polynomial is (1.281a, b):

$$A(ic)^2 + 2Bi(ic) + Ci^2 = -\left(Ac^2 + 2Bc + C\right) = -P_2(c), \tag{1.281a, b}$$

minus (1.275a). It follows that in the non-resonant case (1.282a) when c is not a root of the characteristic polynomial (1.282b) the simplest particular integral of (1.280b) is (1.282c–e):

$$s = 0, \quad c \neq a_{\pm}: \quad \Phi(x,y) = \text{Re}\left\{-\frac{e^{i(y+cx)}}{P_2(c)}\right\} = -\frac{\cos(y + cx)}{Ac^2 + 2Bc + C}$$

$$= -\frac{\cos(y + cx)}{A(c - a_+)(c - a_-)}. \tag{1.282a–e}$$

In the singly resonant case (1.283a) when c is one of the roots (1.283b) of the characteristic polynomial, bearing in mind (1.279c) that the primitive of $\cos \xi$ is $\sin \xi$, the simplest particular integral of (1.280a) is (1.283c–f):

$$s = 1, \quad c = a_\pm: \quad \Phi(x,y) = \frac{x \sin(y + a_\pm x)}{-P'_2(a_\pm)} - \frac{x}{2} \frac{\sin(y + a_\pm x)}{A\, a_\pm + B}$$

$$= \mp \frac{x \sin(y + a_\pm x)}{2\sqrt{D}} = \frac{x}{2A} \frac{\sin(y + a_\pm x)}{a_\mp - a_\pm}, \qquad (1.283a\text{–}f)$$

corresponding to the forced partial differential equation (1.280a: 1.283b)≡(1.283g)

$$A \frac{\partial^2 \Phi}{\partial x^2} + 2B \frac{\partial^2 \Phi}{\partial x \partial y} + C \frac{\partial^2 \Phi}{\partial y^2} = \cos(y + a_\pm x). \qquad (1.283g)$$

A double resonance (1.284a) is possible only if c equals a double root (1.284b–d) of the characteristic polynomial, in which case the p.d.e. (1.280a) reduces to (1.284e) and its simplest particular integral is (1.289f,g):

$$s = 2, \quad c = a_\pm = -\frac{B}{A} = -\sqrt{\frac{C}{A}} \equiv a:$$

$$A \frac{\partial^2 \Phi}{\partial x^2} + 2B \frac{\partial^2 \Phi}{\partial x \partial y} + \frac{B^2}{A} \frac{\partial^2 \Phi}{\partial y^2} = \cos\left(y - \frac{B}{A} x\right), \qquad (1.284a\text{–}e)$$

$$\Phi(x,y) = \frac{x^2}{2A} \cos(y + cx) = \frac{x^2}{2A} \cos\left(y - \frac{B}{A} x\right) \qquad (1.284f,g)$$

It has been shown that *the linear partial differential equation with constant coefficients and derivatives all of the second order with forcing by a circular cosine similarity function (1.280a, b) has simplest particular integral given: (i) by (1.282c–e) in the non-resonant case (1.282a, b); (ii) by (1.283c–f) in the singly resonant case (1.283a, b) corresponding to the forced p.d.e. (1.283g); (iii) by (1.284f,g) in the doubly resonant case (1.284a–d) corresponding to the forced p.d.e. (1.284e).* The general (particular) integral(s) of the unforced (1.203a, b) [forced (1.201a, b)] p.d.e. obtained before [subsection(s) 1.6.8 (1.8.6–1.8.7)] can be added to specify the complete integral (subsection 1.8.8).

1.8.8   GENERAL, PARTICULAR AND COMPLETE INTEGRALS

As for forced linear ordinary differential equations (section IV.1.2) a forced linear partial differential equation (1.11b) ≡ (1.285b) consists of a linear partial differential operator (1.285a) applied to the dependent variable and equated to the forcing function:

$$L\left(\frac{\partial}{\partial x}, \frac{\partial}{\partial y}\right) \equiv \sum_{n=0}^{N} \sum_{m=0}^{n} A_{n,m}(x,y) \frac{\partial^n}{\partial x^m \partial y^{n-m}} : \quad \left\{L\left(\frac{\partial}{\partial x}, \frac{\partial}{\partial y}\right)\right\} \Phi(x,y) = B(x,y). \quad (1.285a, b)$$

As in the case of linear forced ordinary differential equations (section IV.1.2) *the complete integral (1.286a) of a linear forced partial differential equation (1.286c):*

$$\bar{\Phi}(x,y) = \Phi(x,y) + \Phi_*(x,y): \qquad \left\{ L\left(\frac{\partial}{\partial x}, \frac{\partial}{\partial y}\right) \right\} \Phi(x,y) = 0, \quad (1.286a, b)$$

$$\left\{ L\left(\frac{\partial}{\partial x}, \frac{\partial}{\partial y}\right) \right\} \bar{\Phi}, \Phi_*(x,y) = B(x,y) \qquad (1.286c, d)$$

consists of the sum of: (i) the general integral of the unforced differential equation (1.286b) involving N arbitrary functions; (ii) any particular integral of the forced equation (1.286c), the simpler the better since it does not need to include any arbitrary functions. To the particular integral of the forced equation can be added any particular integrals of the unforced equation and the forced equation is still satisfied.

1.8.9 COMPLETE INTEGRALS WITH SIMPLE/DOUBLE RESONANCES

As an example:

$$A\frac{\partial^2 \Phi}{\partial x^2} + 2B\frac{\partial^2 \Phi}{\partial x \partial y} + C\frac{\partial^2 \Phi}{\partial y^2} = E_0 \exp(y+cx) + E_+ \exp(y+a_+x)$$
$$+ E_- \exp(y+a_-x) + F_0 \cos(y+cx) \qquad (1.287)$$
$$+ F_+ \cos(y+a_+x) + F_- \cos(y+a_-x),$$

the linear partial differential equation with constant coefficients and all derivatives of the second order (1.287) has complete integral (1.288b) consisting of the sum of: (i) the general integral (1.226a–d) of the unforced p.d.e. (1.223a, b) involving the roots (1.225a–c) of the characteristic polynomial (1.224a–d); (ii,iii) the non-resonant particular integrals (1.276a–d) [(1.282a–e)] corresponding to non-resonant forcing (1.274) [(1.280a, b)]; (iv,v) the resonant particular integrals (1.277a–b) (1.283a–f)] corresponding to single forcing:

$$c \neq a_\pm: \quad \Phi(x,y) = f(y+a_+x) + g(y+a_-x) + \frac{E_0 e^{y+cx} - F_0 \cos(y+cx)}{A(c-a_+)(c-a_-)} \qquad (1.288a, b)$$
$$+ \frac{x}{A}\frac{E_+ \exp(y+a_+x) - E_- \exp(y+a_-x) - F_+ \sin(y+a_+x) + F_- \sin(y+a_-x)}{a_+ - a_-},$$

where the arbitrary functions in (1.288b) are twice differentiable (1.226b,c).

The particular doubly resonant case corresponds to the partial differential equation without (1.227a–c) [with (1.278a–f; 1.284a–g)] forcing (1.289):

$$A\frac{\partial^2 \Phi_*}{\partial x^2} + 2B\frac{\partial^2 \Phi_*}{\partial x \partial y} + \frac{B^2}{A}\frac{\partial^2 \Phi_*}{\partial y^2} = E \exp\left(y - \frac{B}{A}x\right) + F\cos\left(y - \frac{B}{A}x\right), \qquad (1.289)$$

TABLE 1.5
Linear Partial Equations with All Derivatives of Order N and Constant Coefficients

Forcing	Unforced		Forced	
Section	1.6		1.8	
Subsection	*single root 1.6.1–1.6.3	*root of multiplicity s 1.6.4–1.6.7	non-resonant 1.8.1–1.8.2	resonant 1.8.3–1.8.5
Partial differential equation	(1.203a, b)	(1.213)	(1.264a)	(1.271a)
Characteristic polynomial	(1.202a, b)	(1.212a, b)	(1.202a, b)	(1.202a, b)
General/Particular Integral	(1.208a–c)	(1.222a–d)	(1.264c; 1.265b)	(1.271b)
Examples	1.6.8		1.8.6–1.8.8	

Note: General (Particular) integral of an unforced (forced) linear partial differential equation with constant coefficients and all derivatives of the same order in the cases of single or multiple roots of the characteristic polynomial (non-resonant and resonant cases).

and to the complete integral (1.290) consisting of the sum of the general (1.228d) [particular (1.278f;1.284f,g)] integral(s) of the unforced (forced) partial differential equation (1.289):

$$\Phi(x,y) = f(Ax - By) + (Ax + By)g(Ax - By)$$
$$+ \frac{x^2}{2A}\left[E\exp\left(y - \frac{B}{A}x\right) + F\cos\left(y - \frac{B}{A}x\right)\right] \quad (1.290)$$

where the arbitrary functions in (1.290) are twice differentiable (1.226b,c). The linear partial differential equation with constant coefficients with all derivatives of the same order has been considered (Table 1.5) without (with) forcing [Section 1.6 (1.8)] with the harmonic and biharmonic equations as examples [Section 1.7 (1.9)]. A case of complete solution of a forced linear p.d.e. with constant coefficients and all derivatives of the third order with two variables is considered in the example 10.4.

1.9 FORCED HARMONIC AND BIHARMONIC EQUATIONS

The use of a complex variable and its conjugate as similarity variables applies to both the forced (unforced) Laplace or harmonic [subsections 1.9.1 (1.7.1–1.7.2)] and double Laplace or biharmonic [subsection(s) 1.9.3 (1.7.3–1.7.4)] equations, and two examples of each are given [subsections 1.9.2 (1.9.4)]. The forcing by similarity functions (Section 1.8) is extended to the forcing by arbitrary functions (Section 1.9): (i) starting with the Laplace (biharmonic) equations [subsection 1.9.1–1.9.2 (1.9.3–1.9.4)]; (ii) extending to general linear partial differential equations with constant coefficients and derivatives all of the same order (subsections 1.9.5–1.9.7).

1.9.1 GENERAL FORCING OF THE LAPLACE EQUATION

The Laplace or harmonic equation is considered (1.291a) with general forcing not restricted to similarity functions:

$$\frac{\partial^2 \Phi}{\partial x^2} + \frac{\partial^2 \Phi}{\partial y^2} = B(x,y) \Leftrightarrow 4\frac{\partial^2 \Psi}{\partial z \partial z^*} = A(z,z^*), \qquad (1.291a, b)$$

and the use of complex (1.233a) and conjugate (1.233b) similarity variables (1.235a, b) in the dependent variable (1.239a) and forcing function (1.292) leads (1.238a, b) to (1.291b):

$$B(x,y) = B\left(\frac{z+z^*}{2}, \frac{z+z^*}{2i}\right) = A(z,z^*) = A(x+iy, x-iy). \qquad (1.292)$$

The integration of (1.291b) is immediate (1.293d):

$$f,g \in \mathcal{D}(|C); \quad A \in \mathcal{E}(|C^2): \quad \Psi(z,z^*) = f(z) + g(z^*) + \frac{1}{4}\int^z d\xi \int^{z^*} d\eta A(\xi,\eta), \quad (1.293a–d)$$

involving two arbitrary differentiable functions (1.293a, b) and integrable forcing function (1.293c).

1.9.2 COMPLETE INTEGRAL OF THE HARMONIC EQUATION

It has been shown that *the complete integral (1.293d) of the Laplace or harmonic equation (1.291a) with integrable forcing (1.293c) can be obtained using complex and conjugate (1.223a, b) similarity variables (1.225a, b) both in the dependent variable (1.239a) and in the forcing function (1.292) and consists of the sum of: (i) the general integral (1.240c) ≡ (1.232c) of the unforced equation (1.230e–f) involving two arbitrary differentiable functions (1.240a) ≡ (1.293a, b); (ii) a particular integral of the forced equation (1.291a) ≡ (1.291b) involving two integrations of the forcing function.* Considering only the simplest particular integral of the forced Laplace equation, two examples are: (i) for constant forcing (1.294a) the particular integral (1.294b–d):

$$B(x,y) = b: \qquad \Psi(z+z^*) = b\frac{z^* z}{4} = b\frac{x^2 + y^2}{4} = \Phi(x,y); \qquad (1.294a–d)$$

(ii) for the forcing function (1.295a–c) the particular integral is (1.295d–g):

$$B(x,y) = b(x^2 + y^2) = bz^* z = A(z,z^*):$$

$$\Psi(z,z^*) = b\frac{(z^* z)^2}{4.2^2} = b\frac{(x^2 + y^2)^2}{16} = b\frac{x^4 + y^4}{16} + b\frac{x^2 y^2}{8} = \Phi(x,y). \qquad (1.295a–g)$$

Similar methods apply to the double Laplace or biharmonic equation (subsections 1.9.3–1.9.4).

1.9.3 General Forcing of the Double Laplace Equation

The double Laplace or biharmonic equation is considered (1.296a, b) with general forcing not restricted to similarity functions:

$$B(x,y) = \left(\frac{\partial^2}{\partial x^2} + \frac{\partial^2}{\partial x^2} \right)^2 \Phi(x,y) = \frac{\partial^4 \Phi}{\partial x^4} + 2\frac{\partial^4 \Phi}{\partial x^2 \partial y^2} + \frac{\partial^4 \Phi}{\partial y^4}$$

$$\Leftrightarrow \quad 16 \frac{\partial^4 \Psi}{\partial z^2 \partial y^{*2}} = A(z, z^*); \tag{1.296a–c}$$

the use of complex (1.233a) and conjugate (1.233b) similarity variables (1.235a, b) in the depenent variable (1.239a) and forcing function (1.292) leads (1.291a, b) to (1.296c) whose solution is (1.297f):

$$f, g, h, j, \in \mathcal{D}^2(|C); \quad A \in \mathcal{E}^2(|C^2): \quad \Psi(z, z^*)$$

$$= f(z) + g^*(z) + z^* h(z) + zj(z^*) + \frac{1}{16} \int\limits^z d\xi \int\limits^\xi d\eta \int\limits^{z^*} d\alpha \int\limits^\alpha d\beta B(\eta, \beta), \tag{1.297a–f}$$

involving four arbitrary twice differentiable functions (1.297a–d), and twice integrable forcing (1.297e).

1.9.4 Complete Integral of the Biharmonic Equation

It has been shown *that the complete integral (1.297f) of the double Laplace or biharmonic equation (1.296a, b) with twice integrable forcing (1.297e) can be obtained using complex (1.233a) and conjugate (1.233b) similarity variables (1.235a, b) in the dependent variable (1.239a) and forcing function (1.292) and consists of the sum of: (i) the general integral (1.248e) ≡ (1.249e) of the unforced biharmonic equation, involving four arbitrary twice differentiable functions (1.248a–d) ≡ (1.297a–d); (ii) the simplest particular integral of the forced biharmonic Equation (1.296a–c) involving two double integrations.* Considering only the simplest particular integral of the forced biharmonic equation two examples are: (i) constant forcing (1.294a) ≡ (1.298a) leading to the particular integral (1.298b–e):

$$B(x,y) = b: \quad \Psi(z + z^*) = b\frac{z^2 z^{*2}}{16.2^2} = b\frac{(x^2 + y^2)^2}{64}$$

$$= b\frac{x^4 + y^4}{64} + b\frac{x^2 + y^2}{32} = \Phi(x,y); \tag{1.298a–d}$$

TABLE 1.6

Solutions of the Laplace and Biharmonic Equation

Equation	Laplace		Biharmonic	
Operator	$\nabla^2 = \dfrac{\partial^2}{\partial x^2} + \dfrac{\partial^2}{\partial y^2}$		$\nabla^4 = \dfrac{\partial x^4}{\partial x^4} + 2\dfrac{\partial^4}{\partial x^2 \partial y^2} + \dfrac{\partial^4}{\partial y^4}$	
Forcing	*Unforced**	*Forced***	*Unforced**	*Forced*
Equation	$\nabla^2\Phi = 0$	$\nabla^2\Phi = B$	$\nabla^4\Phi = 0$	$\nabla^4\Phi = B$
Using complex variables	*(1.239b)*	*(1.291a,b)*	*(1.250)*	*(1.296a–c)*
Integral	*(1.240a–c)*	*(1.293a–d)*	*(1.248a–e) ≡ (1.249a–e)*	*(1.297a–f)*
General	*1.7.1*	*1.9.1*	*1.7.3*	*1.9.3*
Examples	*1.7.2*	*1.9.2*	*1.7.4*	*1.9.4*

Note: General (particular) integrals of the unforced (forced) Laplace and biharmonic equations

---

(ii) the forcing (1.295a, b) ≡ (1.299a, b) leads to the particular integral (1.299c–f):

$$B(x,y) = b\left(x^2 + y^2\right) = bz^*z:$$

$$\Psi(z, z_*) = b\frac{z^3 z^{*3}}{16.6^2} = b\frac{\left(x^2 + y^2\right)^3}{576} = b\frac{x^6 + y^6}{576} + b\frac{x^2 y^2 \left(x^2 + y^2\right)}{192} = \Phi(x, y). \qquad (1.299\text{a–e})$$

The general (complete) integrals of the unforced (forced) Laplace and biharmonic equations are compared in Table 1.6.

1.9.5 METHOD OF SIMILARITY VARIABLES FOR ARBITRARY FORCING

The linear partial differential equation with constant coefficients and all derivatives of the same order N and arbitrary forcing was solved: (a) in general (Section 1.8) forced by a similarity function (1.261a, b) of one similarity variable; (b) in the case of the Laplace (1.291a, b) [biharmonic (1.296a–c)] equation [subsections 1.9.1–1.9.2 (1.9.3–1.9.4)] with forcing by arbitrary function (1.239a;1.292) involving two similary variables. There is a common generalization of (a) and (b), as shown next. The linear partial differential equation with constant coefficients and all derivatives of the same order N in terms (1.201a, b) of the original variables, can be rewritten (1.300c) using similarity variables for the dependent variable (1.300a) and forcing function (1.300b):

$$\Phi(x,y) \equiv \Psi(\xi,\eta), \quad B(x,y) = A(\xi,\eta): \qquad C\frac{\partial^N \Psi}{\partial \xi^M \partial \eta^{N-M}} = A(\xi,\eta), \qquad (1.300\text{a–c})$$

where: (i) only two similarity variables (ξ, η) appear; (ii) the constant C arises from the transformation from original variables in (1.201a, b) to similary variables in (1.300c); (iii) the similarity variable ξ (η) corresponds to a root of multiplicity

M (N − M) of the characteristic polynomial (1.202a, b). The complete integral of (1.300a–c) is (1.301c):

$$f_1, \ldots, f_M, g_1, \ldots, g_N \in \mathcal{D}^N \left(|R^2 \right):$$

$$\Psi(\xi, \eta) = \sum_{m=0}^{M-1} \xi^m f_m(\xi) + \sum_{n=0}^{N-1} \eta^n g_n(\eta)$$

$$+ \frac{1}{C} \int^{\xi} dx_1 \int^{x_1} dx_2 \ldots \int^{x_{m-1}} dx_M \int^{\eta} dy_1 \int^{y_1} dy_2 \ldots \int^{y_{N-1}} dy_N A(x_M, y_N),$$

(1.301a–c)

consisting of the sum of: (i/ii) the general integrals of the unforced equation for the similarity variable $\xi(\eta)$ in the first (second) terms on the r.h.s. of (1.301c) involving M (N-M) arbitrary N-times differentiable functions (1.301a) [(1.301b)]; (iii) the particular integral of the forced equation appearing as the third term on the r.h.s. of (1.301c). If there are additional similarity variables, such as ζ appearing in derivatives on the l.h.s. of (1.300c) they lead in the complete integral (1.301b) to terms like the first two on the r.h.s., and imply independent integration of the third term, since only two similarity variables are needed in the forcing function. Thus *the linear partial differential equation with constant coefficients (1.201a, b) and all derivatives of the same order with arbitrary forcing has complete integral (1.301c) where: (i) two similarity variables (ξ, η) replace the original variables (x,y) in the dependent variable (1.300a) and forcing function (1.300b); (ii) if the similarity variable $\xi(\eta)$ is a root of multiplicity M (N − M) of the characteristic polynomial then (1.300c) leads to (1.301c); (iii) any additional similarity variables like ξ that appears as further differential in (1.300c) leads to corresponding terms like the first term on the r.h.s. of (1.301c) and integrations of the last term; (iv) the arbitrary functions $f_1, \ldots, f_M$ and $g_1, \ldots, g_N$ in (1.301b) are N times differentiable (1.301a, b).* An example of a linear particular differential equation with constant coefficients and all derivatives of fourth order follows with two independent variables and three similary variables (subsection 1.6.6).

1.9.6   FOURTH-ORDER EQUATION WITH TWO INDEPENDENT AND THREE SIMILARITY VARIABLES

Consider the linear forced partial differential equation with constant coefficients and all derivatives of order four (1.302a):

$$B(x, y) = \frac{\partial^4 \Phi}{\partial x^4} - \frac{\partial^4 \Phi}{\partial x^2 \partial y^2} = \left\{ \frac{\partial^2}{\partial x^2} \left(\frac{\partial^2}{\partial x^2} - \frac{\partial^2}{\partial y^2} \right) \right\} \Phi(x, y)$$

$$= \left\{ \frac{\partial^2}{\partial x^2} \left(\frac{\partial}{\partial x} - \frac{\partial}{\partial y} \right) \left(\frac{\partial}{\partial x} + \frac{\partial}{\partial y} \right) \right\} \Phi(x, y).$$

(1.302a–c)

The roots of the characteristic polynomial are ±1 (single) and 0 (double), and thus the similarity variables are x with multiplicity two and (1.303a, b) with multiplicity one whose inverses are (1.303c, d):

$$\xi = x + y, \qquad \eta = x - y: \qquad 2x = \xi + \eta, \qquad 2y = \xi - \eta. \qquad (1.303a\text{--}d)$$

The factorization (1.302a–c) and single ξ, η and double x similarity variables show that the general integral of the unforced Equation (1.304b) is (1.304f) where (1.304a–d) are arbitrary four times differentiable functions:

$$f, g, h, j \in \mathcal{D}^4(\mathbb{R}^2), B(x, y) = 0:$$
$$\Phi(x, y) = h(y) + xj(y) + f(x + y) + g(x - y). \qquad (1.304a\text{--}f)$$

The derivatives with regard to original (x,y) and similarity (ξ, η) variables are related (1.303a, b) by (1.305a, b):

$$\frac{\partial}{\partial x} = \frac{\partial \xi}{\partial x} \frac{\partial}{\partial \xi} + \frac{\partial \eta}{\partial x} \frac{\partial}{\partial \eta} = \frac{\partial}{\partial \xi} + \frac{\partial}{\partial \eta}, \quad \frac{\partial}{\partial y} = \frac{\partial \xi}{\partial y} \frac{\partial}{\partial \xi} + \frac{\partial \eta}{\partial y} \frac{\partial}{\partial \eta} = \frac{\partial}{\partial \xi} - \frac{\partial}{\partial \eta}, \qquad (1.305a, b)$$

implying (1.306a, b):

$$\frac{\partial}{\partial x} + \frac{\partial}{\partial y} = 2 \frac{\partial}{\partial \xi}, \quad \frac{\partial}{\partial x} - \frac{\partial}{\partial y} = 2 \frac{\partial}{\partial \eta}: \qquad \frac{\partial^2}{\partial x^2} - \frac{\partial^2}{\partial y^2} = 4 \frac{\partial^2}{\partial \xi \partial \eta}; \qquad (1.306a\text{--}c)$$

thus the differential operator in curved brackets in (1.302b) transforms to (1.306c) in similarity variables, showing that in this case $C = 4$ in (1.300c) with $M = 1 = N$ and no differentiation with regard to x. The forced differential equation (1.302a–c) becomes (1.307e) using the similarity variables for the independent variable (1.307a, b) and forcing function (1.307c, d):

$$\Phi(x, y) = \Phi\left(\frac{\xi + \eta}{2}, \frac{\xi + \eta}{2}\right) \equiv \Psi(\xi, \eta), \qquad (1.307a, b)$$

$$B(x, y) = B\left(\frac{\xi + \eta}{2}, \frac{\xi - \eta}{2}\right) \equiv A(\xi, \eta): \qquad (1.307c, d)$$

$$4 \frac{\partial^4 \Psi}{\partial x^2 \partial \xi \partial \eta} = A(\xi, \eta). \qquad (1.307e)$$

The complete integral of (1.307e) is (1.308):

$$\Psi(\xi; \eta; x) = h(y) + xj(y) + f(\xi) + g(\eta) + \frac{1}{4} \times \frac{x^2}{2} \times \int^{\xi} d\alpha \int^{\eta} d\beta A(\alpha, \beta), \qquad (1.308)$$

that is equivalent to:

$$\Phi(x, y) = h(y) + xj(y) + f(x + y) + g(x - y) + \frac{1}{4} \int^x du \int^u dv \int^{v+y} d\xi \int^{v-y} d\eta A(\xi, \eta). \qquad (1.309)$$

An example with a particular forcing function follows (subsection 1.9.7).

Thus the complete integral of the forced linear partial differential equation with constant coefficients and all derivatives of fourth-order is given (1.302a–c) [(1.307a–e)] in original (1.303c, d) [similarity (1.303a, b)] variables by (1.309) [(1.308)].

1.9.7 TWO METHODS FOR EXPONENTIAL FORCING

Consider the partial differential equation (1.302a) with exponential forcing (1.310):

$$\frac{\partial^4 \Phi}{\partial x^4} - \frac{\partial^4 \Phi}{\partial x^2 \partial y^2} = E e^{2ax+2by}, \tag{1.310}$$

The forcing function (1.311a) can be expressed as (1.311b,c) in terms of the similarity variables (1.303a–d):

$$B(x,y) = E e^{2ax+2by} = E \; e^{(a+b)\xi+(a-b)\eta} = A(\xi,\eta). \tag{1.311a–c}$$

Substituting (1.306c;1.311b) the partial differential equation (1.310) becomes (1.312c) for the dependent variable (1.312a, b) in terms of similarity variables:

$$\Phi(x,y) = \Phi\left(\frac{\xi+\eta}{2},\frac{\xi-\eta}{2}\right) = \Psi(\xi,\eta): \qquad \frac{\partial^4 \Psi}{\partial x^2 \partial \xi \partial \eta} = \frac{E}{4} e^{(a+b)\xi+(a-b)\eta}. \tag{1.312a–c}$$

The general integral of the unforced (1.302c) partial differential equation (1.307e) [(1.312c)] is the same (1.304f) ≡ (1.313a):

$$\Psi(\xi,\zeta) = h(y) + x j(y) + f(\xi) + g(\eta) = \bar{\Psi}(\xi,\eta) - \Psi_*(\xi,\eta); \tag{1.313a, b}$$

for the complete integral (1.313b) only a particular integral of the forced equation (1.312c) is needed leading to (1.314a–e):

$$\begin{aligned}
\Psi_*(\xi,\eta,x) &\equiv \Phi_*(x,y) = \int^x du \int^u dv \int^{v+y} d\xi \int^{v-y} d\eta \; \frac{E}{4} e^{(a+b)\xi+(a-b)\eta} \\
&= \frac{E}{4} \int^x du \int^u dv \; \frac{e^{(a+b)(v+y)}}{a+b} \; \frac{e^{(a-b)(v-y)}}{a-b} \\
&= \frac{E}{4} \frac{e^{2by}}{a^2-b^2} \int^x du \int^u dv \, e^{2av} = \frac{E}{4} \frac{e^{2by}}{a^2-b^2} \frac{e^{2ax}}{4a^2}.
\end{aligned} \tag{1.314a–e}$$

The particular integral of (1.310) could be sought in the form (1.315a):

$$\Phi(x,y) = F e^{2ax+2by}; \quad E = F(2a)^2\left[(2a)^2 - (2b)^2\right]$$

$$= 16a^2(a^2 - b^2); \quad \Phi(x,y) = \frac{E}{16a^2} \frac{e^{2ax+2by}}{a^2 - b^2}, \tag{1.315a–d}$$

substitution of (1.315a) in (1.310) leads to (1.315b,c) omitting the common exponential factor. Substituting (1.315c) in (1.315a) confirms the particular integral (1.315d) ≡ (1.314e). Thus (1.315a–d) was used as a check in *the particular case of exponential forcing (1.310) of the general result (1.301a–c) for arbitrary forcing (1.300a–c).*

NOTE 1.1 SIMILARITY SOLUTIONS OF NON-LINEAR P.D.E.S.

The similarity solutions (1.206a–c) specify the general integral (1.210a, b)[(1.221a–c)] of partial differential equations (1.201a, b)[(1.213)] of a particular type: linear, unforced, with constant coefficients and all derivatives of the same order. A general non-linear partial differential equation (1.8a–c) of order N may also have a similarity solution (1.316a–c) that implies (1.316d):

$$\xi = x + ay: \quad \Phi(x,y) = f(\xi) = f(x + ay):$$
$$\frac{\partial^{n+m}\Phi}{\partial x^n \partial y^m} = a^m \frac{d^{n+m}f}{d\xi^{n+m}} \equiv a^m f^{(n+m)}(\xi), \qquad (1.316a\text{–}d)$$

and thus leads (1.317a, b) to an ordinary differential equation of the same order (1.317b) involving the parameter a:

$$0 = F\left(x + ay; \Phi, \frac{\partial\Phi}{\partial x}, \frac{\partial\Phi}{\partial y}, \frac{\partial^2\Phi}{\partial x^2}, \frac{\partial^2\Phi}{\partial x \partial y}, \frac{\partial^2\Phi}{\partial y^2}, \dots,\right.$$
$$\left.\frac{\partial^N\Phi}{\partial x^N}, \dots, \frac{\partial^N\Phi}{\partial x^n \partial y^{N-m}}, \dots, \frac{\partial^N\Phi}{\partial y^N}\right)$$
$$= F\left(\xi; f, f', af', f'', af'', a^2 f'', \dots, f^{(N)}, \dots, a^{N-n} f^{(N)}, \dots, a^N f^{(N)}\right). \qquad (1.317a, b)$$

The general integral of the ordinary differential equation (1.317b) involves N arbitrary constants of integration (1.318):

$$\Phi(x,y) = f(x + ay, C_1, \dots, C_N), \qquad (1.318)$$

and thus is a particular integral of the partial differential equation (1.317a). Thus *the non-linear vpartial differential equation (1.317a) of order N possibly involving the similarity variable (1.316a) has a particular similarity solution (1.318), involving N arbitrary constants of integration C_1, …, C_N that is the solution of the ordinary differential equation (1.317b) of order N. This shows the contrast (Table 1.7) between the general (a particular) partial differential equation that may (must) be non-linear (linear), with variable (constant) coefficients and derivatives of any order up to N (all of the same order N), that has similarity solutions in the general (a particular) integral involving N arbitrary functions (constants).* The similarity solutions are obtained in the sequel (Notes 1.8–1.16) for the Burgers non-linear convection diffusion equation (Notes 1.2–1.7).

TABLE 1.7
Similarity Solutions of Partial Differential Equation

Partial Differential Equation	General	Particular
Linearity	*Non-linear*	*Linear*
Coefficients	*Variable*	*Constant*
Derivates of order	*Up to N*	*All equal to N*
Similarity solution	*General*	*Particular*
Involves N arbitrary	*Functions*	*Constants*

Note: A general partial differential equations has particular similarity solutions; the similarity solutions provide the general integral in the case of a linear partial differential equation with constant coefficients and all derivatives of the same order.

NOTE 1.2   COMBINATION OF LINEAR DIFFUSION WITH LINEAR/NON-LINEAR CONVECTION

Unsteady heat conduction (Note 1.3) in a medium at rest is described by the diffusion equation for the temperature (Note 1.4), that may be extended to include passive convection in a moving medium (Note 1.5). In the case of the diffusion of the velocity by viscosity there is self-convection leading to the non-linear diffusion equation (Note 1.6). The latter has similarity solutions (Note 1.7) with a similarity function that is not arbitrary, but rather satisfies an ordinary differential equation involving a quadratic term, and leading to three cases: (I) for a double root the similarity function has an isolated singularity moving at constant speed (Note 1.8); (II) for a pair of complex conjugate roots the similarity function corresponds to a smooth wave front that is finite for all distances and times (Note 1.9); (IV) for two real and distinct roots the similarity solution has an denumerable infinite number of singularities (Note 1.11) with the first specifying the closest blow-up distance or time. The three preceding cases (I,II,IV) apply in the presence of viscosity; in the limit of zero viscosity case II of a smooth wave front becomes case III of a finite discontinuity or shock front (Note 1.10). The four cases I to IV can be compared for the velocity (Notes 1.5–1.10) or the rate-of-strain (Note 1.11). The rate-of-strain satisfies a third-order non-linear differential equation (Note 1.12) and specifies the dissipation energy (Note 1.13). There is (are) one (three) non-dissipative (dissipative) solution(s) [Note 1.15 (1.14)].

NOTE 1.3   BALANCE OF THE HEAT DENSITY, FLUX AND SOURCES

The heat conduction in a medium at rest has be considered in the steady case (chapter I.32) and is extended next to unsteady conditions. The heat balance (1.319) states (Figure 1.6) that the rate of increase with time of the heat density q in a domain D of

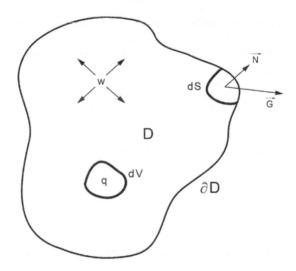

FIGURE 1.6 The heat balance in a domain D, with volume element dV and closed regular boundary surface ∂D, with area element dS and unit normal $\vec{N}$ involves the heat density per unit volume q and the heat sources w in the interior and the heat flux $\vec{G}$ across the boundary.

volume element dV plus the heat flux $\vec{G}$ across the boundary ∂D with area element $d\vec{S}$ equals the output of heat source with volume density w:

$$\frac{d}{dt}\int_D qdV + \int_{\partial D}\vec{G}\cdot d\vec{S} = \int_D wdV. \tag{1.319}$$

Using the divergence theorem (1.320b)≡(III.5.163a, b) for a continuously differentiable vector (1.320a) and (subsection III.5.7.1) a closed regular boundary surface ∂D:

$$\vec{G} \in \mathcal{D}^1\left(|R^3\right): \qquad \int_{\partial D}\vec{G}\cdot d\vec{S} = \int_D \left(\nabla\cdot\vec{G}\right)dV, \tag{1.320a, b}$$

substitution of (1.320b) in (1.319) gives (1.321):

$$\int_D\left(\frac{\partial Q}{\partial t} + \nabla\cdot\vec{G} - w\right)dV = 0. \tag{1.321}$$

Since the domain is arbitrary the integrand in (1.321) must vanish (1.322):

$$\frac{\partial Q}{\partial t} + \nabla\cdot\vec{G} = w, \tag{1.322}$$

leading to the **heat balance equation** *(1.322) stating that the density of heat sources equals the sum of the rate of change with time of the heat density plus the divergence of the heat flux. In the absence of heat sources (1.323a) the heat balance equation (1.322) simplifies to (1.323b):*

$$w = 0: \qquad \frac{\partial Q}{\partial t} + \nabla \cdot \vec{G} = 0; \qquad \frac{\partial q}{\partial t} = 0: \qquad \nabla \cdot \vec{G} = w, \qquad (1.323\text{a–d})$$

in the steady case (1.323c) the heat balance equation simplifies to (1.323d), in agreement with (1.323d)≡(1.32.3). The scalar balance equation (1.322) applies to any conserved scalar quantity, not just the heat, but also, for example, the mass or electric charge densities. A scalar balance equation becomes explicit specifying the density and flux, in the present case of heat (Note 1.3).

NOTE 1.4   UNSTEADY HEAT CONDUCTION IN A MEDIUM AT REST

The **Fourier Law (1818)** states that the heat flux (1.324b) is proportional to the temperature T gradient through the **thermal conductivity** k, that must be positive (1.324a) so that heat flow from the higher to the lower temperatures:

$$k > 0: \qquad \vec{G} = -k\,\nabla T; \qquad \frac{\partial q}{\partial t} = \rho C_V \frac{\partial T}{\partial t}, \qquad (1.324\text{a–c})$$

the change of heat density proportional to the change of temperature (1.324c) through the **mass density** ρ per unit volume and **specific heat at constant volume** C_V. Substituting (1.324b,c) in (1.322) leads to (1.325):

$$\rho C_V \frac{\partial T}{\partial t} = \nabla \cdot \left(k \nabla T \right) + w, \qquad (1.325)$$

that is the **heat conduction equation** *for the temperature. In the case of constant thermal conductivity (1.326a) the temperature satisfies the **diffusion equation** (1.326c) with **thermal diffusivity** (1.326b):*

$$k = \text{const}; \qquad \chi_t \equiv \frac{\kappa}{\rho C_V} > 0: \qquad \frac{\partial T}{\partial t} = \alpha_t \nabla^2 T + \frac{w}{\rho C_V}. \qquad (1.326\text{a–c})$$

In the absence of heat sources (1.327a) the heat equation (1.326c) simplifies to (1.327b) involving only the thermal diffusivity (1.326b):

$$w = 0: \qquad \frac{\partial T}{\partial t} = \chi_t \nabla^2 T; \qquad \frac{\partial T}{\partial t} = 0 \qquad k \nabla^2 T = -w, \qquad (1.327\text{a–d})$$

in the steady case (1.327c) the heat equation (1.326c) simplifies to a Poisson equation (1.327d) involving the thermal conductivity (1.324a), in agreement with (1.328d)≡(1.32.5c). The heat equation (1.310c) applies in a medium at rest, and can be extended to a moving medium (Note 1.4).

NOTE 1.5 PASSIVE THERMAL CONVECTION IN A MOVING MEDIUM

In medium moving with **velocity** (1.328a) the **local time derivative** $\partial/\partial t$ is replaced by the **total or material time derivative** (1.328b):

$$v^i \equiv \frac{dx^i}{dt}: \qquad \frac{d}{dt} = \frac{\partial}{\partial t} + \frac{dx^i}{dt}\frac{\partial}{\partial x^i} = \frac{\partial}{\partial t} + \vec{v}\cdot\nabla, \qquad (1.328a\text{–}c)$$

that is the sum (1.328c) of the local time derivative $\partial/\partial t$ with the **convection by the velocity.** *Since the temperature T is a passive scalar the replacement of the local (1.329a) by the total (1.329b) time derivative:*

$$\frac{\partial T}{\partial t} \to \frac{dT}{dt} = \frac{\partial T}{\partial t} + \vec{v}\cdot\nabla T, \qquad (1.329a, b)$$

*leads from the heat conduction equation (1.325) in a medium at rest to the **convected heat diffusion equation** (1.330) in a medium moving with velocity (1.328a):*

$$\frac{\partial T}{\partial t} + \vec{v}\cdot\nabla T = \chi_t \nabla^2 T + \frac{w}{\rho C_v}, \qquad (1.330)$$

that is linear in the temperature. In the case of diffusion of the velocity by the viscosity there is as self-convection effect (Note 1.6) leading to the non-linear convection-diffusion equation (Note 1.7) as follows from the Navier-Stokes equation (Note N. III.6.6).

NOTE 1.6 NAVIER (1822) – STOKES (1845) EQUATION FOR A VISCOUS FLUID

The balance of momentum for a Newtonian viscous fluid (note N.III.6.6 and subsection 2.4.11) leads to the **Navier (1822) – Stokes (1845) equation** (III.6.390a) = (1.331)≡(2.366):

$$\rho\left[\frac{\partial\vec{v}}{\partial t} + (\vec{v}\cdot\nabla)\vec{v}\right] + \nabla p = \eta\nabla^2\vec{v} + \left(\zeta + \frac{\eta}{3}\right)\nabla(\nabla\cdot\vec{v}) + \vec{f}, \qquad (1.331)$$

where p is the **pressure,** $\vec{f}$ the external **force** per unit volume and η (ζ) are **static shear (bulk) viscosities.** In the one-dimensional case (1.332a) the Navier-Stokes equation simplifies to (1.316b).

$$\vec{v} = \vec{e}_x v: \qquad \rho\left(\frac{\partial v}{\partial t} + v\frac{\partial v}{\partial x}\right) + \frac{\partial p}{\partial x} = \left(\zeta + \frac{4\eta}{3}\right)\frac{\partial^2 v}{\partial x^2} + f. \qquad (1.332a, b)$$

In the absence of a pressure gradient (1.333a) is obtained a **non-linear convection-diffusion equation** (1.333c) for the velocity involving the **total kinematic viscosity** (1.333b) that acts as the **viscous diffusivity:**

$$p = const, \quad \chi_v \equiv \frac{\zeta}{\rho} + \frac{4\eta}{3\rho}: \qquad \frac{\partial v}{\partial t} + v\frac{\partial v}{\partial x} = \chi_v \frac{\partial^2 v}{\partial x^2} + \frac{f}{\rho}. \qquad (1.333a\text{--}c)$$

Thus *the velocity in the one-dimensional (1.332a) flow of a Newtonian viscous fluid (1.331) in the absence of a pressure gradient (1.333a) satisfies a non-linear convection-diffusion equation (1.333c) involving the total kinematic viscosity or viscous diffusivity (1.333b) and balancing: (i) the local acceleration; (ii) self-convection; (iii) viscous stresses; (iv) external forces density per unit mass f/ρ. The effects (i) and (iii) are linear, (ii) is non-linear and (iv) is a forcing term.*

NOTE 1.7 SIMILARITY SOLUTIONS OF THE BURGERS (1948) EQUATION

In the absence of external force (1.334a) the non-linear convection-diffusion equation (1.333c) becomes (1.334b) the **Burgers equation** (1948):

$$f = 0: \qquad \frac{\partial v}{\partial t} + v\frac{\partial v}{\partial x} = \chi \frac{\partial^2 v}{\partial x^2}. \qquad (1.334a, b)$$

A similarity solution (1.335e) is sought for the velocity as a function of position x and time t, through (1.335d) a similarity variable (1.335a) that is constant in (1.335b,c) a reference frame moving at velocity c, and thus represents **permanent waveform** moving in space-time (Figure 1.7) with **propagation speed** c:

$$\xi = x - ct, \quad \left(\frac{dx}{dt}\right)_\xi = -\frac{\partial \xi/\partial x}{\partial \xi/\partial t} = c: \quad v(x,t) = f(\xi) = f(x - ct). \qquad (1.335a\text{--}e)$$

Denoting by prime the derivative with regard to the convected coordinate (1.336a) and noting (1.336b–d):

$$f' \equiv \frac{df}{d\xi}: \qquad \left\{\frac{\partial v}{\partial x}, \frac{\partial v}{\partial t}\right\} = \{1, -c\}f', \qquad \frac{\partial^2 v}{\partial x^2} = f'', \qquad (1.336a\text{--}d)$$

the Burgers equation (1.334b) leads a non-linear second-order ordinary differential equation (1.337b), that is reducible to the first-order (1.337c) using (1.337a):

$$f'' = \frac{d^2 f}{d\xi^2} = \frac{df'}{d\xi} = \frac{df'}{df}\frac{df}{d\xi} = f'\frac{df'}{df}: \qquad f'(f - c) = \chi f'' = \chi f'\frac{df'}{df}. \qquad (1.337a\text{--}c)$$

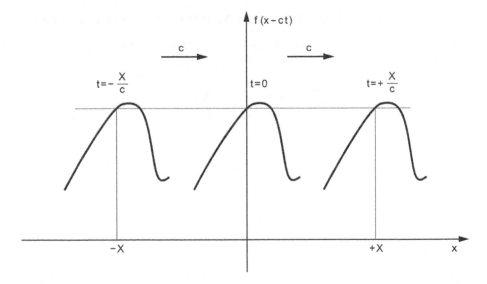

FIGURE 1.7 A linear unforced partial differential equation with constant coefficients and all derivatives of the same order N has a general integral specified by the sum of N similarity functions, N-times differentiable and each involving a similarity variable, that is a linear combination of the independent variables. Taking as example of independent variables time t and a one-dimensional position coordinate x, a similarity variable $\xi = x - ct$ is constant for a constant propagation speed c in the positive x-directions; hence an arbitrary similarly function $f(\xi)$ is a permanent waveform traveling at the same propagation speed; thus it takes the same value at time $t = 0$ at the origin and at time $t \neq 0$ at $x = \pm c\,t$.

The non-linear first-order differential equation (1.337c) has a special (sections IV.5.1-IV.5.3) integral $f'= 0$ that is of no interest since it corresponds to a constant velocity v= const, that is a trivial solution of (1.334b). It is assumed in the sequel that the total kinematic viscosity (1.333b) is constant (1.338a). Suppressing the common factor f in (1.337c) leads to a separable first-order differential equation (1.338b) with general integral is (1.338c) where A is an arbitrary constant of integration:

$$\chi = const; \qquad \chi df' = \left(f - c\right)df: \qquad \chi f' - \frac{A}{2} = \frac{f'^{2}}{2} - cf; \qquad (1.338a\text{--}c)$$

the equivalent differential equation (1.339):

$$\frac{df}{d\xi} = \frac{f^{2} - 2cf + A}{2\chi}, \qquad (1.339)$$

is integrated next (Note 1.8) to specify the velocity (1.335d,e) as a function of position x and time t through the convected coordinate (1.335a–e).

NOTE 1.8   ISOLATED SINGULARITY MOVING AT CONSTANT SPEED

The first-order non-linear differential equation (1.339)≡(1.340b) is solvable by quadratures where B in (1.340a) is another arbitrary constant of integration:

$$\frac{x - ct + B}{2\chi} = \frac{\xi + B}{2\chi} = \int^{\nu} \frac{df}{f^2 - 2cf + A}. \qquad (1.340a, b)$$

The arbitrary constant B may be set to zero (1.341a) choosing the origin $x \to x - B$ along the x-axis, and denoting by (1.341b) the remaining constant A, leads from (1.340b) to (1.341c):

$$B = 0, A = a^2: \qquad \frac{x - ct}{2\chi} = \int^{\nu} \frac{df}{(f - c)^2 + a^2 - c^2} = \int^{\nu} \frac{df}{P_2(f)}. \qquad (1.341a\text{–}d)$$

There are (Table 1.8) three cases of solution of (1.341c) depending on whether the quadratic **characteristic polynomial** (1.342a–c) in the denominator of (1.341c, d) has roots (1.342d):

$$P_2(f) \equiv (f - c)^2 + a^2 - c^2 = f^2 - 2fc + a^2 = (f - f_+)(f - f_-):$$
$$f_\pm = c \pm \sqrt{c^2 - a^2}, \qquad (1.342a\text{–}d)$$

that are: (case I) coincident for c = a (Note 1.8); (case II) complex conjugate for c < a (Note 1.9); (case IV) real and distinct for a < c (Note 1.10). Case I corresponds to the choice of constant of integration (1.343a, b) specifying a double root (1.342d) of the characteristic polynomial (1.342a–c) and leads from (1.341c) to (1.343c, d) and hence the velocity (1.343e):

$$a = c = f_\pm: \qquad \frac{x - ct}{2\chi} = \int^{\nu} \frac{df}{(f - c)^2} = -\frac{1}{\nu - c} \Leftrightarrow v_I(x, t) = c - \frac{2\chi}{x - ct}. \qquad (1.343a\text{–}e)$$

The velocity has an isolated singularity (1.344c) at the central event (1.344b) corresponding (1.344a) to zero convected coordinate (1.344a):

$$\xi = 0: \qquad x = ct: \qquad v_I(ct, t \pm 0) = v_I\left(x \pm 0, \frac{x}{c}\right) = \mp\infty, \qquad (1.344a\text{–}c)$$

with a jump from $+\infty$ at the left to $-\infty$ at right (Figure 1.8) travelling at velocity c the positive x-direction between the same value $v_I = c$ for $\xi \to \pm\infty$.

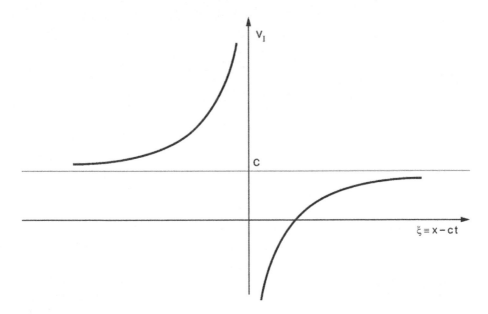

FIGURE 1.8 One case I of a similarity solution of the non-linear diffusion-convection equation in one dimension is an isolated singularity jumping from $+\infty$ before to $-\infty$ after the central event $\xi = 0$ and bounded elsewhere, tending to the same value c asymptotically as $\xi \to \pm \infty$.

NOTE 1.9 SMOOTH MONOTONIC WAVEFORM FOR ALL SPACE-TIME

For a finite velocity at the central event the parameter b in (1.345b) is non-zero, and the change of variable (1.345a, b) leads to the integration of (1.341c)$\equiv$(1.345d) in terms of the inverse hyperbolic tangent (II.7.124b)$\equiv$(1.345e):

$$g \equiv f - c \equiv v - c; b^2 \equiv c^2 - a^2: \frac{x - ct}{2\chi} = \int^{v-c} \frac{dg}{g^2 - b^2} = -\frac{1}{b} arc \tan h\left(\frac{v-c}{b}\right). \quad (1.345a\text{-}e)$$

The similarity solution (1.345e) of the Burgers equation (1.334b) specifies the dependence of the velocity on position and time (1.329):

$$v_{II}(x,t) = c - b \tan h\left[\frac{b(x-ct)}{2\chi}\right], \quad (1.346)$$

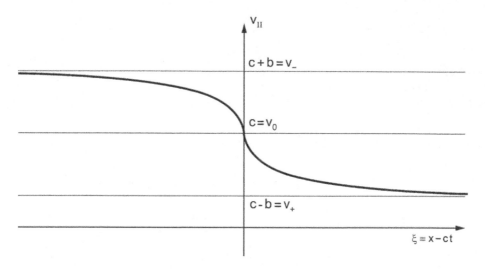

FIGURE 1.9 Another case of similarity solution of the same partial differential equation of Figure 1.7 is a smooth wave front bounded at all positions and times and taking at the central event $\xi = 0$ a value c that is the arithmetic mean of the asymptotic constant values $c \pm b$ for $\xi \to \mp \infty$.

where the constants (b, c) are yet undetermined. Starting with the choice (1.347a) of constant of integration (1.341b) implies (1.347b) and real (1.347c) so that the velocity varies monotonically (Figure 1.9) between the finite values (1.347d,e) at $\xi = \pm \infty$:

$$II: A < c^2: \quad a < c, \ b = |b|: \quad v_{\pm} \equiv f(\pm\infty) = \lim_{x - ct \to \pm\infty} v_{II}(x,t) = c \mp b. \quad (1.347\text{a–e})$$

From (1.347d,e) follows (1.348a) that: (i) the propagation velocity is the arithmetic mean (1.348b) of the velocities at $\pm\infty$ and coincides with the velocity at (1.348c, d) the central event (1.344a, b); (ii) the amplitude (1.348e) is half of the difference of the velocities at $\xi = \pm \infty$:

$$v_+ < v_-: \quad \frac{v_+ + v_-}{2} = c = v_{II}\left(x, \frac{x}{c}\right) = v_{II}(ct,t) \equiv v_0; \quad b = \frac{v_- - v_+}{2}. \quad (1.348\text{a–e})$$

Thus *in case II in (1.347a–c) the similarity solution of the Burger equation (1.334b) is (1.349b) where the velocities* $v_\pm$ *at $\pm\infty$ can be chosen (1.347d,e):*

$$\chi \neq 0: \quad v_{II}(x,t) = \frac{v_+ + v_-}{2} - \frac{v_- - v_+}{2} \tan h\left[\frac{(v_- - v_+)(x - ct)}{4\chi}\right]; \quad (1.349\text{a, b})$$

the continuous solution (1.349b) in all space-time (x, t) assumes non-zero total kine-matic viscosity (1.349a). Case III of zero viscosity is considered next (Section 1.10).

NOTE 1.10   SHOCK DISCONTINUITY FOR ZERO VISCOSITY

In the limit of zero viscous diffusivity (case III) the solution (1.349b) is (Figure 1.10) a discontinuous jump from v_- before (1.350a) to v_+ after (1.350c) the central event (1.344a, b), with the mean value equal to the propagation speed at the central event (1.350b):

$$v_{III}(x,t) \equiv \lim_{\alpha \to 0} v_{II}(x,t) = \begin{cases} c+b=v_- & \text{if} \quad \xi \equiv x-ct < 0, \\ c & \text{if} \quad \xi = 0, \\ c-b=v_+ & \text{if} \quad \xi > 0. \end{cases} \qquad (1.350a\text{--}c)$$

This corresponds to the discontinuous solution (1.351a, b):

$$v_{III}(x,t) \equiv \lim_{\chi \to 0} v_{II}(x,t) = v_- + (v_+ - v_-) H(x-ct). \qquad (1.351a, b)$$

where is used (subsection III.1.2.1) the Heaviside (1876) unit jump is (III.1.28a–c)≡(1.352a–c):

$$H(\xi) = \begin{cases} 0 & \text{if} \quad < 0, \\ \dfrac{1}{2} & \text{if} \quad \xi = 0, \\ 1 & \text{if} \quad \xi > 0, \end{cases} \qquad (1.352a\text{--}c)$$

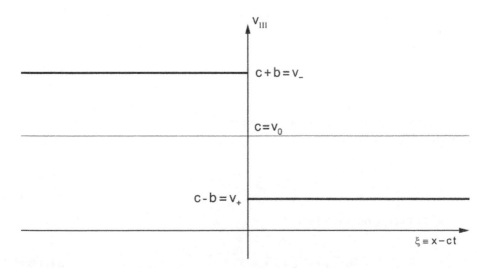

FIGURE 1.10   As the diffusivity decreases, the smooth waveform in case II in Figure 1.9 steepens until, for zero diffusivity, it becomes case III of a discontinuous shock wave jumping between constant values c + b before and c − b, after the central event $\xi = 0$.

and implies a jump of $v_+ - v_-$ across $x = c\,t$. This is also shown by the rate-of-strain, that is the spatial derivative of the velocity (1.353a–d):

$$s_{III}\left(x,t\right) = \lim_{\mu \to 0} s_{II}\left(x,t\right) = \lim_{\mu \to 0} \frac{\partial v_{II}\left(x,t\right)}{\partial x} = \left(v_+ - v_-\right)\delta\left(x - ct\right). \qquad (1.353\text{a–d})$$

where is used (subsection III.1.2.1) the Dirac (1928) unit delta function (III.1.28a–c)≡(1.354a, b):

$$\delta\left(\xi\right) \equiv \frac{dH}{d\xi} \equiv H'\left(\xi\right) = \begin{cases} 0 & if \quad \xi \ne 0, \\ \infty & if \quad \xi = 0. \end{cases} \qquad (354\text{a, b})$$

For non-zero viscous diffusivity the slope (II.7.100b) of (1.349b) specifies the rate-of-strain (1.353a–c) that is negative showing that velocity decreases monotonically in a convected frame:

$$s_{II}\left(x,t\right) \equiv \frac{\partial v_{II}\left(x,t\right)}{\partial x} = -\frac{\left(v_- - v_+\right)}{8\chi} \sec h^2 \left[\frac{\left(v_+ - v_-\right)\left(x - ct\right)}{4\chi}\right] < 0, \qquad (1.355\text{a–c})$$

and has a singularity for zero diffusivity. It remains to consider (Note 1.11) case IV opposite to (1.347a).

NOTE 1.11   BLOW-UP OR DIVERGENCE IN A FINITE SPACE-TIME

The choice of arbitrary constant of integration (1.356a) opposite to (1.347a) leads (1.356b) to (1.356c) imaginary b in (1.346), that is (1.356d):

$$IV \cdot A > c^2 \colon a > c, \; b = i\left|b\right| \colon \qquad v_{IV}\left(x,t\right) = c - i\left|b\right|\tan h\left[\frac{i\left|b\right|\left(x - ct\right)}{2\chi}\right]; \qquad (1.356\text{a–d})$$

this is equivalent to the change (subsection II.5.2.2) from hyperbolic to circular tangent (II.5.26b)≡(1.357):

$$-i\tan h\left(iz\right) \equiv -i\frac{e^{iz} - e^{-iz}}{e^{iz} + e^{-iz}} = \tan z, \qquad (1.357)$$

leading from (1.356d) to (1.358):

$$v_{IV}\left(x,t\right) = c + \left|b\right|\tan\left[\frac{\left|b\right|\left(x - ct\right)}{2\chi}\right]. \qquad (1.358)$$

The same result could be obtained via the change of parameter (1.359a) in the integral (1.345d) leading to (1.359b)≡(1.358) were was used (II.7.112b)≡(1.359c):

$$b \to i|b|: \qquad \frac{x-ct}{2\chi} = \int\limits^{v-c} \frac{dg}{g^2+|b|^2} = \frac{1}{|b|} arc\,tan\left(\frac{v-c}{|b|}\right). \qquad (1.359a\text{–}c)$$

The solution (1.358) of the Burgers equation (1.334b) diverges (1.360c) at the convected coordinates (1.360b) where n is an integer (1.360a):

$$n = 0,\pm1,\pm2,\dots: \qquad \frac{|b|(x-ct)}{2\chi} = \left(n+\frac{1}{2}\right)\pi \Rightarrow v_{IV} = \pm\infty. \qquad (1.360a\text{–}c)$$

The lowest value (1.361b) in (1.360a) shows that the solution (1.358) in case IV *blows-up (Figure 1.11) in a finite time, and a finite solution (1.361a) is constrained to the space-time interval (1.361c):*

$$|v_{IV}(x,t)| < \infty; n = 0: \qquad |x-ct| < \frac{\pi\chi}{|b|}. \qquad (1.361a\text{–}c)$$

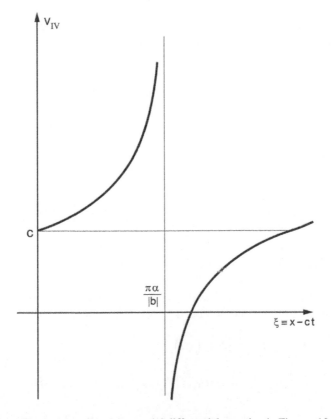

FIGURE 1.11 The last case IV of the partial differential equation in Figures 10.8 to 10.10 is a similarity solution that starts at a finite value c at the central event $\xi = 0$ and blows-up to an infinite value at a finite event after a finite distance or time.

The interpretations (1.348b,e) of the arbitrary constants (b, c) are not valid in the case (1.356a–d) because the limits (1.347d,c) do not exist for (1.358). In the present (1.356a) case c remains the propagation speed in (1.358) and |b| specifies a smaller blow-up interval (1.361c) for larger |b|. The contrast between the two cases II and IV is confirmed by the rate-of-strain (Section 1.12).

NOTE 1.12   COMPARISON OF THE VELOCITIES AND RATES-OF-STRAIN

The rate of strain in the case (1.358) is given (II.7.100a) by (1.362a, b):

$$S_{IV}(x,t) \equiv \frac{\partial v_{IV}(x,t)}{\partial x} = \frac{|b|^2}{2\chi}\sec^2\left[\frac{|b|(x-ct)}{2\chi}\right] > 0. \qquad (1.362a, b)$$

In the cases II (IV) of finite (infinite) velocity (1.349a, b) [(1.358)] the rate-of-strain (1.355a–c) [(1.362a–c)] is negative (positive) showing that the velocity is a monotonically decreasing (increasing) function of the convected coordinate (1.335a), with the difference that: (i) in the case (II) the velocity (1.349b) lies (1.363a–f) between the values (1.347d,e) at infinity (1.346a–f):

$$c+b=v_- = \left[v_{II}(x,t)\right]_{max} = f(-\infty) \geq v_{II}(x,t) \geq f(+\infty)$$
$$= \left[v_{II}(x,t)\right]_{max} = v_+ = c-b; \qquad (1.363a–f)$$

(ii) in the case (IV) the velocity (1.358) equals the propagation speed at the mid-point and increases (1.364a–f) until divergence (1.360c) at (1.361c):

$$c = \left[v_{IV}(x,t)\right]_{max} = v_{IV}\left(x,\frac{x}{c}\right) = v_{IV}(ct,t) \equiv v_0 \leq v(x,t) \leq \left[v_{II}(x,t)\right]_{max}$$
$$= f\left(\frac{\pi\chi}{|b|}\right) = \infty. \qquad (1.364a–f)$$

The rate of strain (1.355a–c) in case II is maximum in modulus, or minimum in value since it is negative, at the mid-point and vanishes at infinity (1.365a–h):

$$0 = \left[s_{II}(x,t)\right]_{max} = f'(\pm\infty) \geq s_{II}(x,t)$$
$$\geq -\frac{(v_+ - v_-)^2}{8\chi} = -\frac{b^2}{2\chi} = \left[s_{II}(x,t)\right]_{min} = s_{II}\left(x,\frac{x}{c}\right) = s_{II}(ct,t) \equiv s_0, \qquad (1.365a–h)$$

whereas in case IV the rate-of-strain (1.362a, b) is minimum and positive at mid-point and diverges towards the blow-up point (1.366a–h):

$$\frac{|b|^2}{2\chi} = \left[s_{II}(x,t)\right]_{min} = s_{IV}\left(x,\frac{x}{c}\right) = s_{IV}(ct,t) \equiv s_0$$

$$\geq s_{IV}(x,t) \leq \left[s_{VI}(x,t)\right]_{max} = s_{IV}\left(ct+\frac{\pi\chi}{|b|}\right) = \infty. \tag{1.366a–h}$$

In the cases I (III) of velocity with an isolated singularly (1.343e) [with a finite shock type discontinuity (1.350a–c)≡(1.351b;1.352a–c)] the rate-of-strain (1.367a) [(1.353a–d;1.354a, b)]:

$$s_I(x,t) = \frac{2\chi}{(x-ct)^2}, \qquad \lim_{x-ct\to\pm\infty} s_{III}(x,t) = 0, \tag{1.367a, b}$$

is infinite at the central event and non-zero elsewhere except asymptotically (1.367b) [zero everywhere else (1.353a–d;1.354a, b)] with zero diffusivity (1.368a). In case II of a shock wave (1.350a–c) the Burgers equation for the velocity (1.346b) is replaced by the conservation of the velocity in a convected frame (1.368b):

$$\chi = 0: \qquad\qquad \frac{\partial v}{\partial t} + v\frac{\partial v}{\partial x} = 0, \tag{1.368a, b}$$

changing from a partial differential equation of the second-order to the first-order. The order is one unit higher, that is two (three) for the differential equation for the rate-of-strain (Note 1.13) in the case(s) II (I, II, IV).

NOTE 1.13 THIRD-ORDER DIFFERENTIAL EQUATION FOR THE RATES-OF-STRAIN

The differential equation satisfied by the rates-of-strain (1.355a–c)/(1.362a, b)/ (1.367a) in the cases II/IV/I can be obtained eliminating the velocity from the Burgers equation (1.334b)≡(1.369b) to isolate the rate-of-strain (1.369a):

$$s(x,t) \equiv \frac{\partial v(x,t)}{\partial x}: \qquad\qquad \frac{\partial v}{\partial t} + vs = \chi\frac{\partial s}{\partial x}. \tag{1.369a, b}$$

Differentiating (1.369b) with regard to x leads to (1.370a) that can be solved (1.370c) for v and differentiated again (1.370b):

$$\frac{\partial s}{\partial t} + s^2 + v\frac{\partial s}{\partial x} = \chi\frac{\partial^2 s}{\partial x^2}: \qquad s \equiv \frac{\partial v}{\partial x} = \frac{\partial}{\partial x}\left(\frac{\chi\dfrac{\partial^2 s}{\partial x^2} - s^2 - \dfrac{\partial s}{\partial t}}{\dfrac{\partial s}{\partial x}}\right); \tag{1.370a–c}$$

rearranging (1.370b) leads to (1.371b):

$$\chi\left[\frac{\partial^3 s}{\partial x^3}\frac{\partial s}{\partial x}-\left(\frac{\partial^2 s}{\partial x^2}\right)^2\right]=3s\left(\frac{\partial s}{\partial x}\right)^2+\frac{\partial s}{\partial x}\frac{\partial^2 s}{\partial x\partial t}-\frac{\partial^2 s}{\partial x^2}\left(s^2+\frac{\partial s}{\partial t}\right).\tag{1.371}$$

Thus *the rate-of-strain is given by (1.355a–c)/(1.362a, b)/(1.367a) in the cases II/ IV/I and satisfies a non-linear third-order partial differential equation (1.371) that simplifies to the second-order (1.372b) in case III setting (1.372a) in (1.371):*

$$\chi=0:\qquad 0=3s\left(\frac{\partial s}{\partial x}\right)^2+\frac{\partial s}{\partial x}\frac{\partial^2 s}{\partial x\partial t}-\frac{\partial^2 s}{\partial x^2}\left(s^2+\frac{\partial s}{\partial t}\right).\tag{1.372a, b}$$

The non-linear second-order partial differential equation (1.372b) can be obtained alternatively: (i) by differentiating (1.368b) with regard to x and using (1.369a) leading to (1.373a):

$$0=\frac{\partial s}{\partial t}+s^2+v\frac{\partial s}{\partial x};\qquad v=-\frac{s^2+\dfrac{\partial s}{\partial t}}{\dfrac{\partial s}{\partial x}},\tag{1.373a, b}$$

(ii) re-arranging (1.373a) as (1.373b) and differentiating again with regard to x leads to (1.372b).

NOTE 1.14   KINETIC ENERGY AND DISSIPATION RATE

The **kinetic energy** (1.374a) [**dissipation rate** (1.374b)] per unit mass equals the square of the velocity (strain rate) multiplied by one-half (the total kinematic viscosity or viscous diffusivity):

$$E_v\left(x,t\right)=\frac{1}{2}\left[v\left(x,t\right)\right]^2,\qquad D\left(x,t\right)=\chi\left[s\left(x,t\right)\right]^2.\tag{1.374a, b}$$

The dissipation rate (1.374b) is given by (1.375a/b/c), respectively, in the cases I/II/ IV of velocity (1.343e)/(1.349b)/(1.358) and strain rate (1.367a)/(1.355c)/(1.362b):

$$D\left(x,t\right)=\begin{cases}\dfrac{4\chi^3}{\left(x-ct\right)^4}\ in\ case\ I,\\[3mm]\dfrac{b^4}{4\chi}sech^4\left[\dfrac{b\left(x-ct\right)}{2\chi}\right]in\ case\ II,\\[3mm]\dfrac{|b|^4}{4\chi}sec^4\left[\dfrac{|b|\left(x-ct\right)}{2\chi}\right]in\ case\ IV.\end{cases}\tag{1.375a–c}$$

In case I of an isolated singularity the dissipation has an infinite maximum at the central event and decays monotonically to zero at infinity (1.376a-f):

$$\infty = D_{I\max} = D_I\left(ct,t\right) \geq D_1\left(x,t\right) \geq 0 = D_{I\max} = D_I\left(x - ct = \pm\infty\right). \quad (1.376a\text{--}f)$$

In case III of a shock wave the dissipation is infinite at the central event and zero elsewhere. The dissipation in case IV of blow-up in a finite space time is always larger than in case II of a smooth waveform for all time, except at the central event where they coincide (1.377a-e):

$$0 = D_{II}\left(x - ct = \pm\infty\right) = D_{II\min} \leq D_{II}\left(x,t\right) \leq D_{II\max}$$

$$= D_{II}\left(ct,t\right) = \frac{|b|^2}{4\chi} = D_{IV}\left(ct,t\right) \qquad\qquad (1.377a\text{--}e)$$

$$= D_{IV\min} \leq D_{IV}\left(x,t\right) \leq D_{IV\max} = D_{IV}\left(x - ct = \frac{\pi}{\alpha|b|}\right) = \infty.$$

Away from the coincident dissipation at the central event the dissipation increases (decreases) monotonically to infinity (zero) in case IV (II) at the blow-up event (1.361c) [at infinity]. The distinction between the three viscous cases I, II and IV in terms of velocity (Notes 1.7–1.11), rates-of-strain (Notes 1.12–1.13) and dissipation (Note 1.14) is due to the choice of constant of integration (Note 1.15).

NOTE 1.15   CHOICE OF CONSTANT OF INTEGRATION AND DISSIPATION VELOCITY

In case II (IV) of smooth waveform (blow-up in finite space-time) the velocities (1.346) [(1.358)] coincide at the central event (1.378a) where they equal the propagation speed (1.378b–d):

$$v_{II}\left(t,ct\right) = v_{IV}\left(t,ct\right) = c \equiv v_0, \qquad (1.378a\text{--}d)$$

and the rate-of-strain (1.355c) [(1.362b)] have opposite negative (1.379a) [positive (1.379b)] sign with the modulus (1.379c, d):

$$-s_{II}\left(ct,t\right) = -\frac{b^2}{2\chi} = -S_{IV}\left(ct,t\right) = \frac{|b|^2}{2\chi} \equiv s_0. \qquad (1.376a\text{--}d)$$

The constant of integration (1.341b) is given (1.345c) by (1.380a–d):

$$A = a^2 = c^2 - b^2 = v_0^2 \mp |b|^2 \, \text{sgn}\left(b^2\right) = v_0^2 \mp 2s_0\chi \, \text{sgn}\left(b^2\right). \qquad (1.380a\text{--}d)$$

*The parameter b can be interpreted as a **dissipation velocity** that is: (i) real (1.381a) in the finite (1.347a–e) case II; (ii) imaginary (1.381c) in the divergent (1.356a–d) case IV; (iii) zero (1.381b) in case I of an isolated singularity (1.343a–e):*

$$b = \begin{cases} \sqrt{2s_0\alpha} & \text{in the finite case II,} \\ 0 & \text{in the case I of isolated singularity,} \\ i\sqrt{2s_0\alpha} & \text{in the divergent case IV.} \end{cases} \qquad (1.381a\text{--}c)$$

Besides the three dissipative cases I, II and IV that are solutions of the non-linear convection-diffusion equation (1.318b) the non-dissipative shock case IV that is a solution of non-linear convection alone (1.351b) are compared next (Note 1.15).

NOTE 1.16   NON-LINEAR CONVECTION WITH OR WITHOUT DIFFUSION

The main features of the four cases are summarized next with reference to Figures 1.7–1.10. The Burgers equation (1.334b) has similarity solutions (1.335a–e) leading to two degenerate and two non-degenerate cases. The degenerate case I (Figure 1.8) is singular at the central event (1.343e), in contrast with the other three cases that are finite at the central event, where the velocity equals the propagation speed. The finite non-degenerate case II corresponds (1.349a, b) to a hyperbolic tangent wave front (Figure 1.9) specifying a monotonic variation between finite velocities at the convected coordinates $\pm\infty$; the second degenerate case III is the limit of zero viscosity that leads (Figure 1.9) to a shock (1.350a–c)≡(1.351a, b) with arbitrary velocities before and after. *The dissipative degenerate I/non-degenerate II/IV cases have velocities (1.343e)/(1.355a–c)/(1.362a, b) [rates-of-strain (1.367a)/(1.355a–c)/(1.362a, b)] that satisfy a non-linear second (third)- order partial differential equation (1.334b) [(1.371)]. The non-dissipative degenerate case III has a discontinuous velocity (1.350a–c)≡(1.351a, b) [singular rate-of-strain (1.353a–d; 1.354a, b)] that satisfies a first (second)-order partial differential equation (1.368b) [(1.372b)].*

Thus in the non-dissipative case the velocity satisfies the non-linear convection equation (1.368b)≡(1.382b) leads to (1.382a):

$$\left(\frac{dx}{dt}\right)_v = -\frac{\partial v/\partial t}{\partial v/\partial x} = v; \quad f \in \mathcal{D}(\mathbb{R}): \quad v(x,t) = f\big(x - tv(x,t)\big), \qquad (1.382a\text{--}c)$$

whose solution (1.382c) is implicit since the velocity coincides with the propagation speed. Choosing arbitrary function in (1.382c) to be the Heaviside unit jump (1.352a–c) leads to the velocity (1.351a, b) in case III with propagation speed c, taking a mean values (1.348b) at the jump (1.348e). The unit jump is not a differentiable (1.354a, b) as an ordinary function, but is infinitely differentiable as a generalized function (chapters III.1 and III.3). The similarity solutions of the non-linear convection-diffusion equation (1.334b) in inviscid limit $\chi \to 0$ do not coincide with the solution (1.382c) of the non-linear convection equation (1.368b) for $\chi = 0$ because (1.382c) is not a similarity solution, but rather an implicit solution involving the velocity both on the l.h.s. and in the argument of the function on the r.h.s.. The similarity solution (1.335a–e) of (1.368b) would satisfy (1.338b) with $\chi = 0$, leading to $f(f - c) = 0$ whose solution is a constant velocity $c = f = v$; this corresponds to the value (1.348b) at the central event of the similarity solution of the Burgers equation in the limit $\chi \to 0$. This limit specifies case III of a shock wave solution (1.352a–c) that can be also be obtained using a different approach (Section 2.7). Thus the similarity solutions of the Burgers equation show that a non-linear partial differential equation with constant coefficients can have smooth, discontinuous or singular solutions in space-time.

1.10   CONCLUSION

A partial differential equation (Table 1.1) involves one dependent variable and several independent variables and also the derivatives of the former with regard to the latter up to some order, and it is linear iff there are no products or powers of derivatives. For example, a linear first-order partial differential equation (Table 1.3) is a sum of partial derivatives with coefficients that may involve the independent variable, and is equivalent to inner product of the vector of coefficients by the gradient of the dependent variable; there may be in addition a forcing term involving only the independent variables. A linear first-order partial differential equation without forcing term and with vector of coefficients $\vec{X}$, has for general integral the family of all surfaces Φ_1, Φ_2, whose intersection (Figure 1.1) is the characteristic curve C, tangent to $\vec{X}$; for example (Figure 1.2), the position vector in three-dimensional space is contained is a family of planes P passing through the origin as well as non-plane surfaces S. The linear first-order partial differential equations are related to first-order differentials. Both are specified by the vector of coefficients. In the three-dimensional case (Figure 1.3) a first-order differential is (i) exact if it excludes rotation (Figure 1.3); (ii) inexact if it includes rotation, in which case it has an integrating factor if the helicity (Figure 1.3) is zero corresponding to rotation in a plane; (iii) if the helicity is non-zero the differential is inexact without integrating factor correspoinding to non-plane rotation.

The three-dimensional vector fields (Figure 1.4) can be represented by scalar potentials: (i) one for an exact differential (Figure 1.4a); (ii) two for an inexact differential with integrating factor (Figure 1.4b); (iii) two choices of three for an inexact differential without integrating factor (Figure 1.4c, d). A vector potential can be used to represent a solenoidal vector field (Figure 1.5a) plus a scalar potential in the non-solenoidal case (Figure 1.5b). These results (Tables 1.2 and 1.4) have a geometrical interpretation. An irrotational vector field, associated with an exact differential has a set of orthogonal hypersurfaces (Figure 1.4a), which are the equipotentials; a Beltrami vector field, with zero helicity, is associated with an inexct integrable differential form and has orthogonal hypersurfaces (Figure 1.4), if its modulus but not direction is changed using a second potential. A vector field with non-zero helicity, associated with an inexact non-integrable differential form involves three potentials (Figure 1.4b) and the subtraction of the gradient of the third potential leads to a Beltrami vector field lying on the intersection of two surfaces; (Figure 1.4c) a solenoidal vector field

TABLE 1.8

Similarity Solutions of the Non-Linear Convection-Diffusion Equation

Case	I	II	III	IV
Section	*N1.8*	*N1.9*	*N1.10*	*N1.11*
Condition	$a = c$	$a < c$	$\chi \to 0$	$a > c$
Meaning	*Isolated singularity*	*Smooth front*	*Shock wave*	*Periodic singularity*
Velocity	*(1.343e)*	*(1.346)≡(1.349a, b)*	*(1.350a–c)≡(1.351a, b;1.352a–c)*	*(1.358)*
Rate-of-strain	*(1.367a)*	*(1.355a–c)*	*(1.353a–d;1.354a, b)*	*(1.362a)*

Note: Four cases of similarity solutions of the burgers equation for one-dimensional unsteady linear diffusion and non-linear self-convection.

relates to a helical geometry through a vector potential (Figure 1.5a); if the vector field is not solenoidal it suffices to add the normal to a surface (Figure 1.5b).

A linear unforced partial differential equation with constant coefficients and all derivatives of the same order (Table 1.5) has similarity solutions specifying the general integral as a sum of arbitrary functions of similarity variables that are linear combination of the independent variables. The Laplace (biharmonic) equations is an example of a linear differential equation with constant coefficients with all derivatives of the second (fourth) order (Table 1.6). The similarity solutions may exist (Table 1.7) for non-linear partial differential equations of any order, but in this case are particular integrals of a specific form, because they lead to a non-linear ordinary differential equation that must be satisfied. For example, the non-linear diffusion-convection equation (Figure 1.6) in one dimension in space time, known as the Burgers equation (Table 1.8), has similarity solutions corresponding to a permanent waveform (Figure 1.7) moving at constant propagation speed: (i) a travelling isolated singularity (Figure 1.8); (ii) a smooth wave front finite at all times and at all positions (Figure 1.9), that may (iii) steepen into a discontinuous shock wave (Figure 1.10) in the limit of zero dissipation; (cv) a divergent solution that blows-up in finite time or distance (Figure 1.11).

2 Thermodynamics and Irreversibility

The work of a force is defined as its projection on a displacement, for example, the work of the: (i) volume forces, like inertia, gravity, electric and magnetic forces; (ii) surface forces, like the pressure and other stresses. The work in general depends not only on the initial and final state, but also on the evolution between them; thus it is specified by a first-order differential in three variables which is generally inexact and non-integrable. This implies (Section 1.5) that: (i) there is another inexact non-integrable differential, the heat, equal to the product of the temperature by the change in entropy; (ii) the sum of the two, i.e. heat plus work, is an exact differential of a function of state, the internal energy. This statement of the first principle of thermodynamics (Section 2.1) shows that the inexact non-integrable first-order differential specifying the work leads to an exact differential by introducing three functions of state: the entropy, the temperature and the internal energy.

In its simplest form, the internal energy: (i) depends on two extensive parameters, the volume (entropy), both of which are additive, that is, add for two systems in equilibrium; (ii) the coefficients are the minus the pressure (temperature), which are intensive parameters, that is, are equal for two systems in equilibrium. Other functions of state can be obtained from the internal energy, e.g.: (i/ii) replacing volume (entropy) by the pressure (temperature) leads to the enthalpy (free energy); (iii) replacing both leads to the free enthalpy. Thus, the first-order derivatives of the functions of state (internal energy, enthalpy, free energy and free enthalpy) are either extensive parameters (volume and entropy) or intensive parameters (pressure and temperature). The second-order derivatives of functions of state are derivatives relating intensive to extensive parameters and specify (Section 2.2) the constitutive properties of matter, such as specific heats, elastic constants, dielectric permittivity and magnetic permeability, etc...

The first principle of thermodynamics (Section 2.1) concerns the states of equilibrium, but gives no indication about their stability. The second principle of thermodynamics identifies the equilibrium states as those corresponding (Section 2.3) to minimum energy or maximum entropy. It implies that the entropy of an isolated system cannot decrease: (i) it is constant in a "reversible" process; (ii) it must increase in an "irreversible" process. The reversible (irreversible) thermodynamic process is (is not) a succession of equilibrium states, for example, a body heated uniformly (non-uniformly), so that there are no (there are) heat fluxes. The second-order differential of the energy (entropy) is a quadratic function of the differentials of extensive or intensive thermodynamic variables (of their gradients or fluxes); the symmetric coefficients specify the constitutive (diffusive) properties of matter [Section 2.2 (2.4)]. The diffusive properties include (Section 2.4) heat and electrical conduction, viscosity, etc... The thermodynamic properties of matter are specified by the constitutive and diffusive properties.

DOI: 10.1201/9781003186595-2

Some of the constitutive properties can be derived from the equation of state (Section 2.5) relating the pressure, volume and temperature. The simplest equations of state apply to ideal or perfect gases, consisting of non-interacting molecules; other equations of state apply to real gases, condensed matter (liquids and solids) and changes of state (vaporization, solidification, etc…). The principles of thermodynamics together with the equation of state describe a variety of thermodynamic processes, for example, isochoric/isobaric/isothermal if, respectively, the volume / pressure/temperature are kept constant; also adiabatic if the entropy is kept constant, that is, there is no heat exchange. The thermodynamic processes can be combined into a cycle, which brings the system back to the original state, such as a piston or jet engine which admits air from the atmosphere, heats it by combustion and ejects it back to the environment. The work performed by an engine, and its efficiency, are determined by its thermodynamic cycle. A thermodynamic process reaching zero absolute temperature requires a third principle of thermodynamics.

An example of a thermodynamic process is the flow of a compressible fluid: it must satisfy the conservation of mass, momentum and energy (Section 2.6). The flow may be: (i) continuous in which case it is a reversible process, if no heat is exchanged; (ii) discontinuous, if there are jumps of flow quantities, such as velocity, density, pressure. There are two basic types of flow discontinuities, where the velocity is: (i) tangential, namely a vortex sheet between flow at the same pressure but different velocities and or densities; (ii) normal, that is, a normal shock (Section 2.7), with a jump in pressure, density and temperature; (iii) an oblique shock (Section 2.8) is a combination of (i) and (ii). A normal or oblique shock is always an irreversible process, leading to entropy production. The second law of thermodynamics, that entropy must increase in an irreversible process, leads to inequalities across a shock: the velocity must decrease from supersonic to subsonic, and the pressure and density must increase. Thus, shock waves are non-linear processes which can occur in a supersonic flow, for example, when it impinges on a body like an aircraft.

The flow in a nozzle, that is, a duct of varying cross-section is an example of the need or not for thermodynamics. For an incompressible flow, at a speed much lower than the sound speed, thermodynamics is not needed: the flow velocity varies inversely with the cross-section, to preserve the volume flux (with rigid impermeable walls). If the speed is not small compared with the sound speed, the flow is compressible and the mass flux must be conserved, rather than the volume flux. The conservation of the mass flux involves the product of the density, velocity and cross-section, which can be traded against each other. If the difference in pressure between the inlet and exhaust is not too large, an adiabatic flow is possible; if the difference in pressure exceeds what can be achieved with an adiabatic flow, a pressure jump or shock wave forms, in order to match the inlet to the exhaust conditions (Section 2.9).

2.1   WORK, HEAT, ENTROPY AND TEMPERATURE

The work of a force in an infinitesimal displacement is defined (subsection 2.1.1) by their inner product. For volume forces the work leads to: (i) the kinetic energy for the inertia force (subsection 2.1.3); (ii) the potential energy for the gravity force (subsection 2.1.4); (iii/iv) the electric (magnetic) energy for the electric (magnetic) force [subsections 2.1.5–2.1.8 (2.1.9–2.1.14)]. The work of the surface forces in a displacement is the double contraction of the stress and strain tensors (subsections 2.1.16–2.1.19), as for the elastic energy; in the particular isotropic case, it reduces to the work performed by the pressure in a volume change (subsection 2.1.15). In addition to the work of volume (surface) forces [subsections 2.1.3–2.1.14 (2.1.15–2.1.19)] there is the chemical work, associated with reaction between several substances (subsections 2.1.20–2.1.21). The sum of the preceding seven forms of work (subsections 2.1.3–2.1.21) is the total work (subsection 2.1.22). The work performed taking a thermodynamic system from an initial to a final state depends on the intermediate states, that is, on the thermodynamic process (subsection 2.1.1); thus, the work is specified by a first-order differential with three variables that is generally inexact and non-integrable (Section 1.5). Thus, it can be written as in terms of (subsection 2.1.2) three functions of state, namely: (i, ii) the temperature and entropy specify the heat, that is also a first-order inexact non-integrable differential; (iii) the sum of the heat and work is the exact differential of a function of state, the internal energy. The statements (i) to (iii) correspond to the first principle of thermodynamics (subsection 2.1.2), that specifies the internal energy of a general thermodynamic system (subsection 2.1.22). The internal energy: (i) is a function of the extensive parameters (subsection 2.1.23), namely entropy, electric and magnetic fields, strain tensor and volume and mole numbers of chemical species, that are additive for two thermodynamic systems; (ii) the coefficients in the exact differential for the internal energy are the intensive parameters (subsection 2.1.24), respectively, the temperature, electric displacement, magnetic induction, stress tensor, pressure and chemical potential, that are equal for two thermodynamic systems in equilibrium (subsections 2.1.25–2.1.26).

2.1.1 WORK OF CONSERVATIVE AND NON-CONSERVATIVE FORCES

The **work** W performed by the force density per unit volume $\vec{f}$ in an infinitesimal displacement $d\vec{x}$ is their inner product, in vector (2.1a) and index (2.1b) notation, and use is made of the **summation convention** that a repeated index i signifies (2.1c) summation over $i=1,2,3$:

$$dW = \vec{f} \cdot d\vec{x} = \sum_{i=1}^{3} f_i dx_i = f_i dx_i; \quad \dot{W} = \frac{dW}{dt} = f_i \frac{dx_i}{dt} = f_i v_i; \qquad (2.1\text{u e})$$

The **power** or **activity** is the work per unit time (2.1d, e) and involves the **velocity** (2.2a) that is the derivative of the position vector with regard to time:

$$\vec{v} \equiv \frac{d\vec{x}}{dt}, \quad W^{12} = \int_{\vec{x}_1}^{\vec{x}_2} \vec{f} \cdot d\vec{x} = \int_{t_1}^{t_2} \left(\vec{f} \cdot \vec{v} \right) dt = \int_{t_1}^{t_2} A \, dt \qquad (2.2\text{a, b})$$

The work performed in a finite displacement (2.2b) generally depends: (i) on the initial $\vec{x}_1$ and final $\vec{x}_2$ positions; (ii) on the path followed between the two positions or on the power used over time. *The work is an exact differential independent of the path, and the power is a single-valued function of time iff (≡ if and only if) the force field is conservative (1.64a–c), that is, equals the gradient of a potential (2.3a):*

$$dW = \nabla\Phi \cdot d\vec{x} = d\Phi, W^{12} = \Phi(\vec{x}_2) - \Phi(\vec{x}_1), \dot{W} = \frac{\partial\Phi}{\partial\vec{x}} \cdot \frac{d\vec{x}}{dt} = \frac{d\Phi}{dt}, \qquad (2.3a\text{–}d)$$

in which case: (i) the work is the exact differential of the potential (2.3b); (ii) the work performed between two states is the difference of the potential at the final and initial state (2.3c), and thus independent of the path; (iii) the power is the derivative of the potential with regard to time (2.4d). The examples of the conservative force fields include the gravity (electrostatic) field [subsections 2.1.4 (2.1.5–2.1.6)].

As an example of a non-conservative force field consider a force of resistance (2.4a) proportional to the velocity and in the opposite direction (2.4b) through the **friction coefficient** k:

$$\vec{f} = -k\vec{v}, k = const > 0: \quad W^{12} = -\int_{\vec{x}_1}^{\vec{x}_2} k\vec{v} \cdot d\vec{x} = -\int_{t1}^{t2} kv^2 dt, \qquad (2.4a\text{–}d)$$

whose work is given by (2.4c, d). If the path is taken at constant velocity (2.4e) the work is proportional to the distance (2.4g) or time (2.4f):

$$|\vec{v}| = const = \frac{L}{t_2 - t_1}: \quad W = -kv^2(t_2 - t_1) = -kvL, \qquad (2.4e\text{–}g)$$

and therefore can be different for different paths between the same initial and final conditions, depending (2.4g) [(2.4f)] on the length (time). Thus, *for a general non-conservative (1.65a–c) helical (1.66a–c) force field, the work (2.1a, b) = (2.5a, b) is an inexact non-integrable first-order differential in three variables that can be represented by (2.5c) in terms of three scalar functions*:

$$dW = \vec{F} \cdot d\vec{x} = F_x dx + F_y dy + F_z dz = dU - TdS. \qquad (2.5a\text{–}c)$$

The last statement corresponds to the first principle of thermodynamics (subsection 2.2.2) by interpreting the three functions as the internal energy U, temperature T and entropy S.

2.1.2   INTERNAL ENERGY AND FIRST PRINCIPLE OF THERMODYNAMICS

The result (2.5c) ≡ (2.6a, b) can be re-stated:

$$dU = dW + dQ, \ dQ = TdS, \qquad (2.6a, b)$$

as the **first principle of thermodynamics**: *(i) although the work (2.1a, b) is not an exact differential, there is another inexact differential, namely the **heat** (2.6b), such* *that the sum (2.6a) is the exact differential of a function of state, namely the **internal energy**; (ii) although the heat is an inexact differential (2.6b), if divided by the **temperature** it becomes an exact differential of another function of state, namely the **entropy**.* The first principle of thermodynamics thus states that: (i) the work is not an exact differential because it is based on external forces and misses the heat as a form of internal energy; (ii) the heat must be an inexact differential because it equals the difference between an exact (inexact) differential, namely the internal energy (work); (iii) there is a function of state related to the heat, namely the entropy, obtained dividing by the temperature. The physical interpretations of U, S, T and the mathematical theorems (2.5a–c) ≡ (2.6a, b) are substantiated by: (i) the internal energy being a function of state with the dimensions of energy and including all forms of energy for matter in equilibrium (Sections 2.1–2.2); (ii) the entropy being associated with the stability of equilibrium states and the evolution between two states by non-equilibrium processes (Sections 2.3–2.4); (iii) the absolute temperature specifying thermal equilibrium (subsections 2.1.24–2.1.25) and being always positive (subsections 2.3.18–2.3.23) for most thermodynamic systems. In order to specify explicitly the internal energy (2.7a) of a general thermodynamic system (subsections 2.1.22–2.1.26) all forms of work must be considered, namely: (i) the work of the inertia force that leads to the kinetic energy (subsection 2.1.3); (ii) the work of the gravity field that leads to the gravity potential energy (subsection 2.1.4); (iii/iv) the work of the electric (magnetic) force that leads to the electric (magnetic) energy [subsections 2.1.5–2.1.8 (2.1.9–2.1.14)]; (v) the work of the pressure in a volume change (subsection 2.1.15) generalized to (vi) the work of the stresses on the strains (subsections 2.1.16–2.1.19); (vii) the work associated with chemical reactions (subsections 2.1.20–2.1.21).

2.1.3 INERTIA FORCE AND KINETIC ENERGY

The **linear momentum** (2.7b) is the product of the velocity (2.2a) by **the mass density** (2.7a) that is the mass per unit volume:

$$\rho = \frac{dm}{dV}: \quad p = \rho \vec{v}. \tag{2.7a, b}$$

The inertia force (2.8a) is the derivative of the linear momentum (2.7b) with regard to time (2.8b) and equals (2.8c):

$$\vec{f}_k \equiv \frac{d\vec{p}}{dt} = \frac{d}{dt}\left(\rho\vec{v}\right) = \rho\frac{d\vec{v}}{dt} + \vec{v}\frac{d\rho}{dt}. \tag{2.8a–c}$$

If the mass density is independent of time (2.9a) the inertia force (2.8c) is (2.9c) its product by the **acceleration** (2.9b) that is the derivative of the velocity with regard to time:

$$\frac{d\rho}{dt} = 0, \quad \vec{a} = \frac{d\vec{v}}{dt}: \quad \vec{f}_k = \rho\frac{d\vec{v}}{dt} = \rho\vec{a}. \tag{2.9a–c}$$

The work (2.1a) of the inertia force (2.8c) is generally given by (2.10a, b):

$$dW_k \equiv \vec{f}_k \cdot d\vec{x} = \frac{d}{dt}\left(\rho\vec{v}\right)\cdot d\vec{x} = \vec{v}\cdot d\left(\rho\vec{v}\right) \qquad (2.10\text{a, b})$$

and in the case of mass density independent of time (2.9a) $\equiv$ (2.11a) equals (2.11b):

$$\frac{d\rho}{dt} = 0: \quad dW_k = \rho\vec{v}\cdot d\vec{v} = d\left(\frac{1}{2}\rho\vec{v}\cdot\vec{v}\right) = dE_k, \quad E_k = \frac{1}{2}\rho v^2 \qquad (2.11\text{a–c})$$

the variation of the **kinetic energy** per unit volume (2.11c) that is half the mass density multiplied by the square of the modulus of the velocity.

It has been shown that *the work (2.1a, b) of the inertia force per unit volume (2.9a–c) is given generally by (2.10a, b). If the mass density does not depend on time (2.9a) $\equiv$ (2.11a), the work (2.11b) of the inertia force (2.9a–c) equals the variation (2.11b) of the kinetic energy (2.11c). In this case (2.12a) the work performed in a finite displacement is a function of state*:

$$\frac{d\rho}{dt} = 0: \quad W_k^{12} = \int_1^2 dE_k = E_{k2} - E_{k1} = \frac{\rho}{2}\left[\left(v_2\right)^2 - \left(v_1\right)^2\right], \qquad (2.12\text{a, b})$$

since it equals (2.12b) the difference of final and initial kinetic energies. The inertia force per unit volume:

$$\vec{f}_k = \vec{f}_g + \vec{f}_e + \vec{f}_h + \vec{f}_u, \qquad (2.13)$$

is the sum of the gravity, electric, magnetic and pressure or stress forces per unit volume, whose work is considered next (respectively, in the subsections 2.1.4/2.1.5–2.1.8/2.1.9–2.1.14/2.1.15–2.1.19).

2.1.4   MASS AND GRAVITY FIELD, POTENTIAL, FORCE AND ENERGY

The gravity field (chapter I.18) is irrotational (2.14a) and derives (2.14b) from a **gravity potential**:

$$\nabla \wedge \vec{g} = 0 \Leftrightarrow \vec{g} = -\nabla\Phi_g. \qquad (2.14\text{a, b})$$

The **gravity force** per unit volume is (I.18.28) $\equiv$ (2.15a) the product of the mass density by the **gravity field** (2.15b) or **acceleration of gravity**:

$$\vec{f}_g = \rho\vec{g} = -\rho\nabla\Phi_g: \quad dW_g \equiv \vec{f}_g \cdot d\vec{x} = -\rho d\vec{x}\cdot\nabla\Phi_g = -\rho d\Phi_g, \qquad (2.15\text{a, b})$$

the work (2.1a, b) of the gravity force (2.15a) relates to the variation of gravity potential (2.14b) through (2.15b). Since the gravity potential is defined by (2.14b) to within an added constant, the latter may be chosen so that it vanishes at infinity (2.16a):

$$0 = \Phi_g(\bar{x} = \infty): \quad E_g = -\int_{\infty}^{\bar{x}} dW_g = \rho \int_{\infty}^{\bar{x}} d\Phi_g = \rho \Phi_g(\bar{x}). \quad (2.16a, b)$$

Thus *the potential energy of the gravity field (2.16b) is minus the work of the gravity force (2.15a) to bring a mass from infinity (2.16a), and equals the mass density multiplied by the gravity potential (2.16a, b). In this case the work of the gravity field displacing a mass between two points and equals the difference of potential energies (2.16d), that equals the mass multiplied by the difference of the gravity potentials (2.16e):*

$$W_g^{12} = \int_1^2 dW_g = E_{g2} - E_{g1} = \rho \left[\Phi_g(\bar{x}_2) - \Phi_g(\bar{x}_1) \right]. \quad (2.16c\text{–}e)$$

*The gravity field is generated by the mass (2.17a), where K_m is the **gravitational constant**, and (2.19b) implies that the gravitational potential satisfies a Poisson equation (2.17b), with a Laplace operator (2.17c) forced by $K_m \rho$:*

$$\nabla \cdot \bar{g} = -\rho K_m: \quad \rho K_m = -\nabla \cdot \left(-\nabla \Phi_g \right) = \nabla^2 \Phi_g. \quad (2.17a\text{–}c)$$

Thus for a given distribution of mass density (2.17a) in space and time, the gravity potential is obtained as the solution of the Poisson equation (2.17c) and specifies the gravity field (2.14b) and force (2.15a) and the gravity energy (2.16b) and work (2.16e). The electrostatic field is irrotational as the gravity field, and thus the electric energy relates to the electrostatic potential (subsections 2.1.5–2.1.8).

2.1.5 ELECTRIC CHARGE, FIELD, SCALAR POTENTIAL AND FORCE

The electrostatic field (2.18a) is irrotational (2.18b):

$$\frac{\partial}{\partial t} = 0: \quad \nabla \wedge \bar{E} = 0 \Leftrightarrow \bar{E} = -\nabla \Phi_e, \quad (2.18a\text{–}c)$$

implying (1.114a, b) that the **electric field** is (2.18c) minus the gradient of a **scalar electric potential**. The **electric force density** per unit volume (I.28.44) ≡ (2.19a) is the product of the electric field by the **electric charge density per unit volume**:

$$\bar{f}_e = q\bar{E}: \quad dW_e = \bar{f}_e \cdot d\bar{x} = q\bar{E} \div d\bar{x} = -q\nabla \Phi_e \cdot d\bar{x} = -qd\Phi_e, \quad (2.19a, b)$$

and the corresponding work (2.1b) is related to the scalar potential by (2.19b) the electric charge. The **electric energy** is (2.20b) the work to bring an electric charge from infinity where the scalar potential is zero (2.20a):

$$0 = \Phi_e(\bar{x} = \infty): \quad E_e = -\int_{\infty}^{\bar{x}} dW_e = q \int_{\infty}^{\bar{x}} d\Phi_e = q\Phi_e(\bar{x}); \quad (2.20a, b)$$

thus *the electric energy (2.20a, b) of the electrostatic field (2.18a–c) equals the electric charge density multiplied by the electrostatic potential, corresponding to the work (2.19b) of the electric force (2.19a) to bring the electric charge from infinity*. An alternative expression for the electric energy in terms of the electric field and displacement is given next (subsection 2.1.6).

2.1.6 ELECTRIC DISPLACEMENT, WORK AND ENERGY

The electric energy (2.20b) can alternatively be expressed in terms of the fields, starting (2.21a) with the variation of the electric charge δq contained in a region D_3 of volume element dV:

$$\delta \bar{E}_e = \int_{D_3} \delta E_e dV = \int_{D_3} \Phi_e \delta q dV; \quad q = \nabla \cdot \vec{D}, \qquad (2.21\text{a, b})$$

the electric charge is related to the **electric displacement** vector through the Maxwell equation (2.21b), that may be substituted in the total electric energy (2.22) of the three-dimensional region or domain D_3 with a closed regular boundary ∂D_3:

$$\delta \bar{E}_e = \int_{D_3} \Phi_e \partial_i (\delta D_i) dV = \int_{D_3} \left[\partial_i (\Phi_e \delta D_i) - \delta D_i (\partial_i \Phi_e) \right] dV. \qquad (2.22\text{a, b})$$

and in (2.22b) was used an integration by parts that is equivalent to the property of the divergence of the product of a scalar by a vector.

The first term of the r.h.s. of (2.22b) may be transformed by the divergence theorem (III.5.163a–c) ≡ (2.23b) into an integral over the surface $d\vec{S}$ of the boundary ∂D_3, where the electric displacement $\vec{D}$ is fixed, so its variation $\delta \vec{D}$ vanishes (2.22a):

$$0 = \delta \vec{D}\Big|_{\partial D_3} : \quad \int_{D_3} \nabla \cdot (\Phi_e \delta \vec{D}) dV = \int_{\partial D_3} \Phi_e \delta \vec{D} \cdot d\vec{S} = 0; \qquad (2.23\text{a–c})$$

thus (2.23c) the first term on the r.h.s. of (2.22b) vanishes, and the remaining second term in (2.22b) involves both the electric field (2.18c) and displacement in (2.24a):

$$\delta \bar{E}_e = \int_{D_3} (\vec{E} \cdot \delta \vec{D}) dV = \int_{D_3} \delta E_e dV: \quad \delta E_e = \vec{E} \cdot \delta \vec{D}, \qquad (2.24\text{a, b})$$

comparison of (2.21a) ≡ (2.24a) specifies an alternate form of the electric energy density (2.24b). Thus *the electric energy density is specified (2.20b) by the electric charge density in the electric force density (2.19a) multiplied by the electrostatic potential (2.18c); the variation of the electric energy (2.24b) equals the electric field (2.18b) projected on the variation of the electric displacement (2.21b)*. The electrostatic energy can be expressed in terms of either the electric field or displacement alone using their constitutive relation through the dielectric permittivity (subsection 2.1.7).

2.1.7 DIELECTRIC PERMITTIVITY TENSOR AND SCALAR

In a **linear dielectric** medium, the electric displacement (2.21b) is proportional to the electric field (2.18b) through (2.25a) the **dielectric permittivity tensor**, implying that they need not be parallel in an **anisotropic medium** like a crystal:

$$dD_i = \varepsilon_{ij} dE_j; \quad dE_e = E_i dD_i = \varepsilon_{ij} E_i dE_j, \qquad (2.25a\text{–}c)$$

the electric energy (2.24b) $\equiv$ (2.25b) is given (2.25a) by (2.25c) that simplifies to (2.26b) in the case of constant dielectric permittivity tensor (2.26a),

$$\varepsilon_{ij} = const: \quad E_e = \frac{1}{2}\varepsilon_{ij}E_iE_j = \frac{1}{2}D_iE_i = \frac{1}{2}\varepsilon_{ij}^{-1}D_iD_j, \qquad (2.26a, b)$$

where ε_{ij}^{-1} is the matrix inverse to ε_{ij}; the dielectric permittivity tensor ε_{ij} must have non-zero determinant, and hence an inverse matrix ε_{ij}^{-1} in order that the constitutive relation (2.25a) be invertible, expressing the electric field in terms of the electric displacement. In the case of an **isotropic medium**: (i) the dielectric permittivity tensor reduces to a **scalar dielectric permittivity** (2.27a) multiplied by the identity matrix; (ii) the electric displacement and field are parallel (2.27b); (iii) the electric energy (2.27c) involves only the modulus of the electric field E or electric displacement D and the dielectric permittivity scalar ε:

$$\varepsilon_{ij} = \varepsilon \delta_{ij}: \quad \vec{D} = \varepsilon \vec{E}, \quad E_e = \frac{\varepsilon E^2}{2} = \frac{\vec{E} \cdot \vec{D}}{2} = \frac{D^2}{2\varepsilon}. \qquad (2.27a\text{–}c)$$

Thus *the electric energy is specified: (i/ii) in general by the electric charge and scalar electrostatic potential in (2.20a, b) or by the electric field and displacement in (2.24b); (iii/iv) in the case of a linear dielectric (2.25a) by (2.25b) leading in the anisotropic (2.26a) [isotropic (2.27a)] case with constant dielectric permittivity tensor (scalar) to (2.26b) [(2.27c)] involving only either the electric field or displacement. From (2.18c; 2.21b; 2.25a) follows that, for a linear dielectric, the scalar electric potential satisfies an* **inhomogeneous anisotropic Poisson equation** *(2.28a):*

$$-q = \partial_i \left(\varepsilon_{ij} \partial_j \Phi \right), \quad \varepsilon_{ij} \partial_i \partial_j \Phi_e = -q, \qquad (2.28a, b)$$

$$\nabla \cdot \left(\varepsilon \nabla \Phi_e \right) = -q, \quad \nabla^2 \Phi_e = -\frac{q}{\varepsilon}, \qquad (2.28c, d)$$

that: (i) for constant dielectric permittivity tensor (2.26a) becomes an anisotropic Poisson equation (2.28b); (ii) for an isotropic dielectric permittivity scalar (2.27a), becomes an inhomogeneous Poisson equation (2.28c) if the dielectric permittivity scalar is not constant; (iii) a constant dielectric permittivity scalar leads to the original Poisson equation (2.28d) $\equiv$ (I.24.5b) forced by $-\frac{q}{\epsilon}$. For a given distribution of electric charge density q as a function of position the electrostatic potential is a solution of (2.28a–d), and specifies the electric field (2.18c) and displacement (2.25a), *and the electric energy in two equivalent forms (2.20b) and (2.24b). The work of the electric force (2.19a) can be expressed in terms of the electric charge and potential (2.19b) or in terms of the electric field and displacement (subsection 2.1.8).*

2.1.8 WORK IN TERMS OF THE ELECTRIC FIELD AND DISPLACEMENT

Using (2.21b), the electric force density (2.19a) per unit volume is given in vector (index) notation by (2.29a) [(2.29b)]

$$\vec{f}_e = \vec{E}(\nabla \cdot \vec{D}) \Leftrightarrow f_{e_i} = E_i(\partial_j D_j): \quad dW_e = E_i(\partial_j D_j)dx_i, \qquad (2.29a\text{–}c)$$

leading to the **electric work** (2.29c) per unit volume. The total electric work in a three-dimensional domain D_3 is given by (2.30a) where is performed an integration by parts (2.30b):

$$d\overline{W}_e \equiv \int_{D_3} dW_e dV = \int_{D_3} \left[\partial_j (E_i D_j dx_i) - D_j \partial_j (E_i dx_i) \right] dV. \qquad (2.30a, b)$$

If the domain D_3 has a closed regular boundary, the divergence theorem (III.5.163a–c) $\equiv$ (2.31) gives:

$$I \equiv \int_{D_3} \partial_j (E_i D_j dx_i) dV = \int_{\partial D_3} E_i dx_i D_j dS_j. \qquad (2.31)$$

Choosing the domain D_3 to be all space, ∂D_3 is the surface at infinity. The surface at infinity lies between two spheres of radii $R_1 \le r \le R_2$ as $R_1 \to \infty$; to prove that (2.21) vanishes on the surface at infinity it is sufficient to show it vanishes on a sphere (2.32a) of infinite radius (2.32b):

$$|\vec{x}| = R, R \to \infty: |d\vec{x}| \sim 0(R), |d\vec{S}| \sim 0(R^2),$$
$$|\vec{E}| \sim 0(R^{-2}) \sim |\vec{D}|, I \sim 0(R^{-1}) \to 0, \qquad (2.32a\text{–}h)$$

in which case: (i/ii) the displacement (area) scale (2.32c) [(2.32d)] on the radius (square of the radius); (iii/iv) the electric field (2.31e) and displacement (2.32f) decay at least (Chapter I.24) as the inverse square of the radius (for a monopole, and faster for a multipole). The product of (2.31) decays (2.32g) like the inverse of the radius, showing that the integral (2.31) vanishes over the surface at infinity (2.32h).

Thus the electric work (2.30a) over all space reduces to the second term on the r.h.s. of (2.30b) $\equiv$ (2.33b), consisting of two terms in (2.33c):

$$d\overline{W}_e = -\int_{D_3} D_j \partial_j (E_i dx_i) dV = -\int_{D_3} D_j (\partial_j E_i) dx_i dV - \int_{D_3} D_j E_i \partial_j (dx_i) dV. \qquad (2.33a\text{–}c)$$

The cross-derivatives of the coordinates (2.34b) are unity (zero) for the same $i = j$ (different $i \neq j$) coordinates and specify (2.34a) the **identity matrix**:

$$\frac{\partial x_i}{\partial x_j} = \delta_{ij} = \begin{cases} 1 \, if \, i = j \\ 0 \, if \, i \neq j \end{cases}; \quad \partial_j (dx_i) = d(\partial_j x_i) = d(\delta_{ij}) = 0. \qquad (2.34a\text{–}c)$$

The identity matrix (2.34a) has constant components, hence its derivative is zero (2.34c) and thus the second term on the r.h.s. of (2.33c) vanishes leaving only the first term (2.35b) where may be substituted (2.18b) $\equiv$ (2.35a) leading to (2.35c).

$$\partial_j E_i = \partial_i E_j : d\bar{W}_e = -\int D_j (\partial_i E_j) dx_i dV = -\int D_j dE_j dV = -\int (\vec{D} \cdot d\vec{E}) dV. \qquad (2.35a\text{–}c)$$

Comparing (2.35c) $\equiv$ (2.30a) it follows that *the work (2.1b, c) of the electric force density (2.29a, b) equals (2.36a) minus the inner vector product of the electric field (2.18aa) by the variation of the electric displacement (2.21b):*

$$dW_e \cong \vec{F}_e \cdot d\vec{x} = -\vec{D} \cdot d\vec{E}; \qquad dE_e - dW_e = \vec{E} \cdot d\vec{D} + \vec{D} \cdot d\vec{E} = d(\vec{E} \cdot \vec{D}) \, (2.36a, b)$$

thus the electric energy (2.24b) minus the electric work (2.36a) is an exact differential (2.36b). In the case of a linear dielectric (2.25a) with constant dielectric permittivity (2.26a) $\equiv$ (2.37a) the electrical energy equals minus the electrical work (2.37b):

$$\varepsilon_{ij} = const: dW_e = -\varepsilon_{ij} E_i dE_j = -dE_e, \qquad \varepsilon_{ij} = \varepsilon_{ji}, \qquad (2.37a\text{–}c)$$

and the quadratic form (2.26b) implies that the dielectric permittivity tensor is symmetric (2.37c). The electric field (magnetic induction) differ [subsections 2.1.5–2.1.8 (2.1.9–2.1.14)] in being irrotational (solenoidal) vector fields [subsection 1.5.11 (1.5.12)].

2.1.9 Electric Current and Magnetic Vector Potential

The magnetic induction is solenoidal (2.38a), implying (1.117a, b) that it equals the curl (2.38b) of a **vector magnetic potential**:

$$\nabla \cdot \vec{B} = 0 \Leftrightarrow \vec{B} = \nabla \wedge \vec{A}. \qquad (2.38a, b)$$

The **magnetic force density** is (I.26.15) $\equiv$ (2.39a) the outer product of the **electric current density** by the magnetic induction divided by the speed of light in vacuo (2.39a):

$$c\vec{f}_h = \vec{J} \wedge \vec{B} = \vec{J} \wedge (\nabla \wedge \vec{A}) \Leftrightarrow cf_{hi} = J_k (\partial_i A_k - \partial_k A_i), \qquad (2.39a, b)$$

and can be written (2.39b) in index notation. The equivalence of (2.39a) ≡ (2.39b) can be seen from the $\ell = 1$ component in Cartesian coordinates:

$$
\begin{aligned}
\left[J \wedge \left(\nabla \wedge \vec{A} \right) \right]_1 &= J_2 \left(\nabla \wedge \vec{A} \right)_3 - J_3 \left(\nabla \wedge \vec{A} \right)_2 = J_2 \left(\partial_1 A_2 - \partial_2 A_1 \right) - J_3 \left(\partial_3 A_1 - \partial_1 A_3 \right) \\
&= J_1 \partial_1 A_1 + J_2 \partial_1 A_2 + J_3 \partial_1 A_3 - J_1 \partial_1 A_1 - J_2 \partial_2 A_1 - J_3 \partial_3 A_1 \\
&= \sum_{i=1}^{3} J_i \left(\partial_1 A_i - \partial_i A_1 \right),
\end{aligned}
\tag{2.40}
$$

and likewise for the components $i = 2, 3$. The electric current density is related to the magnetic field by (2.41b) in the magnetostatic case (2.41a):

$$
\frac{\partial}{\partial t} = 0: \ c\nabla \wedge \vec{H} = \vec{J}; 0 = \nabla \cdot \left(\nabla \wedge \vec{H} \right) = \nabla \cdot \vec{J} = \partial_k J_k,
\tag{2.41a–e}
$$

$$
0 = \int_{D_3} \left(\nabla \cdot \vec{J} \right) dV = \int_{\partial D_3} \vec{J} \cdot d\vec{S},
\tag{2.41f, g}
$$

the divergence of (2.41b) is zero (2.41c) showing that the electric current has zero divergence (2.41d) and thus, by the divergence theorem (2.41f), zero flux across a closed regular surface (2.41g).

Using (2.40) and (2.41c) the magnetic force (2.39b) simplifies to (2.42a–c):

$$
c f_{hi} - J_k \partial_i A_k = -J_k \partial_k A_i = -J_k \partial_k A_i - A_i \partial_k J_k = -\partial_k \left(J_k A_i \right).
\tag{2.42a–c}
$$

Integrating over a region D_3 with volume element dV with a closed regular boundary ∂D_3 with area element $d\vec{S}$, the divergence theorem (III.5.163a–c) ≡ (2.42b) may be applied to the last term on the r.h.s. of (2.42c) leading to (2.43b):

$$
0 = \delta \vec{A} \big|_{\partial D_3} : \ \int_{D_3} \partial_k \left(J_k \delta A_i \right) dV = \int_{D_3} J_k \delta A_i dS_k = \int_{D_3} \left(\vec{J} \cdot d\vec{S} \right) \delta A_i = 0,
\tag{2.43a–c}
$$

that vanishes (2.43c) because the vector potential is fixed on the boundary, that is, has zero variation (2.43a). The total work of the magnetic force (2.42c) is thus given by (2.44a–e):

$$
\begin{aligned}
\delta \overline{W}_m &\equiv \int_{D_3} \delta W_m dV = \int_{D_3} \left(\vec{f}_h \cdot \delta \vec{x} \right) dV = \frac{1}{c} \int_{D_3} J_k \left(\partial_i A_k \right) dx_i dV \\
&= \frac{1}{c} \int_{D_3} J_k dA_k dV = \int_{D_3} \frac{\vec{J} \cdot d\vec{A}}{c} dV,
\end{aligned}
\tag{2.44a–e}
$$

where were used (2.42a–c) and (2.43a–c). The magnetic energy is specified by (2.45b) the work to bring an electric current from infinity, where the vector potential is zero (2.45a):

$$0 = \vec{A}\left(\bar{x} = \infty\right): \quad E_h = \frac{\vec{J}}{c}\int\limits_{\infty}^{\bar{x}} d\vec{A} = \frac{\vec{J}\cdot\vec{A}}{c}, \tag{2.45a–c}$$

and is given (2.45c) by the projection on the vector potential divided by the speed of light in vacuo. It has been shown that *the magnetic energy density defined (2.45b) as the work (2.44a–e) of the electric force (2.39a, b) ≡ (2.42a–c) to bring an electric current (2.41a–f) from infinity, equals (2.45c) its inner vector product by the vector potential (2.38a, b), divided by the speed of light in vacuo, taking the value zero for the vector potential at infinity (2.45a).* An alternative expression for the magnetic energy can be obtained (subsection 2.1.10) in terms of the magnetic field and induction.

2.1.10 MAGNETIC FIELD, INDUCTION AND ENERGY

The magnetic energy can also be expressed in terms of the magnetic field and induction substituting (2.41b) in (2.45c) leading to (2.46a):

$$\delta E_h = \left(\nabla \wedge \vec{H}\right)\cdot\delta\vec{A} = \nabla\cdot\left(\vec{H}\wedge\delta\vec{A}\right) + \vec{H}\cdot\left(\nabla\wedge\delta\vec{A}\right); \tag{2.46a, b}$$

in (2.46b) was used the identity:

$$\nabla\cdot\left(\vec{X}\wedge\vec{Y}\right) = \vec{Y}\cdot\left(\nabla\wedge\vec{X}\right) - \vec{X}\cdot\left(\nabla\wedge\vec{Y}\right), \tag{2.47}$$

that can be proved using Cartesian coordinates:

$$\begin{aligned}
\nabla\cdot\left(\vec{X}\wedge\vec{Y}\right) &= \partial_x\left(\vec{X}\wedge\vec{Y}\right)_x + \partial_y\left(\vec{X}\wedge\vec{Y}\right)_y + \partial_z\left(\vec{X}\wedge\vec{Y}\right)_z \\
&= \partial_x\left(X_yY_z - X_zY_y\right) + \partial_y\left(X_zY_x - X_xY_z\right) + \partial_z\left(X_xY_y - X_yY_x\right) \\
&= Y_x\left(\partial_yX_z - \partial_zX_y\right) + Y_y\left(\partial_zX_x - \partial_xX_z\right) + Y_z\left(\partial_xX_y - \partial_yX_x\right) \\
&\quad -X_x\left(\partial_yY_z - \partial_zY_y\right) - X_y\left(\partial_zY_x - \partial_xY_z\right) - X_z\left(\partial_xY_y - \partial_yY_x\right) \\
&= Y_x\left(\nabla\wedge\vec{X}\right)_x + Y_y\left(\nabla\wedge\vec{X}\right)_y + Y_z\left(\nabla\wedge\vec{X}\right)_z \\
&\quad -X_x\left(\nabla\wedge\vec{Y}\right)_x - X_y\left(\nabla\wedge\vec{Y}\right)_y - X_z\left(\nabla\wedge\vec{Y}\right)_z \\
&= \vec{Y}\cdot\left(\nabla\wedge\vec{X}\right) - \vec{X}\cdot\left(\nabla\wedge\vec{Y}\right).
\end{aligned} \tag{2.48a–e}$$

Integrating (2.46b) over a region, the first term of the r.h.s. leads as for (2.43a–c) by the divergence theorem (2.49b) ≡ (III.5.163a–c) to a vanishing integral (2.49c)

$$\delta\vec{A}\Big|_{\partial D3} = 0: \quad \int\limits_{D_3}\left[\nabla\cdot\left(\vec{H}\wedge\delta\vec{A}\right)\right]dV = \int\limits_{\partial D_3}\left(\vec{H}\wedge\delta\vec{A}\right)\cdot d\vec{S} = 0, \tag{2.49a–c}$$

because the vector potential is fixed (2.49a) on the closed regular boundary ∂D_3. The second term on the r.h.s. of (2.46a) specifies (2.38b) the magnetic energy (2.50a) in terms of the magnetic field and induction (2.50b):

$$\delta E_h = \vec{H} \cdot \left(\nabla \wedge \delta \vec{A} \right) = \vec{H} \cdot \delta \vec{B}. \tag{2.50a, b}$$

Thus, *the magnetic energy per unit volume is specified (2.45c) by the projection of the electric current (2.41a–f) on the vector potential (2.38b) divided by the speed of light in vacuo; its variation (2.50a) equals that of the magnetic induction (2.38a) projected on the magnetic field (2.41b) in steady conditions (2.41a).* The magnetic energy can be expressed in terms of only one of the magnetic field or induction for a linear medium (subsection 2.1.11), using the magnetic permeability.

2.1.11 MAGNETIC PERMEABILITY TENSOR AND SCALAR

In a linear medium the magnetic induction (2.38a) is proportional to the magnetic field (2.41b) through the **magnetic permeability tensor** (2.51a), implying that they are generally not parallel in an anisotropic medium like a crystal:

$$dB_i = \mu_{ij} dH_j; \qquad dE_h = \mu_{ij} H_i dH_j, \tag{2.51a, b}$$

the magnetic energy (2.50b; 2.51a) ≡ (2.51b) in the case of constant magnetic permeability (2.52a) is given by (2.52b):

$$\mu_{ij} = const: \quad E_h = \frac{1}{2} \mu_{ij} H_i H_j = \frac{\vec{H} \cdot \vec{B}}{2} = \frac{1}{2} \mu_{ij}^{-1} B_i B_j \tag{2.52a, b}$$

where μ_{ij}^{-1} is the matrix inverse to μ_{ij}. In the case of an isotropic medium (2.53a) the magnetic induction is parallel to the magnetic field (2.53b) and the magnetic energy depends (2.53c) only on the modulus of the magnetic field H or magnetic induction B and the **magnetic permeability scalar** μ:

$$\mu_{ij} = \mu \delta_{ij}: \vec{B} = \mu \vec{H}, \quad E_h = \frac{\mu^2}{2} = \frac{\vec{H} \cdot \vec{B}}{2} = \frac{B^2}{2\mu}. \tag{2.53a–c}$$

Thus *the magnetic energy is specified by: (i/ii) in general, the inner vector product (2.45a–c) of the vector potential (2.38b) by the electric current (2.41a–f) divided by the speed of light in vacuo or, alternatively, (2.50a, b) by the magnetic field (2.41b) and induction (2.38a); (iii–v) in the case of a linear medium (2.51a), by (2.51b), leading to (2.52b) [(2.53c)] in the anisotropic (2.51a) [isotropic (2.53b)] case with constant magnetic permeability tensor (2.52a) [scalar (2.53a)], involving only the magnetic field or induction.* The electric (magnetic) potential satisfies scalar (vector) inhomogeneous anisotropic Poisson equation [subsection 2.1.7 (2.1.13)]. As a preliminary are needed the outer vector product and curl in index notation, involving the permutation symbol (subsection 2.1.12).

2.1.12 Permutation Symbol, Curl and Outer Vector Product

The **permutation symbol** with three indices in a three-dimensional space (2.54a) is defined by its components as plus (minus) unity for even (odd) permutations (2.54b) [(2.54c)] of (1, 2, 3) and zero (2.54d) otherwise, that is, when there are repeated indices:

$$i, j, k = 1, 2, 3: \quad e_{ijk} = \begin{cases} +1 \, if \, (i,j,k) \, in \, even \\ permutation \, of \, (1,2,3) \\ \\ -1 \, for \, odd \, permutation \\ \\ 0 \, otherwise \end{cases} \tag{2.54a–d}$$

The permutation symbol has $3 \times 3 \times 3 = 27$ components, since each index can take 3 values; since repeated indices lead to zero, there are only six non-zero components (2.55a):

$$e_{123} = e_{231} = e_{312} = +1 = -e_{132} = -e_{213} = -e_{321}. \tag{2.55a}$$

From (2.54a–d) also follows that the permutation symbol is skew-symmetric in all indices:

$$e_{ijk} = e_{jki} = e_{kij} = -e_{ikj} = -e_{jik} = -e_{kji}, \tag{2.55b}$$

since: (i) for distinct indices (2.55b) coincides with (2.55a); (ii) for repeated indices all terms in (2.55b) are zero by (2.54d). The permutation symbols appear in: (i) the outer product of two vectors (2.56a); (ii) the curl of a vector field (2.56b):

$$\left(\vec{A} \wedge \vec{B} \right)_i \equiv e_{ijk} A_j B_k, \quad \left(\nabla \wedge \vec{X} \right)_i = e_{ijk} \partial_j X_k. \tag{2.56a, b}$$

Cartesian coordinates can be used to confirm that (2.56a) $\equiv$ (2.56c) [(2.56b) $\equiv$ (2.56d)] for the first component:

$$\left(\vec{A} \wedge \vec{B} \right)_1 = e_{1jk} A_j B_k = e_{123} A_2 B_3 + e_{132} A_3 B_2 = A_2 B_3 - A_3 B_2, \tag{2.56c}$$

$$\left(\nabla \wedge \vec{X} \right)_1 = e_{1jk} \partial_j X_k = e_{123} \partial_2 X_3 + e_{132} \partial_3 X_2 = \partial_2 X_3 - \partial_3 X_2, \tag{2.56d}$$

and likewise for the other components. The permutation symbol is not (is) needed for the scalar (vector) inhomogeneous anisotropic Poisson equation [subsection 2.1.7 (2.1.13)].

2.1.13　ANISOTROPIC INHOMOGENEOUS POISSON EQUATION

Substituting in (2.41b) ≡ (2.57a) the inverse of (2.51a) leads to (2.57b) and then (2.38b) specifies the **anisotropic inhomogeneous vector Poisson equation** (2.57c) satisfied by the vector magnetic potential forced by the electric current:

$$\frac{1}{c} J_i = e_{ijk}\,\partial_j\,H_k = e_{ijk}\,\mu_{k\ell}^{-1}\,\partial_j\,B_\ell = e_{ijk}\,e_{\ell mn}\,\mu_{k\ell}^{-1}\,\partial_j\,\partial_m\,A_n, \qquad (2.57\text{a–c})$$

where: (i) the inverse dielectric permeability tensor (2.51b) ≡ (2.58a–c) appears outside the differentiation

$$dx_j\left(\partial_j H_k\right) = dH_k = \mu_{k\ell}^{-1} dB_\ell = \mu_{k\ell}^{-1} dx_j\left(\partial_j B_\ell\right); \qquad (2.58\text{a–c})$$

(ii) the permutation symbols are not differentiated because they are constant (2.54a–d). The magnetic permeability tensor need not be constant since it appears outside the differentiations in (2.57a–c, 2.58a–c). In the isotropic case (2.53b) in (2.41b; 2.38b) leads (2.58a) to the **isotropic homogeneous Poisson equation** (2.59a–d):

$$\frac{\vec{J}}{c} = \frac{1}{\mu}\nabla\wedge\vec{B} = \frac{1}{\mu}\nabla\wedge\left(\nabla\wedge\vec{A}\right) = \frac{1}{\mu}\left[\nabla\left(\nabla\cdot\vec{A}\right) - \nabla^2\vec{A}\right]. \qquad (2.59\text{a–d})$$

Imposing the gauge condition (1.120a) ≡ (2.60a) on the vector potential leads to (2.60b) ≡ (I.26.6a, b) the original form of the vector Poisson equation forced by $-\dfrac{\mu\,\vec{j}}{c}$:

$$\nabla\cdot\vec{A} = 0: \quad -\frac{\mu}{c}\vec{J} = \nabla^2\vec{A}. \qquad (2.60\text{a, b})$$

Thus *the magnetic vector potential (2.38a, b) satisfies the anisotropic inhomogeneous vector Poisson equation (2.57c) forced by the electric current (2.41a–f) in a linear medium with (2.51a) inverse magnetic permeability tensor μ_{ij}^{-1}, valid for an homogeneous or inhomogeneous medium. In the case of an isotropic medium (2.59a–d) imposing a gauge condition of zero divergence (2.60a) the magnetic potential satisfies the original vector Poisson equation (2.60b) forced by $-\dfrac{\mu\,\vec{J}}{c}$. For a given distribution of electric current density $\vec{J}$ as a function of position the vector magnetostatic potential is a solution of (2.57a–c) [(2.60a, b)] for a linear anisotropic (isotropic) medium, and specifies the magnetic induction (2.38b) and field (2.51a) and the magnetic energy in two equivalent forms (2.45c) and (2.51b). The work of the magnetic force (2.39a, b) can be expressed in terms of the electric current and vector potential (2.44a–e) or in terms of the magnetic field and induction (Section 2.1.14).*

2.1.14   WORK IN TERMS OF THE MAGNETIC FIELD AND INDUCTION

Substituting (2.41b) in the magnetic force (2.39a) leads to (2.61a) in vector notation, or equivalently (2.39a, b; 2.40) to (2.61b) in index notation:

$$\vec{f}_h = \left(\nabla \wedge \vec{H}\right) \wedge \vec{B} \Leftrightarrow f_{mi} = B_k\left(\partial_k H_i - \partial_i H_k\right). \tag{2.61a, b}$$

Using (2.38a) ≡ (2.62a) from (2.61b) follows (2.62b–d):

$$0 = \partial_k B_k: f_{hi} + B_k \partial_i H_k = B_k \partial_k H_i = \partial_k\left(B_k H_i\right) - H_i \partial_k B_k = \partial_k\left(B_k H_i\right). \tag{2.62a–d}$$

The work (2.1b, c) of the magnetic force (2.62d) in a displacement in a three-dimensional domain is given by (2.63a–c):

$$\begin{aligned}
d\bar{W}_h = \int_{D_3} dW_h dV &= \int_{D_3} f_{hi} dx_i dV \\
&= -\int_{D_3} B_k\left(\partial_i H_k\right) dx_i dV + \int_{D_3}\left[\partial_k\left(B_k H_i\right)\right] dx_i dV.
\end{aligned} \tag{2.63a–c}$$

Using (2.34a–c), the last integral on the r.h.s. of (2.63c) becomes (2.64a):

$$\bar{I} = \int_{D_3}\left[\partial_k\left(B_k H_i\right)\right] dx_i dV = \int_{D_3}\left[\partial_k\left(B_k H_i dx_i\right)\right] dV = \int_{\partial D_3} B_k H_i dx_i dS_k, \tag{2.64a, b}$$

leading to (2.64b) by the divergence theorem. Using (Chapter I.26) the same scalings for the magnetic field and induction as for the electric field (2.32e) and displacement (2.32f), from (2.32a–d) follows that the integral (2.64b) vanishes on the surface at infinity like (2.32g, h). Thus, the magnetic work simplifies to the first term on the r.h.s. of (2.63c) leading to (2.65a–c):

$$d\bar{W}_h = -\int_{D_3} B_k\left(\partial_i H_k\right) dx_i dV = -\int_{D_3} B_k dH_k dV = -\int_{D_3}\left(\vec{B} \cdot d\vec{H}\right) dV. \tag{2.65a–c}$$

Comparing (2.65c) ≡ (2.63a) it follows that *the work (2.1b, c) of the magnetic force (2.39a) ≡ (2.61a, b) equals (2.66a) minus the inner vector product of the magnetic induction (2.38a, b) by the magnetic field (2.41a–f)*:

$$W_h = -\vec{B} \cdot d\vec{H}; \quad dE_h - dW_h = \vec{H} \cdot d\vec{B} + \vec{B} \cdot d\vec{H} = d\left(\vec{B} \cdot \vec{H}\right), \tag{2.66a, b}$$

the difference between the magnetic energy (2.50b) and the magnetic work (2.66a) is an exact differential (2.66b). In the case of a linear medium (2.51a) with constant magnetic permeability tensor (2.52a) the magnetic work equals minus the magnetic energy (2.67a–c):

$$dW_h = -B_i dH_i = -\mu_{ij} H_j dH_i = -dE_h; \mu_{ij} = \mu_{ji}, \tag{2.67a–d}$$

the quadratic forms (2.52b) and (2.67b) imply that the magnetic permeability tensor is symmetric. After the work [subsection 2.1.3 (2.1.4)] of the inertia force (gravity field), and of the electric (magnetic) field [subsections 2.1.5–2.1.8 (2.1.9–2.1.14)] is considered the mechanical work (subsections 2.1.15–2.1.19) associated with surface forces in general (subsections 2.1.16–2.1.19), and the pressure in particular subsection 2.1.5).

2.1.15   Work of the Pressure in a Volume Change

Besides the volume forces (subsections 2.1.3–2.1.14) there are surface forces (subsections 2.1.15–2.1.19). The simplest is the **pressure** corresponding to a force per unit area opposite to the normal (2.68b, c):

$$d\vec{S} = \vec{N}\,dS: \quad d\vec{F}_p = -p\vec{N}dS = -pd\vec{S}, \qquad (2.68\text{a–c})$$

where $\vec{N}$ is the unit outward normal (2.68a) and $d\vec{S}\,(dS)$ the vector area element (its modulus or area scalar). The work (2.1a, b) of the pressure forces

$$dW_p = d\vec{F}_p \cdot d\vec{x} = -p\vec{N} \cdot d\vec{x}dS = -pdSd\ell, \qquad (2.69\text{a–c})$$

involves the product of the area by the normal displacement (2.70a) that specifies (Figure 2.1a) the volume change (2.70b):

$$d\ell = \vec{N} \cdot d\vec{x}, \quad dV = dSd\ell: \quad dW_p = -pdV. \qquad (2.70\text{a–c})$$

Thus *the work of the pressure in a volume change (2.70a, b; 2.68a) is given by (2.70c) and is positive (negative) in (Figure 2.1b) a contraction $dV < 0$ (expansion > 0).* The pressure is a particular case of isotropic, inward stresses and both are associated with volume forces (subsection 2.1.16).

2.1.16   Volume Forces Associated with Surface Pressure or Stresses

The **total pressure force** over a closed regular surface with element (2.68a) is given by (2.71a) the integral of (2.68b):

$$\vec{F}_p = -\int_{\partial D_3} pd\vec{S} = -\int_{D_3} \nabla pdV = \int_{D_3} \vec{f}_pdV: \quad \vec{f}_p \equiv \frac{d\vec{F}_p}{dV} = -\nabla p, \qquad (2.71\text{a–e})$$

the use of the gradient theorem (III.5.170a–c) $\equiv$ (2.71b) *shows that the surface inward pressure (2.68a–c) over a closed regular surface (2.71a–c) is equivalent to a force density per unit volume (2.71d) in the interior equal to (2.71e) minus the gradient of the pressure.* The result (2.71e) may be extended to a **surface stress vector** (2.72a) that may have normal and/or tangential components, for example, associated, respectively, with pressure and/or friction. The linear relation between a stress vector

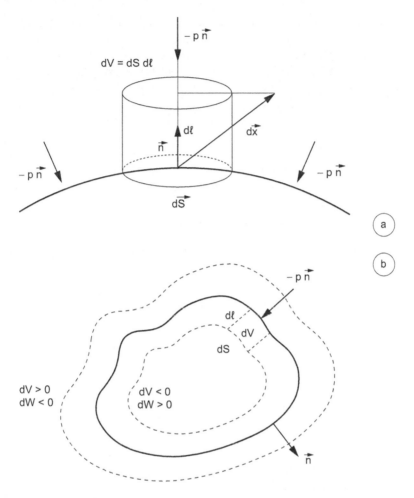

FIGURE 2.1 The pressure acting along the inward normal to the surface of a body performs work in a volume change. Work has to be supplied $dW > 0$ in a compression $dV < 0$ and work can be extracted $dW < 0$ in an expansion $dV > 0$.

T_i with arbitrary direction relative to the unit normal N_j to a surface is specified by the **stress tensor** T_{ij}, leading to a total force (2.72b–d) on a closed regular surface:

$$T_i = T_{ij}N_j: \quad F_{ui} = \int_{\partial D_3} T_i dS = \int_{\partial D_3} T_{ij}N_j dS = \int_{\partial D_3} T_{ij} dS_j; \qquad (2.72\text{a–d})$$

using the divergence theorem (III.5.163a–c) $\equiv$ (2.73a):

$$F_{ui} = \int_{D_3} (\partial_j T_{ij}) dV = \int_{D_3} f_{ui} dV \Leftrightarrow f_{ui} \equiv \frac{dF_{ui}}{dV} = \partial_j (T_{ij}), \qquad (2.73\text{a–d})$$

shows that *a stress surface force (2.72a) over a closed regular surface (2.72b–d) is equivalent (2.73a, b) to a force per unit volume (2.73c) in the interior equal (2.73d) to the divergence of the stress tensor. The particular case of the presence of normal stresses or pressure only (2.74a) leads to an* **isotropic stress tensor** *(2.74b) whose divergence (2.74c) equals minus the gradient of the pressure (2.74d):*

$$\vec{T} = -p\vec{N}: \quad T_{ij} = -p\delta_{ij}, \quad f_{pi} = \partial_j\left(-p\delta_{ij}\right) = -\partial_i p, \qquad (2.74\text{a–d})$$

in agreement with (2.71d, e) ≡ (2.74c, d) as a particular case of the volume stress force (2.73c, d). The work is generalized [subsection 2.1.15 (2.1.17)] from the pressure (to anisotropic stresses) in the volume change (deformation) of a medium.

2.1.17   WORK OF THE SURFACE STRESSES IN A DISPLACEMENT

The work of the stress vector (2.72a) in an infinitesimal displacement δu_i of a closed regular surface is given by (2.75a–c):

$$d\bar{W}_u = \int_{\partial D_3} T_i \delta u_i dS = \int_{\partial D_3} T_{ij} \delta u_i N_j dS = \int_{\partial D_3} T_{ij} \delta u_i dS_j. \qquad (2.75\text{a–c})$$

Applying the divergence theorem (III.5.163a–c) ≡ (2.76a) transforms the surface (2.75c) to a volume integral (2.76b):

$$\partial_j T_{ij} = 0: \quad d\bar{W}_u = \int_{D_3} \partial_j\left(T_{ij}du_j\right)dV = \int_{D_3} T_{ij}\partial_j\left(du_i\right)dV \qquad (2.76\text{a–c})$$

(subsection IV.6.8.4) leading to (2.76c) assuming (2.76a) that there are no external volume forces (2.73d) because the latter have been considered before (subsections 2.1.3–2.1.14). Comparing (2.76c) with (2.77a):

$$\bar{W}_u = \int_{D_3} dW_u dV: \quad dW_u = T_{ij}dD_{ij}, \quad D_{ij} \equiv \partial_i u_j \qquad (2.77\text{a–c})$$

follows that the *work (2.1a, b) of the stress vector (2.72a) in the displacement (2.75a–c; 2.76b, c) of a closed regular surface in absence of external forces (2.76a) equals per unit volume (2.77a) the double inner product or* **contraction** *(2.77b) of the stress tensor with the variation of the* **displacement tensor** *(2.77c) defined by the partial spatial derivatives of the* **displacement vector***.* Besides the balance of surface and volume forces (subsections 2.1.15–2.1.17) the balance of the moment of forces is also considered leading (subsection 2.1.18) to a symmetric stress tensor.

2.1.18   BALANCE OF MOMENT OF FORCES AND SYMMETRIC STRESS TENSOR

The balance of the moment of the stress forces on a closed regular surface is given by (2.78a–d) the outer product of the position vector $\bar{x}$ by the stress vector (2.72a) integrated over the surface.

$$0 = \int_{\partial D_3} \left(\vec{x} \wedge \vec{T} \right)_i dS = \int_{\partial D_3} e_{ijk} x_j T_k dS = \int_{\partial D_3} e_{ijk} x_j T_{k\ell} N_\ell dS = \int_{\partial D_3} e_{ijk} x_j T_{k\ell} dS_\ell. \qquad (2.78\text{a–d})$$

Applying the divergence theorem (III.5.163a, b) transforms the surface (2.78d) to a volume (2.79a) integral:

$$0 = \int_{D_3} \left[\partial_\ell \left(e_{ijk} x_j T_{k\ell} \right) \right] dV: \quad 0 = \partial_\ell \left(e_{ijk} x_j T_{k\ell} \right), \qquad (2.79\text{a, b})$$

and since the domain is arbitrary follows (2.79b). The permutation symbol (2.54a–d) is constant and thus (2.79b) leads to (2.80a):

$$0 = e_{ijk} \left[\left(\partial_\ell x_j \right) T_{k\ell} + x_j \partial_\ell T_{k\ell} \right] = e_{ijk} \delta_{\ell j} T_{k\ell} = e_{ijk} T_{kj}, \qquad (2.80\text{a–c})$$

that simplifies to (2.80b, c), using (2.76a) and (2.34a). From (2.80c) follows (2.81a) and likewise (2.81b, c) proving that the stress tensor is symmetric (2.81d):

$$0 = e_{1jk} T_{kj} = e_{123} T_{32} + e_{132} T_{23} = T_{23} - T_{32} \qquad (2.81\text{a})$$

$$0 = e_{2jk} T_{kj} = T_{31} - T_{13}, 0 = e_{3jk} T_{kj} = T_{21} - T_{12} : T_{ij} = T_{ji}. \qquad (2.81\text{b–d})$$

It has been shown that *the balance of the moment of the stress forces on a surface (2.78a–d) implies that the stress tensor is symmetric (2.81d)*. The symmetry of the stress tensor (2.81d) can be used (subsection 2.1.19) to simplify the mechanical work (2.77b).

2.1.19 DISPLACEMENT AND STRAIN TENSORS AND ROTATION VECTOR

The displacement tensor (2.77c) can be split into the sum (2.82a) of: (i) a skew-symmetric part (2.82b) that corresponds (2.82c) ≡ (1.70c) to the vector angular velocity of rotation:

$$D_{ij} = \Omega_{ij} + S_{ij}: \quad \Omega_{ij} - \frac{1}{2} \left(\partial_i x_j - \partial_j x_i \right) - -\Omega_{ji} \qquad (2.82\text{a, b})$$

$$\Omega_k = e_{ijk} \Omega_{ij}, \quad S_{ij} = \frac{1}{2} \left(\partial_i x_j + \partial_j x_i \right) = S_{ji}; \qquad (2.82\text{c, d})$$

(ii) the symmetric part is the **strain tensor** (2.82d) that separates out the effect of rotation on the displacement, for example, in a rotation in the plane (1.68a–c; 1.69a, b). *The symmetric stress tensor (2.81d) performs no work (2.84a–d) in a rotation (2.83b, c)*

$$T_{ij} d\Omega_{ij} = T_{ji} d\Omega_{ij} = -T_{ji} d\Omega_{ji} = -T_{ij} d\Omega_{ij} = 0; \qquad (2.83\text{a–d})$$

thus the work of the surface stresses in the displacement of a closed regular surface (2.75a–c) corresponds per unit volume (2.77a–c) to (2.84a; 2.83d) ≡ (2.84b)

$$dW_u = T_{ij}d\left(\Omega_{ij} + S_{ij}\right) = T_{ij}dS_{ij}, \qquad (2.84a, b)$$

the double inner product or contraction (2.84b) of the stress tensor (2.72a) with the variation of the strain tensor (2.82d). In the case of isotropic stresses (2.74b) ≡ (2.85a) corresponding to a pressure, the mechanical work (2.84b) involves (2.85b–d) only the volume change, that is, the contraction (2.85f) of the strain tensor:

$$T_{ij} = -p\delta_{ij}: \quad dW_u = -p\delta_{ij}dS_{ij} = -pdS_{ii} = -pdV = dW_p, \quad dV = dS_{ii}, \qquad (2.85a\text{–}f)$$

in agreement with (2.85d) ≡ (2.70c). The only remaining form of work not considered yet is associated with chemical reactions (subsections 2.1.20–2.1.21).

2.1.20   Mole and Avogadro Numbers and Chemical Work, Potential and Affinity

The work of volume (surface) forces [subsections 2.1.3–2.1.14 (2.1.15–2.1.19)] is the total work for a single **chemical substance**. Thermodynamics has some of its most important applications for gaseous or liquid mixtures, or chemical reactions, when several substances $\ell = 1, ..., L$ are present. The quantity of each substance is specified by the **mole number** (2.86a) that is the number of molecules n_ℓ divided by the **Avogadro constant** (2.86b):

$$N_\ell = \frac{n_\ell}{A}, \quad A \equiv \left(6.022169 \pm 0.00040\right) \times 10^{23}. \qquad (2.86a, b)$$

The change in the mole number of substances specifies (2.86c) the **chemical work**:

$$dW_c = \sum_{\ell=1}^{L} \mu_\ell dN_\ell, \qquad (2.86c)$$

with the **molar chemical potentials**, or **simply chemical potentials** μ_ℓ as coefficients.

The mass m_ℓ of each substance (2.87a) is the product of the mole number by the **molecular mass** M_ℓ; the sum of the masses of all substances is the **total mass** (2.87b, c); the mass of each substance divided by the total mass is the **mass fraction** (2.87d, e):

$$m_\ell = M_\ell N_\ell, \quad m = \sum_{\ell=1}^{L} m_\ell = \sum_{\ell=1}^{L} M_\ell N_\ell, \quad \xi_\ell = \frac{m_\ell}{m} = \frac{M_\ell N_\ell}{m}. \qquad (2.87a\text{–}e)$$

The molecular masses are constant (2.87f) and in a chemical non-nuclear reaction there is conservation of the total mass (2.87g) implying that the change in mass fraction (2.87e) is related to the change in mole number by (2.87h):

$$M_\ell = const, \quad m = const: d\xi_\ell = \frac{M_\ell}{m} dN_\ell. \tag{2.87f–h}$$

Substituting (2.87h) in (2.86c) it follows that the chemical work is proportional to the change in mass fractions (2.87i, j):

$$dW_c = m \sum_{\ell=1}^{L} \frac{\mu_\ell}{M_\ell} d\xi_\ell = \sum_{\ell=1}^{L} v_\ell d\xi_\ell, \quad v_\ell \equiv m \frac{\mu_\ell}{M_\ell}, \tag{2.87i–k}$$

through the mass chemical potentials or **affinities** (2.87k). The mole number (2.86a, b) and mass fraction (2.87a–e) are distinct from the **concentration** of a substance N_ℓ/V defined as the mole number per unit volume V.

From (2.87c) together with (2.87f, g) follows (2.88a, b) ≡ (2.88c):

$$0 = dm = \sum_{\ell=1}^{L} M_\ell dN_\ell \Leftrightarrow M_L dN_L = \sum_{\ell=1}^{L-1} M_\ell dN_\ell. \tag{2.88a–c}$$

Substituting (2.88c) in (2.86c) ≡ (2.88d) the chemical work (2.88e) is expressed in terms of the mole numbers of the first $L - 1$ substances:

$$dW_c = \mu_L dN_L + \sum_{\ell=1}^{L-1} \mu_\ell dN_\ell = \sum_{\ell=1}^{L-1} \left(\mu_\ell - \frac{M_\ell}{M_L} \mu_L \right) dN_\ell, \tag{2.88d, e}$$

through (2.88e) ≡ (2.88f) the **relative chemical potentials** (2.88g):

$$dW_c = \sum_{\ell=1}^{L-1} \bar{\mu}_\ell dN_\ell, \quad \bar{\mu}_\ell \equiv \mu_\ell - \frac{M_\ell}{M_L} \mu_L. \tag{2.88f, g}$$

From (2.87a, c, e) follows that the molar fractions add to unity (2.89a, b) and thus their differentials are related by (2.89c):

$$\sum_{\ell=1}^{L} \xi_\ell = \frac{1}{m} \sum_{\ell=1}^{L} M_\ell N_\ell = 1 \Leftrightarrow d\xi_L = -\sum_{\ell=1}^{L-1} d\xi_\ell. \tag{2.89a–c}$$

Substitution of (2.89c) in (2.87i) ≡ (2.89d–f) expresses the chemical work in terms of the changes in mass fractions of $L - 1$ substances:

$$dW_c = \sum_{\ell=1}^{L-1} v_\ell d\xi_\ell + v_L d\xi_L = \sum_{\ell=1}^{L-1} (v_\ell - v_L) d\xi_\ell = \sum_{\ell=1}^{L-1} \bar{v}_\ell d\xi_\ell, \quad \bar{v}_\ell \equiv v_\ell - v_L, \tag{2.89d–g}$$

with the **relative affinities** (2.89g) as coefficients. Thus *the chemical work equals the sum over all substances (2.86c) [(2.87j)] of the change in mole numbers (2.86a, b) [mass fractions (2.87a–e)] multiplied by the chemical potentials μ_ℓ [affinities (2.87k)]. Bearing in mind that the molecular masses are constant (2.87f) and the total mass is conserved (2.87g) in a non-nuclear chemical reaction, the chemical work can be expressed as a sum (2.88f) [(2.89f)] over $L-1$ substances of the change of mole numbers (mass fractions) multiplied by the relative chemical potentials (2.88g) [relative affinities (2.89g)].* As examples (Section 2.1.21) with the smallest possible number of constituents L = 2 (L = 3) are considered for two-phase matter (a combustion reaction).

2.1.21   Matter with Two Phases and a Combustion Reaction

In **two-phase matter** *the conservation of total mass (2.90a, b) leads to (2.90c):*

$$M_1 N_1 + M_2 N_2 = m = const, \quad M_1 dN_1 + M_2 dN_2 = 0. \tag{2.90a–c}$$

The chemical work (2.86c) involves two chemical potentials and mole numbers (2.90d), and reduces (2.90e, f) to one mole number and one relative chemical potential (2.90g):

$$dW_c = \mu_1 dN_1 + \mu_2 dN_2 = \left(\mu_1 - \frac{M_1}{M_2}\mu_2\right) dN_1 = \bar{\mu} dN_1, \quad \bar{\mu} = \mu_1 - \frac{M_1}{M_2}\mu_2. \tag{2.90d–g}$$

The chemical work can be expressed in terms of the change (2.91d) of the mass fraction of one species (2.91a–c):

$$\xi = \frac{M_1 N_1}{m} = \frac{M_1 N_1}{M_1 N_1 + M_2 N_2} = \left(1 + \frac{M_2 N_2}{M_1 N_1}\right)^{-1}, \quad d\xi = \frac{M_1}{m}dN_1; \tag{2.91a–d}$$

using (2.91d) in (2.90e) $\equiv$ (2.91e–g)

$$dW_c = \frac{m\,\bar{\mu}}{M_1}d\xi = \frac{m}{M_1}\left(\mu_1 - \frac{M_1}{M_2}\mu_2\right)d\xi = \bar{v}\,d\xi, \quad \bar{v} \equiv \left(\frac{\mu_1}{M_1} - \frac{\mu_2}{M_2}\right)m, \tag{2.91e–h}$$

with the relative affinity (2.91h) as coefficient.

The simplest chemical reaction involves 3 constituents, for example (2.92d), the combustion of hydrogen (2.92a) with oxygen (2.92b) to produce water (2.92c) vapour:

$$1 \equiv H_2, 2 \equiv O_2, 3 \equiv OH_2: \quad 2H_2 + O_2 \to 2OH_2. \tag{2.92a–d}$$

The mole numbers (2.92e) are taken negative for the **reactants** that are consumed, and positive for the **products** of the reaction; the atomic masses of hydrogen and oxygen (2.92f) lead to the molecular masses (2.92g):

$$N_{1-3} = \{-2, -1, 2\}, \quad \{M_H, M_O\} = \{1, 16\}, \quad M_{1-3} = \{2, 32, 18\}, \quad (2.93\text{a–c})$$

The conservation of total mass (2.94a) is satisfied leading to (2.94b) ≡ (2.88c):

$$2N_1 + 32N_2 + 18N_3 = 0: \quad 9dN_3 = -dN_1 - 16dN_2. \quad (2.94\text{a, b})$$

The chemical work (2.86c) is given by (2.94c, d) [(2.94g)] in terms of the 3 chemical potentials [2 relative chemical potentials (2.94e, f)]:

$$dW_c = \mu_1 dN_1 + \mu_2 dN_2 + \mu_3 dN_3 = \left(\mu_1 - \frac{\mu_3}{9}\right)dN_1 + \left(\mu_1 - \frac{16}{9}\mu_3\right)dN_2, \quad (2.94\text{c, d})$$

$$\bar{\mu}_1 = \mu_1 - \frac{\mu_3}{9}, \quad \bar{\mu}_2 = \mu_1 - \frac{16}{9}\mu_3: \quad dW_c = \bar{\mu}_1 dN_1 + \bar{\mu}_2 dN_2. \quad (2.94\text{e–g})$$

The mass fractions of the three constituents (2.95a, b):

$$\{\xi_1, \xi_2, \xi_3\} \equiv \frac{\{M_1 N_1, M_2 N_2, M_3 N_3\}}{M_1 N_1 + M_2 N_2 + M_3 N_3} = \frac{\{M_1 N_1, M_2 N_2, M_3 N_3\}}{m}, \quad (2.95\text{a, b})$$

have changes (2.95c–e) and satisfy (2.95f, g):

$$d\xi_1 = \frac{M_1}{m}dN_1, \quad d\xi_2 = \frac{M_2}{m}dN_2, \quad d\xi_3 = \frac{M_3}{m}dN_3. \quad (2.95\text{c–e})$$

$$\xi_1 + \xi_2 + \xi_3 = 1 \Rightarrow d\xi_3 = -d\xi_1 - d\xi_2; \quad (2.95\text{f, g})$$

The chemical work (2.94c) ≡ (2.95h):

$$dW_c = m\left(\frac{\mu_1}{M_1}d\xi_1 + \frac{\mu_2}{M_2}d\xi_2 + \frac{\mu_3}{M_3}d\xi_3\right)$$

$$= m\left(\frac{\mu_1}{M_1} - \frac{\mu_3}{M_3}\right)d\xi_1 + m\left(\frac{\mu_2}{M_2} - \frac{\mu_3}{M_3}\right)d\xi_2 = v_1 d\xi_1 + v_2 d\xi_2, \quad (2.95\text{h–j})$$

involves two mass fractions (2.95i, j) and two relative affinities (2.95k, l):

$$\{\bar{v}_1, \bar{v}_2\} = m\left\{\frac{\mu_1}{M_1}, \frac{\mu_2}{M_2}\right\} - m\frac{\mu_3}{M_3}. \quad (2.95\text{k, l})$$

All forms of work and energy associated with the forces (2.13), chemical reactions (2.89c) and heat (2.6b) are considered in the energy balance (subsection 2.1.23) using the augmented (modified) internal energy [subsections 2.1.22 (2.1.23)].

2.1.22   TOTAL WORK AND AUGMENTED INTERNAL ENERGY: THE TOTAL WORK (2.96)

$$dW = -dW_k + dW_g + dW_e + dW_h + dW_u + dW_\ell \qquad (2.96)$$

consists of the sum of: (i/ii) the kinetic (2.10b) and gravity potential (2.15b) energies which involve the mass density (2.7a), velocity (2.2a) and gravity potential (2.14b), with minus sign for the kinetic energy because the inertia force appears in (2.13) with opposite sign to the other forces, since it equals their sum; (iii/iv) the electric (magnetic) work (2.36a) [(2.66a)] which involve the electric (magnetic) field (2.18a–c) [(2.41a–f)] and the electric displacement (2.21b) [magnetic induction (2.38a, b)]; (v/vi) the mechanical work in general (2.84b) [in the isotropic case (2.70c)] involves the stress tensor (2.72a) [pressure (2.68a–c)] and strain tensor (2.82d) [volume change (2.70a, b) ≡ (2.85f)]; (vii) the chemical work of L substances (2.86c) ≡ (2.88f) with mole numbers (2.86a, b) involving the chemical potentials or relative chemical potentials (2.89d):

$$dW = -\vec{v} \cdot d\left(\rho\vec{v}\right) - \rho d\Phi_g - \vec{D} \cdot d\vec{E} - \vec{B} \cdot d\vec{H} + T_{ij}dS_{ij} + \sum_{\ell=1}^{L-1} \bar{\mu}_\ell dN_\ell \qquad (2.97)$$

Adding to the total work (2.97) the heat (2.6b) specifies the **augmented internal energy** (2.98a, b):

$$d\bar{U} \equiv dU + \vec{v} \cdot d\left(\rho\vec{v}\right) + \rho d\Phi_g \qquad (2.98a)$$

$$= TdS - \vec{D} \cdot d\vec{E} - \vec{B} \cdot d\vec{H} + T_{ij}dS_{ij} + \sum_{\ell=1}^{L-1} \bar{\mu}_\ell dN_\ell, \qquad (2.98b)$$

that: (i) includes the kinetic and gravity potential energies (2.99a);

$$\bar{U} = U + E_k + E_g = \bar{U}\left(S, \vec{E}, \vec{H}, S_{ij}, N_\ell\right); \qquad (2.99a, b)$$

(ii) depends on (2.99b) the entropy, electric and magnetic field vectors, strain tensor and mole numbers. The augmented internal energy is modified to lead to the energy balance (subsection 2.1.23).

2.1.23   ENERGY BALANCE AND MODIFIED INTERNAL ENERGY

A Legendre transform (subsection IV.5.5.1 and 2.2.1) is used to relate the augmented (2.99a, b) to the **modified internal energy** (2.100a, b)

$$\tilde{U} \equiv \bar{U} + \vec{D} \cdot \vec{E} + \vec{B} \cdot \vec{H}: \quad d\tilde{U} = d\bar{U} + \vec{D} \cdot d\vec{E} + \vec{E} \cdot d\vec{D} + \vec{B} \cdot d\vec{H} + \vec{H} \cdot d\vec{B}, \qquad (2.100a, b)$$

whose differential (2.100b; 2.98b) ≡ (2.101):

$$d\tilde{U} = TdS + \vec{E} \cdot d\vec{D} + \vec{H} \cdot d\vec{B} + T_{ij}dS_{ij} + \sum_{\ell=1}^{L-1} \bar{\mu}_\ell dN_\ell, \qquad (2.101)$$

shows that it depends (2.102a) on the electric displacement (magnetic induction) instead of the electric (magnetic) field:

$$\tilde{U} = \tilde{U}\left(S, \vec{D}, \vec{B}, S_{ij}, N_\ell\right): \quad d\tilde{U} = dQ + dE_e + dE_m + dW_u + dW_c. \quad (2.102a, b)$$

*The **energy balance** (2.102b) states that the variation of the modified internal energy (2.100a, b; 2.101) that includes the kinetic (2.10b) and gravity potential (2.15b) energies in the augmented internal energy (2.98a, b; 2.99a, b) equals (2.101) ≡ (2.102a, b) the sum of: (i) heat (2.6b) involving the entropy and temperature; (ii) electric energy (2.24b) involving the electric field (2.18a–c) and displacement (2.21b); (iii) the magnetic energy (2.50b) involving the magnetic induction (2.38a, b) and field (2.41a–f); (iv) the mechanical work (2.84b) involving the stress (2.72a) and strain (2.82d) tensors; (v) the chemical work (2.88f) involving the mole numbers (2.86a, b) and relative chemical potentials (2.88g).* Returning to the augmented internal energy (2.98a, b) leads to the extensive and intensive thermodynamic parameters (subsection 2.1.24).

2.1.24   EXTENSIVE AND INTENSIVE THERMODYNAMIC PARAMETERS

*The variation of the augmented internal energy (2.98b) ≡ (2.103a) equals the product of the **intensive parameters** (2.103d) by the variation of the **extensive parameters** (2.103b):*

$$d\bar{U} = \sum_{n=1}^{N} Y_n dX_n: \quad X_n \equiv \left\{S, \vec{E}, \vec{H}, S_{ij}, N_\ell\right\}, \qquad (2.103a, b)$$

$$N \equiv \# X_n = L + 12 = \# Y_n: \quad Y_n \equiv \left\{T, -\vec{D}, -\vec{B}, T_{ij}, \bar{\mu}_\ell\right\} \qquad (2.103c, d)$$

*where: (i) the extensive parameters are **additive** (2.104b) for two thermodynamic systems as the internal energy (2.104a) and are the entropy (2.104c), electric (2.104d) and magnetic (2.104e) fields, strain tensor (2.104f) and mole numbers (2.104g):*

$$\bar{U} = \bar{U}^{(1)} + \bar{U}^{(2)}, X_\alpha = X_\alpha^{(1)} + X_\alpha^{(2)}: \qquad (2.104a, b)$$

$$S = S^{(1)} + S^{(2)}, \quad \vec{E} = \vec{E}^{(1)} + \vec{E}^{(2)}, \qquad (2.104c, d)$$

$$\vec{H} = \vec{H}^{(1)} + \vec{H}^{(2)}, \quad S_{ij} = S_{ij}^{(1)} + S_{ij}^{(2)}, \quad N_\ell = N_\ell^{(1)} + N_\ell^{(2)}; \qquad (2.104e–g)$$

(ii) the intensive parameters are equal (2.105a) for two thermodynamic systems in **equilibrium**, *and thus the thermal/electric (magnetic)/mechanical/chemical equilibria are specified, respectively, by the equality of the temperature (2.105b)/electric displacement (2.105c) [magnetic induction (2.105d)] field/stress tensor (2.105e)/relative chemical potentials:*

$$Y_n^{(1)} = Y_n^{(2)}: \quad T^{(1)} = T^{(2)}, \quad \bar{D}^{(1)} = \bar{D}^{(2)}, \quad \bar{B}^{(1)} = \bar{B}^{(2)}, \quad T_{ij}^{(1)} = T_{ij}^{(2)}, \quad \bar{\mu}_{\ell}^{(1)} = \bar{\mu}_{\ell}^{(2)}; \quad (2.105a\text{–}f)$$

(iii) the extensive (2.103b) [intensive (2.103d)] parameters are the **independent** **(dependent) variables** *in the augmented internal energy (2.103a), since it is a function (2.99a, b) of the extensive parameters, whose partial derivatives (2.106a) specify the intensive parameters:*

$$Y_n = \frac{\partial \bar{U}}{\partial X_n}: \quad T = \frac{\partial \bar{U}}{\partial S}, \quad -\bar{D} = \frac{\partial \bar{U}}{\partial \bar{E}}, \quad -\bar{B} = \frac{\partial \bar{U}}{\partial \bar{H}}, \quad T_{ij} = \frac{\partial \bar{U}}{\partial S_{ij}}, \quad \bar{\mu}_{\ell} = \frac{\partial \bar{U}}{\partial N_{\ell}}, \quad (2.106a\text{–}f)$$

that is the temperature (2.106b)/minus the electric displacement (2.106c) and magnetic induction (2.106d)/the stress tensor (2.106e)/relative chemical potentials (2.106f) are the derivatives of the augmented internal energy with regard to, respectively, the entropy/electric and magnetic field/strain tensor/mole number; (iv) the number of extensive (2.103b) and intensive (2.103d) parameters is the same (2.103c) and consists of 1 thermal, 3 each electrical and magnetic vector, 6 symmetric tensor mechanical and $L - 1$ chemical substances for a total of $1 + 2 \times 3 + 6 + L - 1 = L + 12 \equiv N$ in (2.103c) that is the **dimension of the thermodynamic space**. *The equilibrium conditions (2.105a–f) of equal intensive parameters are proved next (Section 2.1.25).*

2.1.25 THERMAL, ELECTRICAL, MAGNETIC, MECHANICAL AND CHEMICAL EQUILIBRIUM

Consider (Figure 2.2) two sub-systems in equilibrium: (i) the internal energy is additive (2.103a; 2.104a) ≡ (2.107a, b) as all the extensive parameters (2.104b) ≡ (2.107c):

$$d\bar{U} = d\bar{U}^{(1)} + d\bar{U}^{(2)} = \sum_{n=1}^{N} Y_n^{(1)} dX_n^{(1)} + Y_n^{(2)} dX_n^{(2)}: \quad dX_n = dX_n^{(1)} + dX_n^{(2)}; (2.107a\text{–}c)$$

(ii) if the total system is isolated, that is, does not interact with the outside, the total extensive parameters (2.107c) are constant (2.108a) and must have opposite signs (2.108b) for the two sub-systems implying (2.108c) for the total internal energy (2.107b):

$$dX_n = 0: \quad dX_n^{(2)} = -dX_n^{(1)}, \quad d\bar{U} = \sum_{n=1}^{N} \left[Y_n^{(1)} - Y_n^{(2)} \right] dX_n^{(1)}; \quad (2.108a\text{–}c)$$

Isolated Thermodynamic System

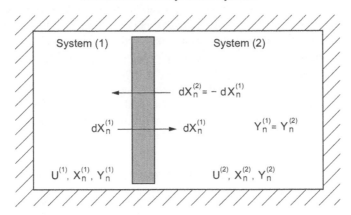

FIGURE 2.2   A thermodynamic system is specified by pairs of extensive and intensive variables, namely thermal, electric, magnetic, and elastic (Table 2.1) plus chemical. The extensive parameters X_n (entropy, electric and magnetic fields, strain tensor and mole numbers of chemical species) are additive for a system consisting of several subsystems. The equilibrium of the subsystems is specified by the equality of the intensive parameters Y_n (temperature, electric displacement, magnetic induction, stress tensor and chemical potentials).

(iii) if the two thermodynamic sub-systems are in **equilibrium**, the internal energy must be stationary (2.109a) and any change in extensive parameters (2.109b) must be accomplished with equal intensive parameters (2.109c):

$$d\bar{U} = 0; dX_n^{(1)} \neq 0: \quad Y_n^{(1)} = Y_n^{(2)}. \tag{2.109a–c}$$

It has been shown that *two interacting thermodynamic systems isolated from their exterior are in **thermal/electrical/magnetic/mechanical/chemical equilibrium** if, respectively, the temperature/electric displacement/magnetic induction/stress tensor/ relative chemical potentials are equal.*

2.1.26   TEMPERATURE, ENTROPY, HEAT AND INTERNAL ENERGY

Besides the mathematical property (1.66a–c) ≡ (2.5a–c) that leads to the first principle of thermodynamics (2.6a, b), it is necessary to justify the physical meaning that has been attributed to the various quantities introduced in this way, namely the internal energy, the heat, the entropy and the temperature. The internal energy as a function of state is additive (2.104a), that is, its value for several systems is the sum of the values for each one, and thus qualifies as an energy. The part of internal energy that is not work is the heat (2.6b). The heat is not a function of state, but the entropy is: like the internal energy it is additive (2.104b). In order to pass between the heat (entropy) which is not (is) a function of state, an integrating factor appears: the temperature in (2.6b). It has been shown that, unlike the internal energy and entropy, that are additive, the temperature is equal for systems in thermal equilibrium (2.105b). In addition, it can be proved from the second principle of thermodynamics (Section 2.3) that $T > 0$ for most substances, so the non-negative temperature T must be the absolute temperature. Before proceeding to prove this last statement, some consequences of the first principle of thermodynamics are explored next (Section 2.2). The consideration of the implications of the first principle of thermodynamics (Section 2.1) can be extended (Section 2.2) in two directions: (i) functions of state alternative to the internal energy using other independent variables (subsections 2.2.1–2.2.4); (ii) adding to the consideration constitutive equations for the electric (2.25a) and magnetic (2.51a) fields, the thermal and mechanical constitutive relations and the interactions among all of them (subsections 2.2.5–2.2.29).

2.2 FUNCTIONS OF STATE AND CONSTITUTIVE PROPERTIES

The most general case of the complete thermodynamic system was used to identify (subsections 2.1.24–2.1.25) the extensive (intensive) parameters which are additive (equal) in equilibrium. There are six pairs: (i) thermal: entropy (temperature); (ii) electric: field (displacement); (iii) magnetic: field (induction); (iv) mechanical: volume (pressure); (v) elastic: strains (stresses); (vi) chemical: mole number (chemical potential). A Legendre transform (subsection IV.5.5.1) can be used (subsection 2.2.1) to exchange each pair of extensive/intensive parameters, leading to new functions of state (subsection 2.2.2) such as the: (i/ii) free energy (enthalpy) that replaces the entropy (strains) in the internal energy by the temperature (stresses) that are easier to measure; (iii) both changes together lead to the free enthalpy (subsection 2.2.3). The first (second) derivatives of the function of state specify [subsection 2.2.4 (2.2.5)] the thermodynamic variables (constitutive parameters). The diagonal (non-diagonal) constitutive parameters specify [subsections 2.2.6 (2.2.7)] the thermal, electric, magnetic and elastic properties in isolation (their couplings, namely thermoelectric, thermomagnetic, thermomechanical, electromagnetic, electromechanical and magnetomechanical). The ensemble of these relations specifies the constitutive properties of matter in the linear case for general anisotropic materials like crystals (subsection 2.2.8). For isotropic matter the constitutive parameters must be independent of direction (subsection 2.2.9) implying that: (i) the electrical and magnetic constitutive relations are separate and independent (subsection 2.2.10); (ii) they are also separate and independent from the thermomechanical constitutive parameters (subsection 2.2.11). The latter remain coupled (subsections 2.2.12–2.2.14) for isotropic matter including the basic thermodynamic system (subsection 2.2.14) for which the stresses (strains) reduce to the pressure (volume change). The total work (subsection 2.1.21) and constitutive relations (subsections 2.2.15–2.2.16) highlight analogies among mechanics, electricity, magnetism and elasticity (2.2.17) including the energies. The conditions that the electric, magnetic and elastic energies of linear, homogeneous materials be positive imply in the isotropic case positive mass density, dielectric permittivity, magnetic permeability and Young modulus and specify the range of values of the Poisson ratio (subsection 2.2.18). In the case of anisotropic matter, the electric and magnetic (elastic) energies are specified by quadratic (biquadratic) forms (subsection 2.2.19), and the conditions that they be positive put constraints on the eigenvalues (components) of [subsections 2.2.20–2.2.24

(2.2.25–2.2.28)] the dielectric permittivity and magnetic permeability tensors (elastic stiffness double tensor). Not all quadratic forms in thermodynamics are positive – definite; for example, the second-order differential of the free energy is indefinite, that is, has no fixed sign (subsection 2.2.29).

2.2.1   DUALITY TRANSFORMATION OF FIRST-ORDER DIFFERENTIALS (LEGENDRE)

Any extensive/intensive parameter can be exchanged within the same pair in the first-order differential for the internal energy (2.103a–d) using (subsection IV.5.5.1) the **Legendre dual transform**:

$$m = 1,\ldots,M < N; \quad s = M+1,\ldots,N: \quad \tilde{U}\left(X_m,Y_s\right) = \bar{U}\left(X_n\right) - \sum_{s=M+1}^{N} X_s Y_s, \quad (2.110a\text{–}c)$$

since: (i) from (2.110c) follows (2.111a):

$$d\tilde{U} = d\bar{U} - \sum_{s=M+1}^{N} \left(X_s dY_s + Y_s dX_s\right), \quad d\tilde{U} = \sum_{m=1}^{M} X_m dY_m - \sum_{s=M+1}^{N} Y_s dX_s; \quad (2.111a, b)$$

(ii) substitution of (2.111a) in (2.103a) proves that N – M extensive parameters (2.103b) that appear as independent variables in the augmented internal energy (2.103a) are replaced by the corresponding intensive parameters (2.103d) as independent variables in the function of state (2.111b) obtained via the Legendre transform. Thus *the **Legendre transform** (2.110a–c) applied to the first-order differential of the augmented internal energy (2.103a–d) leads to the exact differential of another function of state (2.111b) that: (i) retains (2.110a) the first extensive (2.103b) [intensive (2.103d)] parameters as (2.103a) independent and dependent variables (2.112a, b); (ii) concerning (2.110b) the last N – M parameters the independent (dependent) variables are the intensive (minus the extensive) parameters (2.112c, d):*

$$m = 1,\ldots,M: \left\{X_m,Y_m\right\} \rightarrow \left\{X_m,Y_m\right\}; \qquad (2.112a, b)$$

$$s = M+1,\ldots,N: \left\{X_s,Y_s\right\} \rightarrow \left\{Y_s,-X_s\right\}. \qquad (2.112c, d)$$

The Legendre transform was used (subsection 2.1.23) to pass (2.100a) between the first-order differentials of the extended (2.98a, b; 2.99a, b) [modified (2.100b; 2.101)] internal energy exchanging (2.113a) minus the electric (magnetic) field by the electric displacement (magnetic induction), since the former are easier to measure:

$$\left\{-\vec{E},-\vec{H}\right\} \rightarrow \left\{\vec{D},\vec{B}\right\}; \quad \left\{S,S_{ij}\right\} \rightarrow \left\{-T,-T_{ij}\right\}. \qquad (2.113a, b)$$

The same practical motivation suggests (2.113b) replacing the entropy (strains) by the temperature (stresses) that are easier to measure, leading to three new functions of state (subsection 2.2.2). Additional functions of state could be introduced, but they are of little interest, since the mole numbers are easier to measure than the chemical potentials, and these are the remaining extensive (2.103b) and intensive (2.103d) parameters in the augmented internal energy.

2.2.2   FREE ENERGY (HELMHOLTZ) AND ENTHALPY (Gibbs)

The temperature is easier to measure than the entropy, suggesting **(Helmholtz 1882)** the **free energy** (2.114a) as the corresponding Legendre transform of the augmented internal energy (2.98a, b),

$$F\left(T,\vec{E},\vec{H},S_{ij},N_{\ell}\right) \equiv \bar{U}\left(S,\vec{E},\vec{H},S_{ij},N_{\ell}\right) - TS: \quad dF = d\bar{U} - TdS - SdT, \quad (2.114a, b)$$

substituting in the first-order differential (2.114b) of the free energy (2.114a) the first-order differential of the augmented internal energy (2.98b) leads to (2.115):

$$dF = -SdT - \vec{D}\cdot d\vec{E} - \vec{B}\cdot d\vec{H} + T_{ij}dS_{ij} + \sum_{\ell=1}^{L-1}\mu_{\ell}dN_{\ell}; \quad (2.115)$$

confirming in agreement with (2.114a) that: (i) the entropy has been replaced by the temperature as independent variable, and minus the entropy is the dependent variable in (2.115); (ii) the remaining independent and dependent variables are unchanged.

The Legendre transform (2.110a–c) of the augmented internal energy (2.98a, b) with regard to the strain and stress tensors leads **(Gibbs 1873)** to the **enthalpy** (2.116a) as a function of state with exact differential (2.116b),

$$H\left(S,\vec{E},\vec{H},T_{ij},N_{\ell}\right) = \bar{U}\left(S,\vec{E},\vec{H},S_{ij},N_{\ell}\right) - T_{ij}\,S_{ij}: \quad dH = d\bar{U} - T_{ij}dS_{ij} - S_{ij}dT_{ij}; \quad (2.116a, b)$$

substitution of (2.98b) in (2.116b) leads to the first-order differential of the enthalpy (2.117):

$$dH = TdS - \vec{D}\cdot d\vec{E} - \vec{B}\cdot d\vec{H} - S_{ij}dT_{ij} + \sum_{\ell=1}^{L-1}\bar{\mu}_{\ell}dN_{\ell}, \quad (2.117)$$

confirming that the only change is that the stress (minus the strain) tensor becomes the independent (dependent) variable. The free energy and enthalpy are combined in the free enthalpy (subsection 2.2.3) as a fourth thermodynamic function of state.

2.2.3   FREE ENTHALPY AS A FOURTH FUNCTION OF STATE

Combining the Legendre transforms of the augmented internal energy (2.98a, b) with regard to the entropy (strain) and temperature (stress) leads (2.114a) [(2.116a)] to the **free enthalpy** (2.118a–c):

$$G\left(T,\vec{E},\vec{H},T_{ij},N_\ell\right) = \bar{U}\left(S,\vec{E},\vec{H},S_{ij},N_\ell\right) - TS - T_{ij}S_{ij}$$
$$= F\left(T,\vec{E},\vec{H},S_{ij},N_\ell\right) - T_{ij}S_{ij} \qquad (2.118\text{a--c})$$
$$= H\left(S,\vec{E},\vec{H},T_{ij},N_\ell\right) - TS,$$

that has exact differentials, respectively (2.119a–c):

$$dG = d\bar{U} - TdS - SdT - T_{ij}dS_{ij} - S_{ij}dT_{ij}$$
$$= dF - T_{ij}dS_{ij} - S_{ij}dT_{ij} \qquad (2.119\text{a--c})$$
$$= dH - TdS - SdT.$$

Substituting (2.98b)/(2.115)/(2.117), respectively, in (2.119a/b/c) leads to the same exact differential for the free enthalpy (2.120):

$$dG = -SdT - \vec{D}\cdot d\vec{E} - \vec{B}\cdot d\vec{H} - S_{ij}dT_{ij} + \sum_{\ell=1}^{L-1}\bar{\mu}_\ell dN_\ell, \qquad (2.120)$$

confirming that the temperature and stresses (minus the entropy and strains) are independent (dependent) variables. It has been shown that *the Legendre transform (2.110a–c; 2.111a, b) of the augmented internal energy (2.98a, b) leads from its exact first-order differential (2.98b) ≡ (2.103a–d) to those (2.115) [(2.117)] of the free energy (2.114b) [enthalpy (2.116b)] that (2.114a) [(2.116a)] have the: temperature (stresses) as independent variables and minus the entropy (strains) as dependent variables. Both transformations together lead (2.120) to the free enthalpy (2.118a–c) whose exact first-order differential (2.119a–c) involves both changes of independent and dependent variables.* The first (second) order derivatives of the functions of state [subsection(s) 2.2.4 (2.2.5–2.2.7)] specify conjugate thermodynamic variables (constitutive properties of matter).

2.2.4 First-Order Derivatives and Conjugate Thermodynamic Variables

The extensive (2.103b) and intensive (2.103d) thermodynamic variables can be rearranged (2.121a) as six pairs conjugate thermodynamic variables, namely: (i) thermal: entropy and temperature (2.121b); (ii) electric: electric field and minus the displacement (2.121c); (iii) magnetic: magnetic field and minus the induction (2.121d); (iv) elastic: strain and stress tensors (2.121e) including as (v) particular case the pressure and minus the volume (2.121f); (vi) chemical: mole numbers and relative chemical potentials (2.121g):

$$\left\{\bar{X}_n,\bar{Y}_n\right\}: \quad \left\{S,T\right\},\left\{\vec{E},-\vec{D}\right\},\left\{\vec{H},-\vec{B}\right\},\left\{S_{ij},T_{ij}\right\},\left\{p,-V\right\},\left\{N_\ell,\bar{\mu}_\ell\right\}. \quad (2.121\text{a--g})$$

The partial derivatives of the thermodynamic functions of state, such as the augmented internal energy (2.98a, b), free energy (2.115), enthalpy (2.117) and free enthalpy (2.120) always involve conjugate thermodynamic variables, namely:

(i) thermal (2.121b): the temperature (2.122a, b) and entropy (2.122c, d); (ii) electric (2.121c): electric field and displacement (2.123a–e); (iii) magnetic (2.121d): magnetic field and induction (2.124a–e); (iv) elastic (2.121e): stress (2.125a, b) and strain (2.125c, d) tensors; (v) mechanical (2.121f): pressure (2.126a, b) and volume (2.126c, d); (vi) chemical (2.121g): mole numbers and relative chemical potentials (2.127a–e):

$$T = \left(\frac{\partial \bar{U}}{\partial S}\right)_{\bar{E},\bar{H},S_{ij},N_\ell} = \left(\frac{\partial H}{\partial S}\right)_{\bar{E},\bar{H},T_{ij},N_\ell} , \qquad \text{(2.122a, b)}$$

$$S = -\left(\frac{\partial F}{\partial T}\right)_{\bar{E},\bar{H},S_{ij},N_\ell} = -\left(\frac{\partial G}{\partial T}\right)_{\bar{E},\bar{H},T_{ij},N_\ell} , \qquad \text{(2.122c, d)}$$

$$\bar{H},N_\ell = const: \quad -\vec{D} = \left(\frac{\partial \bar{U}}{\partial \vec{E}}\right)_{S,S_{ij}} = \left(\frac{\partial F}{\partial \vec{E}}\right)_{T,S_{ij}} = \left(\frac{\partial H}{\partial \vec{E}}\right)_{S,T_{ij}} = \left(\frac{\partial G}{\partial \vec{E}}\right)_{T,T_{ij}} , \quad \text{(2.123a–e)}$$

$$\bar{E},N_\ell = const: \quad -\vec{B} = \left(\frac{\partial \bar{U}}{\partial \vec{H}}\right)_{S,S_{ij}} = \left(\frac{\partial F}{\partial \vec{H}}\right)_{T,S_{ij}} = \left(\frac{\partial H}{\partial \vec{H}}\right)_{S,T_{ij}} = \left(\frac{\partial G}{\partial \vec{H}}\right)_{T,T_{ij}} , \quad \text{(2.124a–e)}$$

$$T_{ij} = \left(\frac{\partial U}{\partial S_{ij}}\right)_{S,\bar{E},\bar{H}} = \left(\frac{\partial F}{\partial S_{ij}}\right)_{T,\bar{E},\bar{H}} , \qquad \text{(2.125a, b)}$$

$$S_{ij} = -\left(\frac{\partial H}{\partial T_{ij}}\right)_{S,\bar{E},\bar{H}} = -\left(\frac{\partial G}{\partial T_{ij}}\right)_{T,\bar{E},\bar{H}} , \qquad \text{(2.125c, d)}$$

$$-p = \left(\frac{\partial \bar{U}}{\partial V}\right)_{S,\bar{E},\bar{H},N_\ell} = \left(\frac{\partial F}{\partial V}\right)_{T,\bar{E},\bar{H},N_\ell} , \qquad \text{(2.126a, b)}$$

$$V = \left(\frac{\partial H}{\partial p}\right)_{S,\bar{E},\bar{H},N_\ell} = \left(\frac{\partial G}{\partial p}\right)_{T,\bar{E},\bar{H},N_\ell} , \qquad \text{(2.126c, d)}$$

$$\bar{E},\bar{H} = const: \quad \bar{\mu}_\ell = \left(\frac{\partial \bar{U}}{\partial N_\ell}\right)_{S,S_{ij}} = \left(\frac{\partial F}{\partial N_\ell}\right)_{T,S_{ij}} = \left(\frac{\partial H}{\partial N_\ell}\right)_{S,T_{ij}} = \left(\frac{\partial G}{\partial N_\ell}\right)_{T,T_{ij}} ; \quad \text{(2.127a–e)}$$

in all the partial derivatives of a function of state with regard to a thermodynamic variable (2.122a–d; 2.123a–e; 2.124a–e; 2.125a–d; 2.126a–d; 2.127a–e) all other thermodynamic variables are kept constant. Any thermodynamic function of state specifies through its second-order derivatives the constitutive properties of matter, and most of the available experimental data refers to the thermodynamic variables in the free energy, that is used in the sequel (subsections 2.2.6–2.2.8).

2.2.5 SECOND-ORDER DERIVATIVES AND CONSTITUTIVE PROPERTIES

*The free energy (2.115) ≡ (2.128a) has for **independent variables** (2.128b) the temperature, electric and magnetic fields, strain tensor and mole numbers:*

$$dF = \sum_{n=1}^{N} \bar{Y}_n d\bar{X}_n; \quad \bar{X}_\alpha = \left\{ T, \vec{E}, \vec{H}, S_{ij}, N_\ell \right\}; \quad (2.128\text{a, b})$$

*and the partial derivatives (2.129a) specify the **dependent variables** (2.129b), respectively (2.129c) minus the entropy, minus the electric displacement, minus the magnetic induction, the stress tensor and the relative chemical potentials:*

$$\bar{Y}_n = \frac{\partial F}{\partial \bar{X}_n} = \left\{ \frac{\partial F}{\partial T}, \frac{\partial F}{\partial \vec{E}}, \frac{\partial F}{\partial \vec{H}}, \frac{\partial F}{\partial S_{ij}}, \frac{\partial F}{\partial N_\ell} \right\} = \left\{ -S, -\vec{D}, -\vec{B}, T_{ij}, \bar{\mu}_\ell \right\} = \bar{Y}_n \left(\bar{X}_n \right). (2.129\text{a–d})$$

If the free energy has continuous second-order derivatives (2.130a) they are symmetric (2.130b):

$$F \in C^2: \quad Z_{nm} = \frac{\partial Y_n}{\partial \bar{X}_m} = \frac{\partial^2 F}{\partial \bar{X}_m \partial \bar{X}_n} = \frac{\partial^2 F}{\partial \bar{X}_n \partial \bar{X}_m} = \frac{\partial \bar{Y}_m}{\partial \bar{X}_n} = Z_{mn}, \quad (2.130\text{a, b})$$

*and specify the **constitutive coefficients** (2.130b) that appear in (2.131b) the second-order differential for the free energy (2.131a):*

$$d^2 F = \sum_{n,m=1}^{N} \frac{\partial^2 F}{\partial \bar{X}_n \partial \bar{X}_m} d\bar{X}_n d\bar{X}_m = \sum_{n,m=1}^{N} Z_{nm} d\bar{X}_n d\bar{X}_m. \quad (2.131\text{a, b})$$

The existence of continuous second-order derivatives for the free energy (2.130a) implies (2.129c) the dependent variables (2.130b, c) are continuously differentiable functions of the independent variables (2.132a) and in their differentials (2.132b):

$$\bar{Y}_n \left(\bar{X}_m \right) \in C^1 \left(IR^N \right): \quad d\bar{Y}_n = \sum_{m=1}^{N} \frac{\partial F}{\partial \bar{X}_m} d\bar{X}_m = \sum_{m=1}^{N} Z_{nm} dX_m \quad (2.132\text{a–c})$$

appear (2.132c) the constitutive coefficients (2.130b).

Substituting the independent (2.128b) and dependent (2.129c) variables in (2.132c) leads to *the **constitutive relations for linear anisotropic matter** specifying the entropy (2.133a), electric displacement (2.133b), magnetic induction (2.133c) and stress tensor (2.133d):*

$$dS = \frac{C_V}{T} dT + f_i dE_i + h_i dH_i + \alpha_{ij} dS_{ij}, \quad (2.133\text{a})$$

$$dD_i = f_i dT + \varepsilon_{ij} dE_j + \vartheta_{ij} dH_j + p_{ijk} dS_{jk}, \qquad (2.133b)$$

$$dB_i = h_i dT + \vartheta_{ji} dE_j + \mu_{ij} dH_j + q_{ijk} dS_{jk}, \qquad (2.133c)$$

$$dT_{ij} = -\alpha_{ij} dT - p_{kij} dE_k - q_{kij} dH_k + C_{ijkl} dS_{kl}, \qquad (2.133d)$$

as functions of the temperature, electric and magnetic fields and strain tensors for a **single chemical species,** *so that the mole numbers and relative chemical potentials do not appear.* For multiple chemical species the constitutive coefficients in (2.133a–d) would appear for each distinct species or the aggregate (subsection 2.3.30–2.3.31). The 10 constitutive coefficients in (2.133a–d) can be divided into two groups [Table 2.1 (2.2)] namely the **single (coupling) constitutive coefficients** for the 4 effects: thermal, electric, magnetic and mechanical [$\binom{4}{2} = 4 \times \dfrac{3}{2} = 6$ com-binations: thermoelectric, thermomagnetic, thermomechanical, electromagnetic, piezo-electric and piezomagnetic] considered next [subsection 2.2.6 (2.2.7)] and indicated in Table 2.1 (2.2).

2.2.6   Specific Heat, Dielectric Permittivity, Magnetic Permeability and Elastic Stiffness

The constitutive parameters can be defined for any function of state, that is any Legendre transform of the augmented internal energy (2.98b); the augmented internal energy (2.98a) shows that the kinetic energy and gravity potential, being exact differentials, do not give rise to constitutive properties. The choice of the free energy (2.114a) was dictated by the most convenient independent variables (2.128b) in (2.115) ≡ (2.128a). The corresponding dependent variables (2.129a–c) are similar for each chemical species corresponding, respectively, to the thermal (2.121b), electric (2.121c), magnetic (2.121d) and elastic (2.121e) pairs of conjugate thermo-dynamic variables (Table 2.1). Starting with the **thermal properties** of matter they are specified (2.134a, b):

$$dQ = TdS = C_V dT: \quad C_V = T \left(\frac{\partial S}{\partial T} \right)_{\bar{E}, \bar{H}, S_{ij}} = -T \left(\frac{\partial^2 F}{\partial T^2} \right)_{\bar{E}, \bar{H}, S_{ij}} \qquad (2.134a\text{–}d)$$

by the **specific heat** (2.134c, d) that is the heat (2.6b) needed to increase the tempera-ture by one degree; the specific heat (2.134c, d) is calculated with all other indepen-dent variables $\left(\bar{E}, \bar{B}, S_{ij} \right)$ constant, and corresponds to the specific heat at constant volume for a basic thermodynamic system.

The **electrical properties** of matter are specified by the **dielectric permittivity tensor** (2.135a) ≡ (2.25a):

$$dD_i = \varepsilon_{ij} dE_j: \quad \varepsilon_{ij} \equiv \frac{\partial D_i}{\partial E_j} = -\frac{\partial^2 F}{\partial E_j \partial E_i} = -\frac{\partial^2 F}{\partial E_i \partial E_j} = \frac{\partial D_j}{\partial E_i} = \varepsilon_{ji} \qquad (2.135a\text{–}c)$$

TABLE 2.1
Four Constitutive Properties

property		Thermal	Electric	Magnetic	Elastic
Variable	Extensive	Entropy:S	Field: E_i	Field H_i	Strain tensor S_{ij}
	Intensive	Temperature: T	Displacement: D_i	Induction: B_i	Stress tensor: T_{ij}
Constitutive tensor	Anisotropic	-	$\varepsilon_{ij} \equiv -\partial^2 F/\partial E_i \partial E_j$ dielectric permittivity	$\mu_{ij} \equiv -\partial^2 F/\partial E_i \partial E_j$ magnetic permeability	$C_{ijka} = \partial^2 F/\partial S_{ij} \partial S_{km}$ elastic stiffness tensor
	Isotropic	$C_v = T\left(\dfrac{\partial S}{\partial T}\right)_v$	$\varepsilon_{ij} = \varepsilon\,\delta_{ij}$	$\mu_{ij} = \mu\,\delta_{ij}$	Lame moduli $c_{ijka} = \lambda\,\delta_{ik}\delta_{in} + \mu\,\delta_{ij}\delta_{km}$
Constitutive relation	Anisotropic	$dS = C_v\,dT$	$D_i = \varepsilon_{ij}E_j$	$B_i = \mu_{ij}H_j$	$T_{ij} = C_{ijkm}S_{km}$
	Isotropic	-	$\bar{D} = \varepsilon\bar{E}$	$\bar{B} = \mu\bar{H}$	$T_{ij} = \lambda S_{ij} + \mu S_{kk}\delta_{ij}$
Energy density	Anisotropic	-	$2E_e = \varepsilon_{ij}E_iE_j$	$2E_n = H_iH_j$	$2E_n = C_{ijkm}S_{ij}S_{kn}$
	Isotropic	Heat $dQ = C_v dT$	$2E_e = \varepsilon E^2 = D^2/\varepsilon$	$2E_n = \mu H^2 = B^2/\mu$	$2E_n = \lambda(S_{kn})^2 + \mu(S_{kk})^2$

Note: The four constitutive properties of matter are: thermal, electric, magnetic and elastic; for each there is an extensive parameter that is additive for several systems and an intensive parameter that is equal at equilibrium; the extensive and intensive parameters are connected by a constitutive relation and have an associated energy that may be simplified passing from anisotropic to isotropic matter.

that is symmetric (2.135b, c) $\equiv$ (2.37c) and relates linearly the electric displacement and field. The **magnetic properties** of matter are specified by the **magnetic permeability tensor** (2.136a) $\equiv$ (2.51a):

$$dB_i = \mu_{ij}dH_j: \quad \mu_{ij} \equiv \frac{\partial B_i}{\partial H_j} = -\frac{\partial^2 F}{\partial H_j \partial H_i} = -\frac{\partial^2 F}{\partial H_i \partial H_j} = \frac{\partial B_j}{\partial H_i} = \mu_{ji}, \quad (2.136\text{a–c})$$

and is symmetric (2.136b, c) $\equiv$ (2.67d) and relates linearly the magnetic induction and field.

The fourth constitutive diagonal concerns the **elastic properties** of matter specified by the **stiffness double tensor** (2.137a):

$$dT_{ij} = C_{ijkl}dS_{kl}: \quad C_{ijkl} \equiv \frac{\partial T_{ij}}{\partial S_{kl}} = \frac{\partial^2 F}{\partial S_{kl} \partial S_{ij}} = \frac{\partial^2 F}{\partial S_{ij} \partial S_{kl}} = C_{klij} = C_{jikl} = C_{ijlk}, \quad (2.137\text{a–g})$$

relating the stress and strain tensors that is symmetric: (i) in the two pairs of indices (2.137b–e); (ii/iii) in the first (2.137f) [second (2.137g)] pair of indices because the stress (2.81d) [strain (2.82d)] tensors are symmetric. *The **deformation energy** (2.138b) $\equiv$ (2.84b) equals the work of deformation, and for a linear material (2.137a) corresponds (2.138c) to the **elastic energy** (2.138d, e) for a constant stiffness double tensor (2.138a):*

$$C_{ijkl} = const: \quad dE_u = dW_u = C_{ijkl}S_{kl}dS_{ij}, \quad E_u = \frac{1}{2}C_{ijkl}S_{ij}S_{kl} = \frac{1}{2}T_{ij}S_{ij}. \quad (2.138\text{a–e})$$

The four **isolated effects** (Table 2.1), namely thermal (2.134a–d), electric (2.135a–c), magnetic (2.136a–c) and mechanical elastic (2.137a–g), when occurring simultaneously (Table 2.2) lead to $\binom{4}{2} = 4 \times \frac{3}{2} = 6$ couplings (subsection 2.2.7).

2.2.7   PYROELECTRIC/MAGNETIC VECTORS, ELECTROMAGNETIC COUPLING, THERMOELASTIC, AND PIEZOELECTRIC/MAGNETIC TENSORS

There are 4 sets of properties which at order 2 can be coupled in $\binom{5}{2} = \frac{4!}{2!2!} = 6$ symmetric ways (Table 2.2). Starting with the **thermoelectric (thermomagnetic) properties** they are specified by the **pyroelectric (pyromagnetic) vector** (2.139a–d) [(2.140a–d)]:

$$f_i \equiv \frac{\partial D_i}{\partial T} = -\frac{\partial^2 F}{\partial T \partial E_i} = -\frac{\partial^2 F}{\partial E_i \partial T} = \frac{\partial S}{\partial E_i}: \quad dS = f_i dE_i, \quad dD_i = f_i dT, \quad (2.139\text{a–f})$$

$$h_i \equiv \frac{\partial B_i}{\partial T} = -\frac{\partial^2 F}{\partial T \partial H_i} = -\frac{\partial^2 F}{\partial H_i \partial T} = \frac{\partial S}{\partial H_i}: \quad dS = h_i dH_i, \quad dB_i = h_i dT, \quad (2.140\text{a–f})$$

TABLE 2.2
Six Constitutive Couplings

Coupling	Constitutive Tensor: Anisotropic (Isotropic)	Thermodynamics Variables: Independent (Dependent)	Constitutive Relation: Anisotropic (Isotropic)
Thermo-electric			
Pyroelectric	$\vec{f}=\partial\vec{D}/\partial T$ $=\partial S/\partial\vec{E}$ $(\vec{f}=0)$	Temperature T (Electric displacement $\vec{D}$)	$d\vec{D}=f\,dT$ $(d\vec{D}=0)$
		Electric field $\vec{E}$ (Entropy S)	$dS=f.\vec{E}$ $(dS=0)$
Thermom-magnetic		Temperature T (Magnetic induction $\vec{B}$)	$dS=\vec{h}.d\vec{H}$ $(dS=0)$
Pyromagnetic	$\vec{h}=\partial\vec{B}/\partial T$ $=\partial S/\partial\vec{H}$ $(\vec{h}=0)$	Magnetic field $\vec{H}$ (Entropy S)	$dS=\vec{h}.d\vec{H}$ $(dS=0)$
Thermo-elastic		Strain tensor S_{jk} (Entropy S)	$dS=\alpha_{ij}dS_{ij}$ $(dS=0)$
Thermal stresses	$\alpha_{ij}=\partial S/\partial S_{ij}$ $=-\partial T_{ij}/\partial T$ $(\alpha_{ij}=\alpha\delta_{ij})$	Stress tensor T_{ij} (Temperature T)	$dT_{ij}=-\alpha_{ij}dT$ $(dp=\alpha dT)$
Electro-elastic		Strain tensor S_{jk} (Electric displacement D_i)	$dD_i=p_{ijk}dS_{jk}$ $(d\vec{D}=0)$
Piezomagnetic	$q_{ijk}=\partial B_i/\partial S_{jk}$ $=-\partial T_{jk}/\partial H_i$ $(q_{ijk}=0)$	Electric field E_i (Stress tensor T_{jk})	$dT_{jk}=-p_{ijk}dE_i$ $(dT_{jk}=0)$

(Continued)

TABLE 2.2 (Continued)

Coupling	Constitutive Tensor: Anisotropic (Isotropic)	Thermodynamics Variables: Independent (Dependent)	Constitutive Relation: Anisotropic (Isotropic)
Magneto-elastic Piezomagnetic	$q_{ijk} = \partial B_i/\partial S_{jk}$ $= -\partial T_{jk}/\partial H_i$ $(q_{ijk} = 0)$	Strain tensor S_{jk} (Magnetic induction B_i) Magnetic field H_i (Stress tensor T_{jk})	$dB_i = -q_{ijk}\, dS_{jk}$ $(d\vec{B} = 0)$ $dT_{jk} = -p_{ijk}\, dH_i$ $(dT_{jk} = 0)$
Electro-magnetic Coupling	$\vartheta_{ij} = \partial D_i/\partial H_j$ $= \partial B_j/\partial E_i$ $(\vartheta_{ij} = 0)$	Magnetic field H_j (Magnetic displacement D_j) Electric field E_i (Magnetic induction B_j)	$dD_i = \vartheta_{ij}\, dH_j$ $(d\vec{D} = 0)$ $dB_i = \vartheta_{ji}\, dE_j$ $(d\vec{B} = 0)$

Note: The four constitutive properties (thermal, electric, magnetic and elastic) give rise to: (i) four constitutive properties in Table 2.1; (ii) six constitutive couplings in Table 2.2; in all cases (i) and (ii) is involved one extensive and one intensive parameter with: (i) one choice; (ii) two choices; in all cases (i) and (ii) the constitutive relation may simplify from anisotropic to isotropic matter, and in some cases the coupling ceases to exist.

that relates: (i) the electric displacement (2.139f) [magnetic induction (2.140f)] to the temperature; (ii) the entropy or heat exchanged to the electric (magnetic) field (2.139e) [(2.140e)].

The third possible thermal coupling specifies the **thermoelastic effect** through the **thermal stress tensor** (2.141a–d),

$$\alpha_{ij} \equiv \frac{\partial S}{\partial S_{ij}} = -\frac{\partial^2 F}{\partial S_{ij} \partial T} = -\frac{\partial^2 F}{\partial T \partial S_{ij}} = -\frac{\partial T_{ij}}{\partial T} = \alpha_{ji} :$$

$$dS = \alpha_{ij} dS_{ij}, \qquad dT_{ij} = -\alpha_{ij} dT, \qquad (2.141a\text{–}g)$$

that is symmetric (2.141e) and relates: (i) the stresses to the temperature (2.141g); (ii) the entropy or heat released to the strain (2.141f).

The **electro(magneto)elastic effect** is specified by the **piezoelectric (piezomagnetic) tensor** (2.142a–d) [(2.143a–d)]:

$$p_{ijk} \equiv \frac{\partial D_i}{\partial S_{jk}} = -\frac{\partial^2 F}{\partial S_{jk} \partial E_i} = -\frac{\partial^2 F}{\partial E_i \partial S_{jk}} = -\frac{\partial T_{jk}}{\partial E_i} = p_{ikj} :$$

$$dD_i = p_{ijk} dS_{jk}, \quad dT_{ij} = -p_{kij} dE_k, \qquad (2.142a\text{–}g)$$

$$q_{ijk} \equiv \frac{\partial B_i}{\partial S_{jk}} = -\frac{\partial^2 F}{\partial S_{jk} \partial H_i} = -\frac{\partial^2 F}{\partial H_i \partial S_{jk}} = -\frac{\partial T_{jk}}{\partial H_i} = q_{ikj} :$$

$$dB_i = q_{ijk} dS_{jk}, \quad dT_{ij} = -q_{kij} dH_k, \qquad (2.143a\text{–}g)$$

that is symmetric in two indices (2.142e) [(2.143e)] and relates: (i) the electric displacement (2.142f) [magnetic induction (2.143f)] to the strains; (ii) the stresses to the electric (magnetic) field (2.142g) [(2.143g)].

The sixth coupling effect concerns **electromagnetic coupling** and is specified by the **electromagnetic coupling tensor** (2.144a–d):

$$\vartheta_{ij} \equiv \frac{\partial D_i}{\partial H_j} = -\frac{\partial^2 F}{\partial H_j \partial E_i} = -\frac{\partial^2 F}{\partial E_i \partial H_j} = \frac{\partial B_j}{\partial E_i} = \vartheta_{ji} :$$

$$dD_i = \vartheta_{ij} dH_j, \quad dB_j = \vartheta_{ji} dE_i, \qquad (2.144a\text{–}g)$$

that is symmetric (2.144e) and relates: (i) the electric displacement to the magnetic field (2.144f); (ii) the magnetic induction to the electric field (2.144g). The constitutive properties (2.133a–d) apply to the complete thermodynamic system (2.128a, b; 2.129a–d: 2.130a, b; 2.131a, b; 2.132a–c) for anisotropic matter and the maximum number of independent constitutive parameters is calculated next (subsection 2.2.8).

2.2.8 MAXIMUM NUMBER OF CONSTITUTIVE COEFFICIENTS FOR ANISOTROPIC MATTER

*The aggregate of 4 single (6 coupled) diagonal (off-diagonal) constitutive properties [subsections 2.2.6 (2.2.7)] leads to **four anisotropic constitutive relations** (2.133a–d), involving 10 **constitutive coefficients** that specify the maximum number of independent components for **anisotropic matter**, specifically: (i) one (2.145a) for the specific heat (2.134c, d) specifying the thermal constitutive relation (2.134a, b); (ii/iii) three (2.145b) [(2.145c)] for the pyroelectric (2.139a–d) [pyromagnetic (2.140a)] vector specifying thermoelectric (2.139e, f) [thermomagnetic (2.140e, f)] coupling;(iv/v) six (2.145d) [(2.145e)] for the dielectric permittivity (2.135b) [magnetic permeability (2.136b)], specifying the electric (2.135a) [magnetic (2.136a)] constitutive relation because it is a symmetric tensor (2.135c) [(2.136c)]; (vi) also six (2.145f) for the electromagnetic coupling tensor (2.144a–d) specifying electromagnetic coupling (2.144f, g) that is symmetric (2.144e); (vii) also six (2.145g) for the thermal stress tensor (2.141a–d) that specifies the thermomechanical coupling (2.141f, g) and is symmetric (2.141c); (viii/ix) eighteen (2.145h) [(2.145i)] for the piezoelectric (2.142a–d) [piezomagnetic (2.143a–d)] tensor specifying the electromechanical (2.142f, g) [magnetomechanical (2.143f, g)]coupling, that is symmetric in two indices (2.142e) [(2.143e)] and thus has $6 \times 3 = 18$ independent components; (x) twenty-one (2.145j) for the stiffness tensor (2.137b–d) specifying the elastic constitutive relation (2.137a), that is a double tensor, thus equivalent to a symmetric 6×6 matrix, with $6 \times \dfrac{7}{2} = 21$ independent components:*

$$\# C_V = 1 \equiv \bar{n}_1, \# f_i = \# h_i = 3 \equiv \bar{n}_2, \# \varepsilon_{ij} = \# \mu_{ij} = \# \vartheta_{ij} = \# \alpha_{ij} = 6 \equiv \bar{n}_3, \quad (2.145a\text{–}g)$$

$$\# p_{ijk} = \# q_{ijk} = 18 \equiv \bar{n}_4, \# C_{ijkl} = 21 \equiv \bar{n}_5:$$
$$\bar{n} = \bar{n}_1 + 2\bar{n}_2 + 4\bar{n}_3 + 2\bar{n}_4 + \bar{n}_5 = 88. \quad (2.145h\text{–}k)$$

*Thus **the total number of constitutive parameters for linear anisotropic matter** cannot exceed 88 in the absence of symmetries. In the case of crystals or other materials with **geometrical symmetries** the number of distinct constitutive parameters is less than 88 and cannot be lower than 7 for isotropic matter (subsections 2.2.9–2.2.10).*

2.2.9 CONSTITUTIVE PROPERTIES OF ANISOTROPIC AND ISOTROPIC MATTER

The constitutive tensors (2.145a–j) for isotropic matter must be independent of direction and thus can depend only on scalars, the identity matrix (2.34a) and the permutation symbol (2.54a–d). It follows that: (i) the specific heat (2.134a–d) is a scalar and exists for isotropic matter (2.146a); (ii/iii) the pyroelectric (2.139a–f) [pyromagnetic (2.140a–f)] vectors cannot be independent of direction unless they are zero (2.147a) [(2.147b)] implying that an electric (magnetic) field cannot cause heating, and a temperature change cannot generate an electric (magnetic) field; (iv/v/vi) the dielectric permittivity (2.135a–c), magnetic permeability (2.136a–c) and thermal stress (2.141a–g) tensors are symmetric and relate polar vectors, and in isotropic matter are proportional to the identity matrix (2.34a), respectively, through the scalar dielectric permittivity (2.148a) ≡ (2.27a), magnetic permeability (2.148b) ≡ (2.53a) and thermal pressure (2.148c):

$$\text{isotropic matter:} \# C_V = \# \varepsilon = \# \mu = \# \alpha = 1 = \underline{n_1}, \quad \# C_{ijkl} = 2 \equiv \underline{n_2}; (2.146\text{a–f})$$

$$\text{isotropic matter:} f_i = 0 = h_i, \quad \vartheta_{ij} = 0, \quad p_{ijk} = 0 = q_{ijk}, \quad (2.147\text{a–e})$$

$$\text{isotropic matter:} \varepsilon_{ij} = \varepsilon\delta_{ij}, \quad \mu_{ij} = \mu\delta_{ij}, \quad \alpha_{ij} = \alpha\delta_{ij}. \quad (2.148\text{a–c})$$

and thus have one independent component (2.146b–d). The remaining results (2.146e, f; 2.147c–e) are proved next (subsections 2.2.10–2.2.12).

2.2.10 ABSENCE OF ELECTROMAGNETIC COUPLING IN ISOTROPIC MATTER

Although the electromagnetic coupling tensor (2.144a–g) is symmetric it relates linearly (2.144g) the electric field (magnetic induction) that is an irrotational (2.18b) [solenoidal (2.38a)] vector, which is not possible in isotropic matter, and thus must vanish. This can be proved considering an **inversion relative to the origin** (2.149a) that changes the sign of all coordinates and hence: (i) changes the sign (2.149b) of the electric field (2.18c); (ii) does not change the sign of the magnetic induction (2.149c) bearing in mind the unsteady Maxwell equation (2.149d):

$$\vec{x} \to -\vec{x}: \quad \{\vec{E}, \vec{B}\} \to \{-\vec{E}, \vec{B}\}, \quad \nabla \wedge \vec{E} = -\frac{1}{c}\frac{\partial\vec{B}}{\partial t}. \quad (2.149\text{a–d})$$

Applying the transformation (2.149b, c) to the electromagnetic coupling constitutive relation (2.144g) ≡ (2.150a) leads to (2.150b):

$$\vartheta_{ji}dE_i = dB_i = -\vartheta_{ji}dE_i \quad \Rightarrow \vartheta_{ji} = 0, \quad (2.150\text{a–c})$$

implying that there is no electromagnetic coupling (2.150c) in isotropic matter. Thus *a steady (2.18a) electric field (2.18b, c) cannot generate a magnetic induction (2.38a) and vice-versa in isotropic matter.*

2.2.11 ABSENCE OF ELECTRO/MAGNETOELASTIC INTERACTION IN ISOTROPIC MATTER

The electro(magneto)mechanical coupling is specified by the piezoelectric (2.142a–g) [piezomagnetic (2.143a–g)] tensor that (a) is symmetric in two indices. In isotropic matter it would have to be proportional (2.151a) [(2.151b)] to the permutation symbol (2.54a–d) that (b) is skew-symmetric in all three indices. The properties (a) and (b) are incompatible unless the piezoelectric (piezomagnetic) tensor is zero (2.151c) [(2.151d)]:

$$p_{ijk} = \bar{p}e_{ijk}, \quad q_{ijk} = \bar{q}e_{ijk}: \quad \bar{p} = 0 = \bar{q}. \quad (2.151\text{a–d})$$

The proof of (2.151c, d) is similar and is made for the piezoelectric tensor: from (2.142e) follows (2.152a) implying (2.152b, c) and hence (2.152d) and (2.152e, f);

$$0 = p_{ijk} - p_{ikj} = \bar{p}\left(e_{ijk} - e_{ikj}\right) = 2\bar{p}e_{ijk} \implies \bar{p} = 0 \implies p_{ijk} = 0 = q_{ijk}. \quad (2.152a\text{–}f)$$

Thus the strains cannot generate an electric (magnetic) field, and the electric (magnetic) field cannot cause stresses in isotropic matter for which the piezoelectric (piezomagnetic) effect does not exist. Combining (2.147a, b) with (2.147c) ≡ (2.150c) and (2.147d, e) ≡ (2.152e, f) it follows that in isotropic matter the steady electric (magnetic) fields are decoupled from each other (2.150c) and from thermal (2.147a) [(2.147b)] and elastic (2.152e) [(2.152f)] effects, and thus the general constitutive relation is (2.133a; 2.148a) ≡ (2.153a) ≡ (2.27b) [(2.133c; 2.148b) ≡ (2.153b) ≡ (2.53b)] implying that the dielectric displacement (magnetic induction) is parallel to the electric (magnetic) field (2.153a) [(2.153b)]:

$$\text{isotropic matter:} \quad d\vec{D} = \varepsilon d\vec{E}, \quad d\vec{B} = \mu d\vec{H}. \quad (2.153a, b)$$

Among the results (2.146a–f; 2.147a–e; 2.148a–c) it remains to prove that the elastic stiffness tensor (2.137a–g) has two independent components (2.146e) in isotropic matter (subsection 2.2.12) leading to a total of 7 constitutive coefficients (2.146f).

2.2.12   LAMÉ MODULI, YOUNG MODULUS AND POISSON RATIO

In isotropic matter the elastic stiffness tensor with symmetries (2.137e–g) cannot depend on the permutation symbol (2.54a–d) and can depend only on two combinations (2.154) of the identity matrix (2.34a) with the **Lamé moduli of elasticity** as coefficients:

$$C_{ijkl} = \lambda\delta_{ik}\delta_{jl} + \nu\delta_{ij}\delta_{kl}; \quad (2.154)$$

the constitutive relation between the stress and strain tensors (2.137a) for isotropic matter (2.154) involves (2.155a) only two elastic moduli in agreement with (2.146e) and simplifies to (2.155b):

$$dT_{ij} = \lambda\delta_{ik}\delta_{jl}dS_{kl} + \nu\delta_{ij}\delta_{kl}dS_{kl} = \lambda dS_{ij} + \nu\delta_{ij}dS_{kk} = \lambda dS_{ij} + \nu\delta_{ij}dV, \quad (2.155a\text{–}c)$$

showing (2.155c) that: (i) the shear $i \neq j$ (normal $i = j$) strains are proportional to the shear (normal) stresses through the first Lamé modulus λ; (ii) the volume changes (2.85f) cause only normal stresses through the second Lamé modulus ν.

*The **Hooke law of elasticity** (2.155c) ≡ (II.4.76a–c) ≡ (2.156c) can be restated using instead of the Lamé moduli the **Young modulus** (2.156a) and **Poisson ratio** (2.156b):*

$$\lambda = \frac{E}{1+\sigma}, \quad \nu = \frac{\lambda\sigma}{1-2\sigma}: \quad dT_{ij} = \frac{E}{1+\sigma}\left[dS_{ij} + \frac{\sigma}{1-2\sigma}\delta_{ij}dS_{kk}\right]. \quad (2.156a\text{–}c)$$

In the case of constant elastic moduli (2.157a, b) the elastic energy (2.138e) is given by (2.157c) [(2.157d)] in terms of the Lamé elastic moduli (2.155b) [Young modulus (2.156a) and Poisson ratio (2.156b)]:

$$\lambda, v = const = E, \sigma: \quad 2E_e = \lambda S_{ij} S_{ij} + v S_{ii} S_{jj} = \frac{E}{1+\sigma}\left[S_{ij} S_{ij} + \frac{\sigma}{1-2\sigma}(S_{ii})^2 \right]. \quad (2.157a\text{--}d)$$

The non-zero constitutive tensors (2.146a; 2.148a–c; 2.154) specify the constitutive properties of isotropic matter (subsection 2.2.13).

2.2.13 CONSTITUTIVE RELATIONS FOR ISOTROPIC MATTER

Substituting (2.147a, b; 2.148c) valid for isotropic matter in (2.133a) leads to (2.158a) for the entropy:

$$\text{isotropic matter:} \quad dS = \frac{C_V}{T} dT + \alpha dS_{ii}, \quad dQ = C_V dT + \alpha T dV, \quad (2.158a, b)$$

implying that heat is exchanged (2.158b) through temperature changes and changes of volume (2.85f). Substituting (2.152e, f; 2.148c; 2.156c) for isotropic matter in (2.133d) leads to the stresses (2.159):

$$\text{isotropic matter:} \quad dT_{ij} = -\alpha \delta_{ij} dT + \frac{E}{1+\sigma}\left(dS_{ij} + \frac{\sigma}{1-2\sigma} \delta_{ij} dV \right), \quad (2.159)$$

that consist of: (i) isotropic **thermal stresses** due to temperature changes; (ii) elastic stresses as in (2.156a–c). The scalar α in (2.148c), (2.158a, b) and (2.159) corresponds to isotropic stresses (2.85a) $\equiv$ (2.160a) and may be designated the **thermal pressure coefficient** (2.160c, d):

$$T_{ij} = -p\delta_{ij}, dS_{ii} = dV: \quad \alpha = \left(\frac{\partial p}{\partial T}\right)_S = \left(\frac{\partial S}{\partial V}\right)_T, \quad (2.160a\text{--}d)$$

because it specifies: (i) the rate of change of pressure with temperature at constant entropy (2.141d; 2.160a) $\equiv$ (2.160c); (ii) the rate of change of entropy with volume at constant temperature (2.158b; 2.160b) $\equiv$ (2.160d).

The electric (2.153a) [magnetic (2.153b)] field need not be constant in (2.160c, d) for isotropic matter because they are decoupled from thermomechanical effects. Thus *the constitutive relations for **linear isotropic matter** involve (2.146f) seven constitutive parameters, namely: (i/ii) the dielectric permittivity (2.161a) [magnetic permeability (2.161c) for the dielectric displacement (magnetic induction) parallel and proportional to the electric (magnetic) field (2.161b) [(2.161d)]:*

$$\varepsilon \equiv \frac{\partial \vec{D}}{\partial \vec{E}}: \quad d\vec{D} = \varepsilon d\vec{E}, \quad \mu \equiv \frac{\partial \vec{B}}{\partial \vec{H}}: \quad d\vec{B} = \mu d\vec{H}, \quad (2.161a\text{--}d)$$

that are decoupled from each other and thermoelastic effects; (iii/iv) the entropy (2.158a) and stress tensor (2.159) are coupled to the temperature and strain tensor through the specific heat at constant volume (2.134a–d) and thermal pressure coefficient (2.160a–d). A further simplification arises for isotropic stresses corresponding to a pressure (2.85a) ≡ (2.160a) leading to the basic thermodynamic system (subsection 2.2.14) involving only five constitutive parameters.

2.2.14 BASIC THERMODYNAMIC SYSTEM WITH THERMOMECHANICAL COUPLING

The basic thermodynamic system is linear (subsections 2.2.5–2.2.8), isotropic (subsections 2.2.9–2.2.13) and subject only to isotropic stresses corresponding to a pressure (2.85a) ≡ (2.160a) ≡ (2.162):

$$\text{basic thermodynamic system:} \quad \text{isotropic,} \quad T_{ij} = -p\delta_{ij}, \qquad (2.162)$$

so that the constitutive relation (2.159) simplifies to (2.163a–c):

$$-3dp = dT_{ii} = -\alpha\delta_{ii}dT + \frac{E}{1+\sigma}\left(dS_{ii} + \frac{\sigma}{1-2\sigma}\delta_{ii}dV \right) \qquad (2.163a\text{–}c)$$

Thus *for a basic thermodynamic system (2.162) consisting of isotropic matter subject to isotropic stresses (2.85a) equivalent to an inward pressure: (i) the pressure is related to the temperature and volume by (2.163c) ≡ (2.164d):*

$$\beta = \frac{E}{3(1-2\sigma)} = v + \frac{\lambda}{3} = -\left(\frac{\partial p}{\partial V}\right)_T : \quad dp = \alpha dT - \beta dV, \qquad (2.164a\text{–}d)$$

*where the **coefficient of isothermal expansion** at constant temperature (2.164c) is related by (2.164a) [(2.164b)] to the Young modulus and Poisson ratio [to the sum of the two Lamé moduli of elasticity (2.156a, b)]; (ii) the thermal pressure coefficient (2.160c, d) appears both in the constitutive equation for the pressure (2.164d) and for the entropy (2.158a) or heat (2.158b); (iii) the latter involves also the specific heat at constant volume (2.134a–d) that is unaffected by electric (2.161a, b) and magnetic (2.161c, d) fields that are decoupled among themselves and from thermomechanical effects.* The constitutive relations (2.133a–d) have been written using differentials so that they apply to general non-linear, anisotropic, inhomogeneous and unsteady matter (subsection 2.2.15) including 24 particular cases (subsection 2.2.16).

2.2.15 NON-LINEAR, ANISOTROPIC, INHOMOGENEOUS AND UNSTEADY MATTER

The constitutive relations (2.133a–d) for the complete thermodynamic system (2.128a, b; 2.129a–d) have been written in terms of differentials (2.130a, b; 2.131a, b; 2.132a–c) to apply to **general non-linear, anisotropic, inhomogeneous and**

unsteady matter for which the constitutive tensors may depend on the independent variables (2.128b), coordinates and time (2.165a–c)

$$m,n,v = 1,\ldots,N; i,j,k = 1,2,3: \quad dY_n = \sum_{m=1}^{N} Z_{nm}\left(X_v,x_i,t\right)dX_m. \quad (2.165\text{a–c})$$

For a **linear material** the constitutive tensors do not depend on the independent variables (2.166a) and the differentials in (2.165a) may be omitted by a simple integration leading to (2.166a):

$$\text{linear: } Z_{nm} = Z_{n,m}\left(x_i,t\right): \quad Y_n - Y_{0n} = \sum_{m=1}^{N} Z_{nm}\left(x_i,t\right)\left(X_m - X_{0m}\right). \quad (2.166\text{a, b})$$

The material is **homogeneous (steady)** iff the constitutive coefficients do not depend on position (2.167a) [time (2.167b)]:

$$\text{homogeneous: } \frac{\partial Z_{nm}}{\partial x_i} = 0; \quad \text{steady: } \frac{\partial Z_{nm}}{\partial t} = 0. \quad (2.167\text{a, b})$$

Matter is **isotropic** (subsections 2.2.9–2.2.14) if the constitutive tensors (2.168a) are independent of direction at each point and time, and thus involve only scalars, the identity matrix (2.34a) and the permutation symbols (2.54a–d):

$$\text{isotropic: } Z_{nm} = Z_{nm}\left(Z,\delta_{ij},e_{ijk}\right); \quad \text{basic: } T_{ij} = -p\delta_{ij}. \quad (2.168\text{a, b})$$

for a **basic thermodynamic system** besides isotropic matter (2.168a) the stresses must also be isotropic (2.168b) ≡ (2.160a) that is reduce to a normal inward pressure. These four criteria of linearity (2.166a, b), homogeneity (2.167a), steadiness (2.167b) and isotropy (2.168a, b) lead to a classification of constitutive relations (2.165a–c) into 24 cases (subsection 2.2.16).

2.2.16 TWENTY-FOUR CASES OF CONSTITUTIVE RELATIONS

The general constitutive relation (2.165a–c) may be classified according to four criteria: (i) non-linear (linear) if the constitutive coefficients do (do not) depend on the independent variables or fields (2.165a–c) ≡ (2.169a) [(2.166a, b) ≡ 2.169b)]; (ii) inhomogeneous (homogeneous) if at least one (all) constitutive tensors depends (2.169c) [do not depend (2.167a) ≡ (2.169d)] on position; (iii) unsteady (steady) if at least one (all) constitutive tensors depends (2.169e) [do not depend (2.167b) ≡ (2.169f)] on time; (iv) anisotropic (isotropic) if at least one (all) constitutive tensors depends (2.165c) ≡ (2.169g) [do not depend (2.168a) ≡ (2.169h)] on direction, and basic (2.168b) ≡ (2.169i) if in addition the stresses are isotropic:

$$\text{constitutive relations}\begin{cases} linearity\begin{cases} non-linear: Z_{nm}\left(X_v,x_i,t\right), \\ linear: Z_{nm} = Z_{nm}\left(x_i,t\right), \end{cases} \\[2em] homogeneity\begin{cases} inhomogeneous: some\ \dfrac{\partial Z_{nm}}{\partial x_i} \neq 0, \\[1em] homogeneous: all\ \dfrac{\partial Z_{nm}}{\partial x_i} = 0, \end{cases} \\[2em] steadiness\begin{cases} unsteady: some\ \dfrac{\partial Z_{n,m}}{\partial t} \neq 0, \\[1em] steady: all\ \dfrac{\partial Z_{n,m}}{\partial t} = 0, \end{cases} \\[2em] isotropy\begin{cases} anisotropic: some\ Z_{nm} = Z_{nm}\left(X_v,x_i,t\right), \\ isotropic: all\ Z_{nm} = Z_{nm}\left(Z,\delta_{ij},e_{ijk}\right), \\ basic:\ T_{ij} = -p\,\delta_{ij}. \end{cases} \end{cases} \qquad (2.169a\text{--}i)$$

There are 2 (linearity) × 2 (homogeneity) × 2 (steadiness) ×3 (isotropy) = 24 cases or combinations of constitutive relations for matter, of which: (i) the most general is (2.169a, c, e, g) non-linear, anisotropic, inhomogeneous, unsteady matter, to which applies (2.133a–d) the general theory of constitutive properties (subsections 2.2.5–2.2.8); (ii) the simplest is (2.169b, d, f, i) linear, homogeneous, steady, basic matter (subsection 2.2.14). An example is a bar or a beam that is: (i) non-linear (linear) if the stress–strain relation is non-linear (linear); (ii) inhomogeneous (homogeneous) if it has variable (constant) cross-section or (and) constitutive coefficients are dependent on (independent of) position; (iii) unsteady (steady) if it is a collapsible (rigid) tube or (and) the constitutive coefficients depend on (are independent of) time; (iv) anisotropic (isotropic) if the properties are dependent (independent) of direction in the cross-section, such as elliptic (circular) cross-section, and basic if the stresses in the cross-section reduce to a pressure. The first principle of thermodynamics in general form (Section 2.1) and its implications as concerns the constitutive relations (subsections 2.2.1–2.2.16) highlight a number of analogies among mechanics, electricity, magnetism and elasticity (subsection 2.2.17).

2.2.17 ANALOGIES AMONG MECHANICS, ELECTRICITY, MAGNETISM AND ELASTICITY

The total work (2.96) of the forces per unit volume (2.13) in the first principle of thermodynamics (2.6a, b), and the resulting constitutive relations (2.169a–i) for matter (Tables 2.1 and 2.2) highlight a number of analogies (Table 2.3) among mechanics

TABLE 2.3
Four Physical Analogies

Field	Mechanics	Electric	Magnetic	Elasticity
Primary field $\vec{X}$	Velocity $\vec{V} = \dfrac{d\vec{X}}{dt}$	Electric field $\vec{E}$	Magnetic field $\vec{H}$	Strain tensor $2S_{ij} = \partial_i u_j + \partial_j u_i$
Material property ξ	Mass density $\rho = \dfrac{dm}{dV}$	Dielectric permittivity $\varepsilon = \dfrac{\partial\vec{D}}{\partial\vec{E}}$	Magnetic permeability $\mu = \dfrac{\partial\vec{B}}{\partial\vec{H}}$	Elastic stiffness tensor $C_{ijk\ell} = \dfrac{\partial T_{ij}}{\partial S_{k\ell}} = \dfrac{E}{1+\sigma}\left(\delta_{ik}\delta_{j\ell} + \dfrac{\sigma}{1-2\sigma}\delta_{ij}\delta_{k\ell}\right)$
Secondary field $\vec{Y} = \xi\vec{X}$	Linear momentum $\vec{p} = \rho\vec{V}$	Electric displacement $\vec{D} = \varepsilon\vec{E}$	Magnetic induction $\vec{B} = \mu\vec{H}$	Stress tensor $T_{ij} = \dfrac{E}{1+\sigma}\left(S_{ij} + \dfrac{\sigma}{1-2\sigma}S_{kk}\delta_{ij}\right)$
Energy density $2E = \vec{Y}.\vec{X}$	Kinetic (of translation) $2E_k = \vec{p}.\vec{V}$	Electric $2E_e = \vec{D}.\vec{E}$	Magnetic $2E_n = \vec{H}.\vec{B}$	Elastic $2E_u = T_{ij}S_{ij}$
Quadratic form	$2E_k = \rho V^2$	$2E_e = \varepsilon E^2$	$2E_n = \mu H^2$	$2E_u = \dfrac{E}{1+\sigma}\left(S_{ij}S_{ij} + \dfrac{\sigma}{1-2\sigma}S_{ii}S_{jj}\right)$
Anistropic	$2E_t = I_{ij}\Omega_i\Omega_j$	$2E_e = \varepsilon_{ij}E_iE_j$	$2E_n = \mu_{ij}H_iH_j$	$2E_u = C_{ijk\ell}S_{ij}S_{k\ell}$

Kinetic energy: of translation E_k; of rotation E_t

Note: The comparison of the four constitutive properties (Table 2.1) has some relation with the set of four physical analogies among mechanics (of particles or rigid bodies), electricity, magnetism and elasticity; the analogies include: (i–iii) a primary field that multiplied by a material property specifies a secondary field; (iv–vi) the product (in scalar, inner vector or contracted tensor form) of the primary and secondary variables specifies an energy, that simplifies from the anisotropic to the isotropic case.

(subsection 2.1.3–2.1.4), electricity (subsections 2.1.5–2.1.8 and 2.2.6–2.2.11), magnetism (subsections 2.1.9–2.1.14 and 2.2.6–2.2.11) and elasticity (subsections 2.1.15–2.1.19 and 2.2.6–2.2.12), some of which are highlighted next. *The **analogies among mechanics, electricity, magnetism and elasticity** include: (i) the **primary fields**, respectively, the velocity (2.2a) ≡ (2.170a), electric (2.18a–c) ≡ (2.170b) and magnetic (2.41a–f) ≡ (2.170c) fields and strain tensor (2.82d) ≡ (2.170d):*

$$\vec{v} = \frac{d\vec{x}}{dt} \leftrightarrow \vec{E} \leftrightarrow \vec{H} \leftrightarrow S_{ij} = \frac{1}{2}\left(\partial_i u_j + \partial_j u_i\right); \qquad (2.170a–d)$$

*(ii) the **material properties**, respectively, the mass density (2.8a) ≡ (2.171a), the dielectric permittivity (2.161a) ≡ (2.171b), the magnetic permeability (2.161c) ≡ (2.171c) and the elastic stiffness tensor (2.154; 2.156a–c) ≡ (2.171d), all in the case of isotropic matter:*

$$\rho = \frac{dm}{dV} \leftrightarrow \varepsilon = \frac{\partial \vec{D}}{\partial \vec{E}} \leftrightarrow \mu = \frac{\partial \vec{B}}{\partial \vec{H}}, \quad C_{ijkl} = \frac{E}{1+\sigma}\left(\delta_{ik}\delta_{jl} + \frac{\sigma}{1-2\sigma}\delta_{ij}\delta_{kl}\right); \quad (2.171a–d)$$

*(iii) the **secondary fields** that are the product of the primary field (2.170a–d) by the material property (2.171a–d) leading respectively to the linear momentum (2.8b) ≡ (2.172a), electric displacement (2.27b) ≡ (2.172b), magnetic induction (2.53b) ≡ (2.172c) and stress tensor (2.156c) ≡ (2.172d), the latter three (2.172b–d) in the case of isotropic matter:*

$$\vec{p} = \rho\vec{v} \leftrightarrow \vec{D} = \varepsilon\vec{E} \leftrightarrow \vec{B} = \mu\vec{H} \leftrightarrow T_{ij} = C_{ijkl}S_{kl} = \frac{E}{1+\sigma}\left(S_{ij} + \frac{\sigma}{1-2\sigma}S_{kk}\delta_{ij}\right); \quad (2.172a–d)$$

(iv) the energy is one half of the inner vector or tensor product of the primary (2.170a–d) and secondary (2.172a–d) variables, leading, respectively, to the kinetic (2.11c) ≡ (2.173a), electric (2.27c) ≡ (2.173b), magnetic (2.53c) ≡ (2.173c) and elastic (2.138d, e) ≡ (2.157d) ≡ (2.173d) energies:

$$E_k = \frac{\vec{p}\cdot\vec{v}}{2}, \quad E_e = \frac{\vec{D}\cdot\vec{E}}{2}, \quad E_h = \frac{\vec{B}\cdot\vec{H}}{2}, \quad E_u = \frac{1}{2}T_{ij}S_{ij}, \qquad (2.173a–d)$$

that equals (2.174a–d) one half of the material property (2.171a–d) multiplied by the square of the primary variable (2.170a–d)

$$E_k = \frac{1}{2}\rho v^2, \quad E_e = \frac{1}{2}\varepsilon E^2, \quad E_h = \frac{1}{2}\mu H^2, \quad E_u = \frac{E}{2(1+\sigma)}\left(S_{ij}S_{ij} + \frac{\sigma}{1-2\sigma}S_{ii}S_{jj}\right). \quad (2.174a–d)$$

The condition that the energy must be positive leads to inequalities for the isotropic (anisotropic) constitutive parameters [Section(s) 2.2.18 (2.2.19–2.2.28)].

2.2.18   INEQUALITIES FOR THE ISOTROPIC CONSTITUTIVE COEFFICIENTS

The condition of positive energy for homogeneous isotropic matter applied to the kinetic (2.174a) ≡ (2.175a), electric (2.174b) ≡ (2.175c), magnetic (2.174c) ≡ (2.175e) and elastic (2.174d) ≡ (2.175g) energies implies, respectively, that the mass density (2.175b), electric permittivity (2.175d) and magnetic permeability (2.175f) are positive:

$$E_k > 0 \Rightarrow \rho > 0; E_e > 0 \Rightarrow \varepsilon > 0; E_h > 0 \Rightarrow \mu > 0;$$

$$E_u > 0 \Rightarrow E > 0 \wedge -1 < \sigma < \frac{1}{2}. \qquad \text{(2.175a–j)}$$

and that the Young modulus (Poisson ratio) satisfy (2.175h) [(2.176i, j)]. The proof that (2.175g) implies (2.175h–j) can be made as follows: (i) in order for the elastic energy (2.174d) to be positive (2.175g) must be satisfied (2.176a, b):

$$\frac{E}{1+\sigma} > 0 < \frac{\sigma}{1-2\sigma}; \qquad \text{(2.176a, b)}$$

(ii) if $E < 0$ then (2.176a) implies $\sigma < -1$ and (2.176b) cannot hold, thus (2.177a) ≡ (2.175h) must be true:

$$E > 0; \quad \sigma > -1, \quad \sigma < \frac{1}{2}, \qquad \text{(2.177a–c)}$$

(iii) from (2.176a) and (2.177a) follows that (2.177b) ≡ (2.175i); (iv) if $\sigma > \frac{1}{2}$ then (2.176b) would not hold, so (2.177c) ≡ (2.175j) must be true. QED. The inequalities for the constitutive parameters can be extended to anisotropic media (Sections 2.2.19 and 2.2.28) bearing in mind that the electric and magnetic (elastic) energies become quadratic (biquadratic) forms (subsections 2.2.20–2.2.27).

2.2.19   VECTOR AND TENSOR QUADRATIC FORMS FOR ENERGIES

*In the case of linear steady homogeneous anisotropic matter, twice the density per unit volume of the electric (magnetic) energy is specified by a **vector quadratic form** (2.26b) ≡ (2.178a) [(2.52b) ≡ (2.178b)] and the elastic energy by a **tensor quadratic form** (2.138d) ≡ (2.178c):*

$$2E_e = \varepsilon_{ij} E_i E_j > 0 < 2E_h = \mu_{ij} H_i H_j; 2E_u = C_{ijkl} S_{ij} S_{kl} > 0; 2E_r = I_{ij}\Omega_i\Omega_j; \qquad \text{(2.178a–d)}$$

the analogous anisotropic mechanical energy is the kinetic energy of rotation (2.178d) involving the inertia tensor I_{ij} and angular velocity Ω_r. These energies must all be positive, leading to constraints, respectively, the dielectric permittivity (2.135b, c) [magnetic permeability (2.136b, c)] tensors and elastic stiffness (2.137b–g) double tensor.

The tensor quadratic form (2.178c) can be transformed to a vector quadratic form by separating the pairs (2.179b) of Cartesian indices (2.179a):

$$i, j = 1,2,3 \equiv x, y, z: \quad \alpha, \beta \equiv 1,2,3,4,5,6 = xx, yy, zz, xy, xz, yz, \quad \text{(2.179a, b)}$$

where: (i) the strain (stress) tensors are symmetric (2.82d) $\equiv$ (2.180a) [(2.81d) $\equiv$ (2.180b)]:

$$T_\alpha \equiv \{T_{xx}, T_{yy}, T_{zz}, T_{xy}, T_{xz}, T_{yz}\}, S_\beta \equiv \{S_{xx}, S_{yy}, S_{zz}, S_{xy}, S_{xz}, S_{yz}\} : T_\alpha = C_{\alpha\beta} S_\beta, \text{(2.180a–c)}$$

with the first (last) three terms corresponding to the normal (shear) components; (ii) the stress (2.179a) and strain (2.179b) in vector form are related by the **elastic stiffness matrix** (2.180c) that contains (2.181) all the distinct components of the elastic stiffness double tensor (2.137a–g):

$$
\begin{bmatrix} T_{xx} \\ T_{yy} \\ T_{zz} \\ T_{xy} \\ T_{xz} \\ T_{yz} \end{bmatrix} =
\begin{bmatrix}
C_{1111} & C_{1122} & C_{1133} & C_{1112} & C_{1113} & C_{1123} \\
C_{1122} & C_{2222} & C_{2233} & C_{2212} & C_{2213} & C_{2223} \\
C_{1133} & C_{2233} & C_{3333} & C_{3312} & C_{3313} & C_{3323} \\
C_{1112} & C_{2212} & C_{3312} & C_{1212} & C_{1213} & C_{1223} \\
C_{1113} & C_{2213} & C_{3313} & C_{1213} & C_{1313} & C_{1323} \\
C_{1123} & C_{2223} & C_{3323} & C_{1223} & C_{1323} & C_{2323}
\end{bmatrix}
\begin{bmatrix} S_{xx} \\ S_{yy} \\ S_{zz} \\ S_{xy} \\ S_{xz} \\ S_{yz} \end{bmatrix}
\quad \text{(2.181)}
$$

*The condition that the electric (2.178a)/magnetic (2.178b)/elastic (2.178c) $\equiv$ (2.179a, b; 2.180a–c) $\equiv$ (2.181) energy be positive requires that the respective quadratic forms be **positive-definite**, leading to a classification of quadratic forms (subsection 2.2.20).*

2.2.20   DEFINITE, SEMI-DEFINITE AND INDEFINITE QUADRATIC FORMS

Taking as example the second-order differential for the free energy, a **quadratic form** (2.182b) in N variables (2.182a) with matrix of coefficients A_{nm}:

$$n, m = 1, \ldots, N: \quad Z = \sum_{n,m=1}^{N} A_{nm} X_n X_m \equiv A_{nm} X_n X_m \quad \text{(2.182a, b)}$$

can be classified in five cases:

$$
(X_1, \ldots, X_N) \neq (0, \ldots, 0) : Z
\begin{cases}
> 0 : positive-definite, \\
< 0 : negative-definite, \\
\geq 0 : positive-semidefinite, \\
\leq 0 : negative-semidefinite, \\
\geq, \leq : indefinite,
\end{cases}
\quad \text{(2.183a–f)}
$$

assuming non-zero variables (2.183a): (i/ii) **positive (negative)-definite** iff it is positive (negative) in all cases (2.183b) [(2.183c)]; (iii/iv) **positive (negative)-semidefinite** if it is non-negative (non-positive), that is positive (negative) or zero (2.183d) [(2.183e)]; (v) **indefinite** if it has no fixed sign, that is can be positive or negative, and possibly but not necessarily, also zero (2.183f). *The matrix of a quadratic form (2.184a–d) can always be replaced by its **symmetric part** (2.184e, f):*

$$Z \equiv A_{nm}X_nX_m = A_{mn}X_mX_n = \frac{1}{2}\left(A_{nm} + A_{mn}\right)X_nX_m = \bar{A}_{nm}X_nX_m:$$

$$\bar{A}_{nm} \equiv \frac{1}{2}\left(A_{nm} + A_{mn}\right) = \bar{A}_{mn}.$$

(2.184a–f)

The conditions for positive energy [subsections 2.2.23–2.2.24 (2.2.29)] are specified by positive-definite quadratic forms [subsections 2.2.21–2.2.22 (2.2.25–2.2.27)] in terms of eigenvalues (principal determinants).

2.2.21 EIGENVALUES, EIGENVECTORS, DIAGONALIZATION AND SUM OF SQUARES (SYLVESTER)

The **eigenvector** of a matrix multiplied by the matrix is parallel to itself through a scalar **eigenvalue** (2.185a):

$$A_{nm}B_n = \lambda B_n; \quad \left(A_{nm} - \lambda\delta_{nm}\right)B_n = 0,$$

(2.185a, b)

rewriting (2.185a) in the form (2.185b) it follows that *non-zero eigenvectors (2.186a) exist for eigenvalues that are roots of the **characteristic determinant** (2.186b):*

$$\left(B_1,...,B_N\right) \neq \left(0,...,0\right): \quad 0 = \det\left(A_{nm} - \lambda\delta_{nm}\right) = P_N\left(\lambda\right) = (-)^N \prod_{n=1}^{N}\left(\lambda - \lambda_n\right),$$

(2.186a–d)

that is a polynomial of degree N in the eigenvalues (2.186c) with N roots (2.186d). To each eigenvalue λ (λ) corresponds an eigenvector (2.187a) [(2.187b)]:

$$A_{nm} \underset{r}{B_m} = \underset{r}{\lambda} \underset{r}{B_n}, \quad A_{nm} \underset{s}{B_m} = \underset{s}{\lambda} \underset{s}{B_n}.$$

(2.187a, b)

From (2.187a, b) using the symmetry of the matrix (2.184f) follows that the inner product of eigenvectors satisfies (2.187c–g):

$$\underset{r}{\lambda}\left(\underset{r}{\vec{B}} \cdot \underset{s}{\vec{B}}\right) = \underset{r}{\lambda} \underset{r}{B_n} \underset{s}{B_n} = A_{nm} \underset{r}{B_m} \underset{s}{B_n} = A_{mn} \underset{r}{B_m} \underset{s}{B_n} = \underset{s}{\lambda} \underset{r}{B_m} \underset{s}{B_m} = \underset{s}{\lambda}\left(\underset{r}{\vec{B}} \cdot \underset{s}{\vec{B}}\right).$$

(2.187c–g)

Thus two distinct eigenvalues (2.188a) correspond to orthogonal eigenvectors (2.188b):

$$\lambda_r \neq \lambda_s \Rightarrow \left(\vec{B}_r \cdot \vec{B}_s\right) = 0; \quad X_n = B_n \, Y_i \Rightarrow Z = C_{ij} Y_i Y_j, \qquad (2.188a\text{--}d)$$

*the change of coordinates X_n to the **reference frame** Y_i of eigenvectors (2.188c) transforms the quadratic form (2.182a, b) with symmetric matrix (2.184a–f) to the quadratic form (2.188d) whose matrix is diagonal (2.189a–d):*

$$C_{ij} \equiv A_{nm} \, B_n \, B_m = \lambda \, B_m \, B_m = \lambda \left(\vec{B}_i \cdot \vec{B}_j\right) = \lambda \left|\vec{B}_i\right|^2 \delta_{ij}. \qquad (2.189a\text{--}d)$$

The eigenvectors (2.185a) are defined to within a multiplying constant, that can be chosen so that the eigenvectors have unit modulus (2.190a), the diagonal terms of the matrix (2.189d) then coincide (2.190b) with the eigenvalues:

$$\left|\vec{B}_i\right| = 1: \quad C_{ij} = \lambda_i \delta_{ij}, \quad Z = \lambda_i \delta_{ij} Y_i Y_j = \lambda_i \left(Y_i\right)^2, \qquad (2.190a\text{--}d)$$

*and the quadratic form (2.189d) becomes a sum of squares with the eigenvalues as coefficients (2.190c, d), corresponding to the **Sylvester (1852) inertia theorem**.* Thus follow necessary and sufficient conditions [subsection 2.2.22 (2.2.23)] for the classification of quadratic forms (2.183a–f) [for the energies (2.178a–c) to be positive].

2.2.22 CLASSIFICATION OF QUADRATIC FORMS BY THE EIGENVALUES OF THE MATRIX

The application of the Sylvester inertia theorem (2.190a–d) to quadratic forms (2.183a, b) provides their classification (2.183a–f) in terms of eigenvalues (2.186a–d). Thus *the eigenvalues (2.186a–d) of the matrix of coefficients of the quadratic form (2.182a, b) $\equiv$ (2.190a–d) lead to the classification (2.191a–e):*

$$\lambda_1 \geq \lambda_2 \geq \cdots \geq \lambda_N \begin{cases} \lambda_N > 0 : positive-definite, \\ \lambda_1 < 0 : negative-definite, \\ \lambda_N = 0 : positive-semidefinite, \\ \lambda_1 = 0 : negative-semidefinite, \\ \lambda_1 > 0 > \lambda_N : indefinite, \end{cases} \qquad (2.191a\text{--}f)$$

with eigenvalues ordered by magnitude (2.191a): (i/ii) if the smallest (largest) eigenvalue is positive (negative), then all eigenvalues are positive (negative), and the matrix of coefficients and quadratic form are positive(negative)-definite (2.191b) [(2.191c)]; (iii/iv) if the smallest (largest) eigenvalue is zero, the non-zero eigenvalues are positive (negative), and the matrix of coefficients and quadratic form are positive(negative)-semidefinite (2.191d) [(2.191e)]; (v) if the largest (smallest) eigenvalue is positive (negative) then, regardless of whether zero eigenvalues occur or not, the quadratic form is indefinite (2.191f). In the case (subsection 2.2.23) of the electric (2.178a), magnetic (2.178b) and elastic (2.180c) energies the quadratic forms must be positive-definite (subsection 2.2.23).

2.2.23 Conditions for Positive Electric, Magnetic and Elastic Energies

The dielectric permittivity (magnetic permeability) is a symmetric tensor (2.135b, c) [(2.136b, c)] and its eigenvalues (2.192a–c):

$$0 = Det\left(\varepsilon_{ij} - \varepsilon\delta_{ij}\right) = \begin{vmatrix} \varepsilon_{11} - \varepsilon & \varepsilon_{12} & \varepsilon_{13} \\ \varepsilon_{12} & \varepsilon_{22} - \varepsilon & \varepsilon_{23} \\ \varepsilon_{13} & \varepsilon_{23} & \varepsilon_{33} - \varepsilon \end{vmatrix} = P_3(\varepsilon) = -\prod_{i=1}^{3}(\varepsilon - \varepsilon_i) \quad (2.192\text{a–c})$$

transform the quadratic form for the electric (2.178a) [magnetic (2.178b)] energy into (2.190a–c) a sum of squares (2.193a) [(2.194a)]:

$$2E_e = \varepsilon_1\left(E_1\right)^2 + \varepsilon_2\left(E_2\right)^2 + \varepsilon_3\left(E_3\right)^2 > 0 \Leftrightarrow \varepsilon_1 \geq \varepsilon_2 \geq \varepsilon_3 > 0, \quad (2.193\text{a–c})$$

$$2E_m = \mu_1\left(H_1\right)^2 + \mu_2\left(H_2\right)^2 + \mu_3\left(H_3\right)^2 > 0 \Leftrightarrow \mu_1 \geq \mu_2 \geq \mu_3 > 0. \quad (2.194\text{a–c})$$

Thus *the electric energy (2.178a) ≡ (2.193a) [magnetic energy (2.178b) ≡ (2.194a)] is positive (2.193b) [(2.194b)] iff the eigenvalues (2.179a, b) of the dielectric permittivity (2.135a–c) [magnetic permeability (2.136a–c)] are all positive (2.193c) [(2.194c)]. Likewise the elastic energy (2.180c) ≡ (2.195a; 2.190a–d) is positive (2.195b) iff the elastic stiffness matrix in (2.181) has all eigenvalues (2.195d–f) positive (2.195c):*

$$2E_e = C_{\alpha\beta}S_\alpha S_\beta > 0 \Leftrightarrow C_1 \geq C_2 \geq \cdots \geq C_6 > 0:$$

$$0 = Det\left(C_{\alpha\beta} - C\delta_{\alpha\beta}\right) \equiv P_6(C) = \prod_{\alpha=1}^{6}(C - C_\alpha). \quad (2.195\text{a-f})$$

The eigenvalues of a tensor can be used to classify the material as biaxial, uniaxial or isotropic (subsection 2.2.24).

2.2.24 Biaxial, Uniaxial and Isotropic Materials

The eigenvalues apply to real symmetric tensors, not only to diagonal constitutive tensors like the dielectric permittivity (2.192a–c) but also to non-diagonal (2.196a–c) like the thermal stress tensor (2.141a, b):

$$0 = Det\left(\alpha_{ij} - \alpha\delta_{ij}\right) \equiv P_3(\alpha) = -\prod_{i=1}^{3}(\alpha - \alpha_i). \quad (2.196\text{a–c})$$

The three eigenvalues of a constitutive tensor (2.192a–c) can be used to **classify the material** as: (i) **biaxial** if they are all distinct (2.197a); (ii/iii) **uniaxial prolate (oblate)** if two are equal and the distinct one is the largest (2.197b) [smallest (2.197c)]; (iv) **isotropic** if all eigenvalues are equal (2.197d):

$$\text{material:} \begin{cases} \varepsilon_1 > \varepsilon_2 > \varepsilon_3 : biaxial, \\ \varepsilon_1 > \varepsilon_2 = \varepsilon_3 : uniaxial\ prolate, \\ \varepsilon_1 = \varepsilon_2 > \varepsilon_3 : uniaxial\ oblate, \\ \varepsilon_1 = \varepsilon_2 = \varepsilon_3 \equiv \varepsilon : isotropic. \end{cases} \qquad (2.197a\text{–}d)$$

*Considering the space with axis along the electric field, the **surface with constant electric energy** (2.193a) $\equiv$ (2.198a) is: (i) in the biaxial case (2.198a) an **ellipsoid** with half-axis (2.198b–d):*

$$1 = \frac{(E_1)^2}{a^2} + \frac{(E_2)^2}{b^2} + \frac{(E_3)^2}{c^2} : \quad 0 < a \equiv \sqrt{2\frac{E_e}{\varepsilon_1}} \le b \equiv \sqrt{2\frac{E_e}{\varepsilon_2}} \le c \equiv \sqrt{2\frac{E_e}{\varepsilon_3}}; \quad (2.198a\text{–}d)$$

*(ii/iii) in the uniaxial prolate (2.197b) [oblate (2.197c)] case a **prolate (oblate) ellipsoid of revolution** $c > a = b$ ($c = b > a$) like a rugby ball (a flying saucer); (iv) in the isotropic case (2.197d) a **sphere** of radius $a = b = c$.* The criteria (2.191a–f) [(2.193a–c; 2.194a–c; 2.195a–f)] for the classification (2.183a–f) of quadratic forms (2.182a, b) [positiveness of, respectively, electric (2.178a)/magnetic (2.178b)/elastic (2.178c) $\equiv$ (2.195a) energies] depend on the calculation of eigenvalues. Alternative criteria that do not require calculation of eigenvalues as a preliminary step and use directly the matrix of coefficients of the quadratic form (the components of the constitutive tensor) are obtained next [Sections 2.2.25–2.2.27 (2.2.28)].

2.2.25 PRINCIPAL SUBMATRICES AND SUBDETERMINANTS OF A SQUARE MATRIX

Given a $N \times N$ **square matrix** (2.199a, b)

$$n = 1,\ldots,N: \quad A_{nn} = \begin{bmatrix} A_{11} & A_{12} & A_{13} & \cdots & A_{1N} \\ A_{21} & A_{22} & A_{23} & \cdots & A_{2N} \\ A_{31} & A_{32} & A_{33} & \cdots & A_{3N} \\ \vdots & \vdots & \vdots & \ddots & \vdots \\ A_{N1} & A_{N2} & A_{N3} & \cdots & A_{NN} \end{bmatrix}, \qquad (2.199a, b)$$

the **principal submatrices** (2.199d) are the smaller $r \times r$ submatrices at the top left corner for (2.199c) and specify the **principal determinants** (2.199e):

$$n = 1,\ldots,r: \quad A_{nm}^r = \begin{bmatrix} A_{11} & A_{12} & \cdots & A_{1r} \\ A_{21} & A_{22} & \cdots & A_{2r} \\ \vdots & \vdots & \ddots & \vdots \\ A_{r1} & A_{r2} & \cdots & A_{rr} \end{bmatrix}, \quad A_r = \det\left(A_{nn}^r\right). \quad \text{(2.199c–e)}$$

For example: (i) the first and smallest principal submatrix of (2.199a, b) is the top left component (2.200a) that equals its principal determinant (2.200b, c),

$$A_{nn}^1 = A_{11} = Det\left(A_{nn}^1\right) = A_1; \qquad\qquad \text{(2.200a–c)}$$

$$A_{nn}^2 = \begin{bmatrix} A_{11} & A_{12} \\ A_{21} & A_{22} \end{bmatrix}, \quad A_2 = Det\left(A_{nn}^2\right) = A_{11}A_{22} - A_{12}A_{21}, \qquad \text{(2.200d–f)}$$

the second and next smallest principal submatrix of (2.199a, b) is the 2×2 top left corner (2.200d) and the corresponding second principal determinant is (2.200e, f); (iii) and so on for $r = 3, \ldots, N$.

Considering the quadratic form (2.182a, b) with only one non-zero variable (2.201a) it is positive (2.201b) iff the corresponding diagonal element is positive (2.201c).

$$X_1 \neq 0 = X_2 = \cdots = X_N: \quad Z_1 = A_{11}\left(X_1\right)^2 > 0 \Rightarrow A_{11} > 0. \qquad \text{(2.201a–c)}$$

Appling the same reasoning to all X_n one by one it follows that *a necessary but not sufficient condition for the quadratic form to be positive-definite (2.202a) is that all diagonal elements be positive (2.202b)*:

$$Z \equiv A_{nm}X_nX_m > 0 \Rightarrow A_{11}, A_{22} > 0,\ldots, A_{NN} > 0. \qquad \text{(2.202a, b)}$$

To see that the conditions (2.202b) are not sufficient consider the case with two variables (2.203a) leading (2.182a, b) to (2.203b, c):

$$X_1 \neq 0 \neq X_2 \neq 0 = X_3 = \cdots = X_N:$$
$$Z_2 \equiv A_{11}\left(X_1\right)^2 + 2A_{12}X_1X_2 + A_{22}\left(X_2\right)^2 = A_{11}\left(X_1 + \frac{A_{12}}{A_{11}}X_2\right)^2 + \left[A_{22} - \frac{\left(A_{12}\right)^2}{A_{11}}\right]\left(X_2\right)^2; \quad \text{(2.203a–c)}$$

even if $A_{11} > 0 < A_{22}$ choosing large enough $|A_{12}|^2 > A_{11}A_{22}$ follows $Z < 0$ in (2.203c) contradicting (2.202a). However (2.203a–c) suggests a set of necessary and sufficient conditions for positive-definiteness of a quadratic form (Section 2.2.26).

2.2.26 SUBDETERMINANTS AND POSITIVE/NEGATIVE DEFINITENESS

From (2.203a–c) it follows that the quadratic form of rank two (2.203b) $\equiv$ (2.204a) is positive-definite (2.204b) iff the first two subdeterminants of the matrix of coefficients are positive (2.204c, d):

$$Z = \sum_{n,m}^{2} A_{nm} X_n X_m > 0 \Leftrightarrow A_1 \equiv A_{11} > 0, \quad A_2 \equiv A_{11} A_{22} - A_{12} A_{21} > 0, \quad (2.204\text{a–d})$$

providing a necessary and sufficient condition in the 2×2 case. This suggests the extension to the $N \times N$ case leading to the **Sylvester subdeterminant theorem**: *the quadratic form (2.182a, b) is positive-definite (2.205a) iff all principal subdeterminants (2.199a–e; 2.200a–f) are positive (2.205b)*:

$$Z = A_{nm} X_n X_m > 0 \Leftrightarrow A_1 > 0, \ldots, A_n > 0, \ldots, A_N > 0. \qquad (2.205\text{a, b})$$

If the sign of the matrix of coefficients is reversed (2.206a) the quadratic form (2.182a, b) becomes negative-definite (2.206b) and the principal determinants have the same modulus with the same (opposite) sign for even (odd) rank (2.206c):

$$A_{nm} \rightarrow -A_{nm}, \quad Z \rightarrow -Z, \quad A_n \rightarrow (-)^n A_n. \qquad (2.206\text{a–c})$$

Thus *the necessary (2.207a, b) [and sufficient (2.208a–c)] conditions that the quadratic form (2.182a, b) is negative-definite are that all diagonal elements are negative (2.207b) [the principal determinants have alternating signs (2.208b, b) starting with minus]*:

$$Z = A_{nm} X_n X_m < 0 \Rightarrow A_{11} < 0, A_{22} < 0, \ldots, A_{NN} < 0, \qquad (2.207\text{a, b})$$

$$Z = A_{nm} X_n X_m < 0 \Leftrightarrow sign(A_n) = (-)^n \Leftrightarrow A_1 < 0, A_2 > 0, A_3 < 0, \ldots \ (2.208\text{a–c})$$

The proof of the necessary condition $\Leftarrow$ in (2.205a, b), that (2.205a) implies (2.205b) can be made as follows: (i) consider the vector with all coefficients beyond X_r zero (2.209a) so that the quadratic form (2.182a, b) of rank N reduces to rank r in (2.209b):

$$0 = X_{r+1} = X_{r+2} = \cdots = X_N: \quad Z = \sum_{n,m=1}^{r} A_{nm} X_n X_m > 0. \qquad (2.209\text{a–c})$$

Since (2.209b) is positive (2.209c) definite its eigenvalues (2.209d) are positive (2.209e), and the determinant of the sub-matrix is the product of the eigenvalues (2.209f) and thus positive (2.209g):

$$Det\left(A_{nm}^r - \lambda A_{n,m}^r\right) = (-)^r \prod_{s=1}^{r} (\lambda - \lambda_s), \lambda_s > 0: \qquad (2.209\text{d, e})$$

$$A_r = Det\left(A_{nm}^r\right) = \sum_{s=1}^{r} \lambda_s > 0, \qquad (2.209\text{f–h})$$

proving (2.209h) ≡ (2.205b). QED. The proof of the sufficient condition ⇐ in (2.205a, b) that (2.205b) implies (2.205a) is less simple and is made next (subsection 2.2.27).

2.2.27   NECESSARY AND SUFFICIENT CONDITIONS FOR POSITIVE/NEGATIVE DEFINITENESS

The proof of the sufficient condition ⇐ in (2.205a, b) follows from the following remarks: (i) for a quadratic form of rank one (2.202a, b) [two (2.204a–d)] proves (2.205a, b); (ii) thus it is sufficient to show that *if the quadratic form of rank r is positive-definite (2.210b) and the subdeterminant of rank r + 1 is positive (2.210c) then the quadratic form of rank r + 1 is positive-definite (2.210d)*:

$$r = 1,\ldots,N: \quad Z_r > 0 \wedge A_{r+1} > 0 \Rightarrow Z_{r+1} > 0; \qquad (2.210a\text{–}d)$$

(iii) applying (2.210b–d) for (2.210a) proves the sufficient condition that (2.205b) implies (2.205a). The proof is made by "reductio ad absurdum" showing that if (2.210d) is false and (2.210c) is true then the hypothesis (2.210b) is false. Thus assume that (2.210d) is false (2.211a) so that A_{nm}^{r+1} must have at least one negative eigenvalue; since Z_r is the product of all eigenvalues and is positive there must be at least two negative eigenvalues (2.211b, c) with eigenvectors (2.186a) ≡ (2.211d, e):

$$Z_{r+1} < 0: \quad \underset{1}{\lambda} < 0 > \underset{2}{\lambda}, \quad A_{nm}^r \underset{1}{B_m} = \underset{1}{\lambda} \underset{1}{B_n}, \quad A_{nm}^r \underset{2}{B_m} = \underset{2}{\lambda} \underset{2}{B_n}. \qquad (2.211a\text{–}e)$$

Let (2.212a) be a linear combination of the eigenvectors with constants (α, β) such that the component $r + 1$ is zero (2.212b) and thus the quadratic form (2.212c) reduces from rank $r + 1$ to rank r and is given by (2.212d-g):

$$X_n = \alpha \underset{1}{B_n} + \beta \underset{2}{B_n}, \quad X_{r+1} = 0:$$

$$\begin{aligned}
Z_{r+1} &= \sum_{n,m=1}^{r+1} A_{nm} X_n X_m = \sum_{n,m=1}^{r} A_{nm} X_n X_m \\
&= \sum_{n,m=1}^{r} A_{nm} \left(\alpha \underset{1}{B_n} + \beta \underset{2}{B_n} \right) \left(\alpha \underset{1}{B_m} + \beta \underset{2}{B_m} \right) \qquad (2.212a\text{–}g) \\
&= \sum_{m=1}^{r} \left(\alpha \underset{1}{\lambda} \underset{1}{B_m} + \beta \underset{2}{\lambda} \underset{2}{B_m} \right) \left(\alpha \underset{1}{B_m} + \beta \underset{2}{B_m} \right) \\
&= \alpha^2 \underset{1}{\lambda} \left| \underset{1}{\vec{B}} \right|^2 + \beta^2 \underset{2}{\lambda} \left| \underset{2}{\vec{B}} \right|^2 + \alpha\beta \left(\underset{1}{\lambda} + \underset{2}{\lambda} \right) \left(\underset{1}{\vec{B}} \cdot \underset{2}{\vec{B}} \right).
\end{aligned}$$

If the eigenvalues are distinct (2.213a) ≡ (2.189) the eigenvectors are orthogonal (2.213b) ≡ (2.189b) and the quadratic form (2.212g) ≡ (2.213c) is negative (2.213d) from (2.211b, c):

$$\lambda_1 \neq \lambda_2: \quad \left(\underset{1}{\vec{B}} \cdot \underset{2}{\vec{B}}\right) = 0, \qquad Z_{r+1} = \lambda_1 \alpha^2 \left|\underset{1}{\vec{B}}\right|^2 + \lambda_2 \beta^2 \left|\underset{2}{\vec{B}}\right|^2 < 0. \qquad (2.213a\text{--}d)$$

If the eigenvalues are equal (2.214a), using the property (2.214b) that the inner product of two vectors cannot be less than the product of their modulus (2.214b) it follows (2.214c–e):

$$\underset{1}{\lambda} = \underset{2}{\lambda} \equiv \lambda, \qquad \left(\underset{1}{\vec{B}} \cdot \underset{2}{\vec{B}}\right) \geq -\left|\underset{1}{\vec{B}}\right|\left|\underset{2}{\vec{B}}\right|:$$

$$\alpha^2 \left|\underset{1}{\vec{B}}\right|^2 + \beta^2 \left|\underset{2}{\vec{B}}\right|^2 + 2\alpha\beta\left(\underset{1}{\vec{B}} \cdot \underset{2}{\vec{B}}\right)$$

$$\geq \alpha^2 \left|\underset{1}{\vec{B}}\right|^2 + \beta^2 \left|\underset{2}{\vec{B}}\right|^2 - 2\alpha\beta\left|\underset{1}{\vec{B}}\right|\left|\underset{2}{\vec{B}}\right| \qquad (2.214a\text{--}b)$$

$$= \left(\alpha\left|\underset{1}{\vec{B}}\right| - \beta\left|\underset{2}{\vec{B}}\right|\right)^2 > 0,$$

And since the eigenvalue is negative (2.214f) the quadratic form (2.212g) is again negative (2.214g)

$$\lambda = 0: \quad \lambda\left[\alpha\left|\underset{1}{\vec{B}}\right|^2 + \beta\left|\underset{2}{\vec{B}}\right|^2 + 2\alpha\beta\left(\underset{1}{\vec{B}} \cdot \underset{2}{\vec{B}}\right)\right] < 0. \qquad (2.214f,g)$$

In both cases of distinct (2.213a) [equal (2.214a)] eigenvalues the quadratic form (2.213d) [(2.214e)] is negative contradicting the hypothesis (2.210b); thus the assumption (2.211a) must be false, implying that (2.210c) is true. QED.

2.2.28   INEQUALITIES FOR CONSTITUTIVE TENSORS ENSURING POSITIVE ENERGIES

The quadratic form for the electric (2.178a) ≡ (2.215a) [magnetic (2.178b) ≡ (2.216a)] energy in a linear homogeneous anisotropic medium shows that it is positive (2.215b) [(2.216b)] iff the principal determinants of the dielectric permittivity (2.135a–c) [magnetic permeability (2.136a–c)] tensor are (2.205a, b) all positive (2.215c–e) [(2.216c–e)]:

$$2E_e = \varepsilon_{ij} E_i E_j > 0 \Leftrightarrow \varepsilon_{11} > 0, \begin{vmatrix} \varepsilon_{11} & \varepsilon_{12} \\ \varepsilon_{12} & \varepsilon_{22} \end{vmatrix} > 0, \begin{vmatrix} \varepsilon_{11} & \varepsilon_{12} & \varepsilon_{13} \\ \varepsilon_{12} & \varepsilon_{22} & \varepsilon_{23} \\ \varepsilon_{13} & \varepsilon_{23} & \varepsilon_{33} \end{vmatrix} > 0, \qquad (2.215a\text{--}e)$$

$$2E_m = \mu_{ij} H_i H_j > 0 \Leftrightarrow \mu_{11} > 0, \begin{vmatrix} \mu_{11} & \mu_{12} \\ \mu_{12} & \mu_{22} \end{vmatrix} > 0, \begin{vmatrix} \mu_{11} & \mu_{12} & \mu_{13} \\ \mu_{12} & \mu_{22} & \mu_{23} \\ \mu_{13} & \mu_{23} & \mu_{33} \end{vmatrix} > 0. \qquad (2.216a\text{--}e)$$

Likewise *the elastic energy (2.178c) ≡ (2.217a) in a linear homogeneous anisotropic medium is positive (2.217b) iff all principal determinants of the elastic stiffness matrix (2.181) are positive (2.217c, d)*:

$$2E_u = C_{ijkl}S_{ij}S_{kl} > 0 \Leftrightarrow r = 1,\ldots,6: \quad C_r \equiv Det\left(A_{\alpha\beta}^r\right) > 0. \qquad (2.217a–d)$$

An example of a quadratic form that is indefinite is the second-order differential (2.131a, b) of the free energy (subsection 2.2.29).

2.2.29 INDEFINITE SECOND-ORDER DIFFERENTIAL OF THE FREE ENERGY

The constitutive relations (2.133a–d) for general matter can be written in matrix form (2.218):

$$\begin{bmatrix} dS \\ dD_i \\ dB_i \\ dT_{ij} \end{bmatrix} = \begin{bmatrix} \dfrac{C_V}{T} & f_k & h_k & \alpha_{kl} \\ f_i & \varepsilon_{ik} & \vartheta_{ik} & p_{ikl} \\ h_i & \vartheta_{ki} & \mu_{ik} & q_{ikl} \\ -\alpha_{ij} & -p_{kij} & -q_{kij} & C_{ijkl} \end{bmatrix} \begin{bmatrix} dT \\ dE_k \\ dH_k \\ dS_{kl} \end{bmatrix}, \qquad (2.218)$$

relating linearly the differentials of the dependent variables (2.129a–d) to the differentials of the independent variables (2.128a, b) through the **constitutive matrix** (2.218) that appears also in the second-order differential of the free energy (2.131a, b) ≡ (2.219):

$$d^2F = \frac{C_V}{T}\left(dT\right)^2 - \varepsilon_{ij}dE_idE_j - \mu_{ij}dH_idH_j - C_{ijkl}dS_{ij}dS_{kl} - 2f_idTdE_i$$
$$-2h_idTdH_i - 2\vartheta_{ij}dE_idH_j - 2p_{ijk}dE_idS_{jk} - 2q_{ijk}dH_idS_{jk} \qquad (2.219)$$

and is indefinite (2.220e), as shown next. To prove that (2.219) is indefinite it is sufficient to show that the necessary conditions (2.202a, b) are violated, that is diagonal elements with opposite signs exist; for example, consider the second-order derivatives of the free energy with regard to the first component of the electric field (2.135c) ≡ (2.220a) [strain tensor (2.137b) ≡ (2.220c)] is negative (2.220b) [positive (2.220d)]:

$$\frac{\partial^2 F}{\partial E_1 \partial E_1} = -\varepsilon_{11} < 0, \frac{\partial^2 F}{\partial S_{11} \partial S_{11}} = C_{1111} > 0: \quad d^2F \gtrless 0, \qquad (2.220a–e)$$

it follows that the necessary conditions (2.202b) for positive-definiteness, or with reversed sign for negative-definiteness are violated by the *quadratic form for the second-order differential of the free energy (2.131a, b), that thus is indefinite (2.220e), that is, it is neither maximum nor minimum.* The first principle of thermodynamics: (i) specifies the conditions of equilibrium (Section 2.1); (ii) leads to the constitutive properties of matter (Section 2.2). The second principle of thermodynamics is needed to: (i) specify the evolution of a thermodynamic system between two equilibrium states, when the intermediate states are not in equilibrium, that is the process is irreversible (Section 2.3); (ii) leads to the diffusive coefficients for an irreversible process (Section 2.4). The second principle of thermodynamics states (Section 2.3) that the internal energy (entropy) is minimum (maximum) and thus its second-order differential is positive (negative) definite, unlike other functions of state, that may be indefinite, for example, the free energy (subsection 2.2.29).

2.3   THREE PRINCIPLES AND FOUR PROCESSES OF THERMODYNAMICS

The first principle of thermodynamics (Section 2.1) specifies the constitutive properties of matter (Section 2.2) in the most general case of the complete thermodynamic system involving besides heat also seven forms of work. The kinetic energy and gravity potential energy can be included in the augmented internal energy. For a single chemical species, in the case of isotropic matter, the electric and magnetic fields decouple, leaving only thermomechanical coupling. In the case of isotropic stresses equivalent to a pressure this leads to the basic thermodynamic system for which the internal energy is the sum of the heat plus the work of the pressure in a volume change. The inclusion of other forms of work would not change the formulation of the second and third principles of thermodynamics, that are considered (Section 2.3) for the basic thermodynamic system.

For a basic thermodynamic system (subsection 2.3.1) there are only four thermodynamic variables, two extensive (entropy and volume) and two intensive (temperature and pressure) leading to: (i) besides the entropy and internal energy three other functions of state (subsection 2.3.2), namely the free energy, enthalpy and free enthalpy; (ii) four thermodynamic processes (subsection 2.3.3) namely isochoric/adiabatic/isobaric/isothermal, respectively (subsections 2.3.4/2.3.5/2.3.6/2.3.7) when are constant the volume, entropy, pressure or temperature. The first (second) order derivatives of the four functions of state specify (subsection 2.3.8) the 4 thermodynamic variables (4 relations among thermodynamic derivatives). There are 12 thermodynamic derivatives (subsection 2.3.14) that can be related (subsections 2.3.10–2.3.13) through the properties of two-dimensional Jacobians (subsection 2.3.9) to just three (subsection 2.3.15) of them, namely: (i, ii) the specific heats at constant volume and pressure; (iii) the coefficient of thermal expansion.

Whereas the first principle of thermodynamics (Sections 2.1–2.2) specifies that functions of state must be stationary at equilibrium, the second principle of thermodynamics (Sections 2.3–2.4) requires (subsections 2.3.16–2.3.17) that the internal energy (entropy) be minimum (maximum). The conditions of [subsections 2.3.18–2.3.19 (2.3.20–2.3.22)] minimum internal energy (maximum entropy) lead to thermodynamic inequalities for general matter. The principle of stability in terms of minimum internal energy: (i) implies maximum entropy, but not the reverse; (ii) implies also that the absolute temperature is positive. Concerning the other three functions of state (subsection 2.3.23) the enthalpy and free energy have no extremals and the free enthalpy is maximum. The increase in entropy in an irreversible thermodynamic process is exemplified [subsections 2.3.24–2.3.25 (2.3.26–2.3.27)] by the interaction of two reservoirs (a body and the surrounding environment). The limiting case of zero absolute temperature leads to the third principle of thermodynamics (subsections 2.3.28–2.3.29) concerning the properties of matter at the zero temperature. The second and third principles of thermodynamics are derived most simply for a basic thermodynamic system, and apply as well to a general thermodynamic system with multiple chemical species and several force fields (subsections 2.3.30–2.3.31).

2.3.1 COMPLETE AND BASIC THERMODYNAMIC SYSTEM

The first principle of thermodynamics in its most general form for arbitrary matter (2.98a, b) adds: (i) the heat (2.6a, b); (ii) seven forms of work (2.96) or energy. Besides the work of chemical reactions (2.88a–d) the work is associated with five forces: (i/ii) the inertia force (gravity field) lead [subsection 2.1.3 (2.1.4)] to the kinetic (potential) energy that can be incorporated into the augmented internal energy (2.98a, b); (iii/iv) the electric (magnetic) force leads [subsections 2.1.5–2.1.8 (2.1.9–2.1.14)] to a decoupling in the case of isotropic matter (2.153a, b); (v) the work of the stresses on the strains (subsections 2.1.15–2.1.18) remains coupled to the thermal effects (2.158a; 2.159) even in the particular case (2.158a; 2.164d) of isotropic stresses or pressure, corresponding to a basic thermodynamic system. The internal energy (2.221d) of a basic thermodynamic system:

$$X_{1,2} = \{S, V\}, Y_{1,2} = \{T, -p\}: \quad dU = \sum_{n=1}^{2} Y_n dX_n = TdS - pdV, \quad (2.221\text{a–d})$$

consists (2.6a) of: (i) the general form of the heat (2.6b) involving the temperature and entropy; (ii) the simplest form of work, namely (2.70c) of the pressure in a change of volume. The basic thermodynamic system involves (2.221c) two extensive (2.221a) and two intensive (2.221b) variables, and can be used to consider the second and third principles of thermodynamics, since the inclusion of other forms of work would not change the essential arguments.

2.3.2 FIVE FUNCTIONS OF STATE

For a basic thermodynamic system the internal energy (2.221d) depends only on (2.221a) entropy and volume and there are only three possible Legendre transforms (subsection 2.2.1) namely: (i) exchanging entropy by temperature (2.222a) leads to (2.222b) the free energy [compare with (2.114a; 2.115)]; (ii) exchanging volume by pressure (2.223a) leads to (2.223b) the enthalpy [compare with (2.116a, b; 2.117)]; (iii) exchanging both entropy (volume) for temperature (pressure) leads (2.224a–c) to (2.224d) the free enthalpy [compare with 2.118a–c; 2.120)]:

$$F(T,V) = U(S,U) - TS: \quad dF = -SdT - pdV, \quad (2.222\text{a, b})$$

$$H(S,p) = U(S,V) + pV: \quad dH = TdS + Vdp, \qquad (2.223a, b)$$

$$G(T,p) = U(S,V) - TS + pV = F(T,V) + pV = H(S,p) - TS: \qquad (2.224a\text{--}d)$$
$$dG = -SdT + Vdp.$$

The entropy is also a function of state specified by the internal energy (2.221d) ≡ (2.225a) [enthalpy (2.223b) ≡ (2.225b)]:

$$\frac{dU}{T} + \frac{p}{T}dV = dS = \frac{dH}{T} - \frac{V}{T}dp. \qquad (2.225a, b)$$

Thus *a basic thermodynamic system has five functions of state, namely: (i) the internal energy (2.221a–c) from the first principle of thermodynamics; (ii/iii/iv) the free energy/enthalpy/free enthalpy, respectively (2.222a, b)/(2.223a, b)/(2.224a–d) as Legendre transforms of the internal energy: (v) the entropy (2.225a) [(2.225b)] as a rearrangement of the internal energy (2.221d) [enthalpy (2.223b)].* Although the work (heat) are generally inexact differentials [subsection 2.1.1 (2.1.2)] they become exact differentials for particular thermodynamic processes for which they coincide with a function of state (subsection 2.3.3).

2.3.3   CASES WHEN WORK AND HEAT ARE FUNCTIONS OF STATE

Since a basic thermodynamic system has 4 variables (2.221a, b) there are 4 particular **thermodynamic processes** in which one of them is constant, namely: (i) **isochoric** process if the volume is constant (2.226a), thus no work is done (2.226b), simplifying the internal energy (2.221d) [free energy (2.222b)] to (2.226c, d) [(2.226e)]; (ii) **adiabatic** process in which the entropy is constant (2.227a), and thus no heat is exchanged (2.227b), simplifying the internal energy (2.221d) [enthalpy (2.223b)] to (2.227c, d) [(2.227e)]; (iii) **isobaric** process if the pressure is constant (2.228a), simplifying the enthalpy (2.223b) [free enthalpy (2.224d)] to (2.228b, c) [(2.228d)]; (iv) **isothermal** process if the temperature is constant (2.229a), simplifying the free energy (2.222b) [free enthalpy (2.224d)] to (2.229b, c) [(2.229d)]:

$$dV = 0: \quad dW = 0, \quad dU = TdS = dQ, \quad dF = -SdT; \qquad (2.226a\text{--}e)$$

$$dS = 0: \quad dQ = 0, \quad dU = -pdV = dW, \quad dH = Vdp; \qquad (2.227a\text{--}e)$$

$$dp = 0: \quad dH = TdS = dQ, \quad dG = -SdT; \qquad (2.228a\text{--}d)$$

$$dT = 0: \quad dF = -pdV = dW, \quad dG = Vdp. \qquad (2.229a\text{--}d)$$

Thus *for a basic thermodynamic system (2.221a–d) there are 4 thermodynamic processes, namely isochoric (2.226a–e)/adiabatic (2.227a–e)/isobaric (2.228a–d)\V isothermal (2.229a–d) for which are constant, respectively, the volume/entropy/pressure/temperature. Although the work is generally an inexact differential (subsection 2.1.1) it becomes an exact differential for an adiabatic (2.227a) [isothermal (2.229a)]*

process when it coincides with the internal energy (2.227d) [free energy (2.229c)].
Although the heat is generally an inexact differential (subsection 2.1.2) it becomes an
exact differential in an isochoric (2.226a) [isobaric (2.228a)] process when it coin-
cides with the internal energy (2.226d) [enthalpy (2.228c)]. The isochoric/adiabatic/
isobaric/isothermal thermodynamic processes are considered in more detail in the
sequel (subsection, respectively, 2.3.4/2.3.5/2.3.6/2.3.7).

2.3.4 ADIABATIC PROCESS AND PRESSURE EQUILIBRIUM

Consider two adiabatic systems, that do not exchange heat, so that: (i) their entropies
are constant (2.230a, b):

$$dS_1 = 0 = dS_2: \quad dU_1 = -p_1 dV_1 = dW_1, \quad dU_2 = -p_2 dV_2 = dW_2; \quad (2.230a\text{--}d)$$

(ii) their internal energies (2.230c, d) coincide with the work. The total volume is
conserved (2.231a, b), but there is a siding partition (Figure 2.3a) with no friction that
allows the volumes to vary, thus varying the total internal energy (2.231c–e):

$$0 = dV = dV_1 + dV_2: \quad dU = dU_1 + dU_2 = (p_1 - p_2)dV_2 = (p_2 - p_1)dV_1. \quad (2.231a\text{--}e)$$

The condition of equilibrium that the total internal energy be stationary (2.232a)
implies that the pressure be equal (2.232b):

$$dU = 0 \Leftrightarrow p_1 = p_2. \qquad (2.232a, b)$$

This result confirms that *the volume (pressure) is an extensive (2.221b) [intensive*
(2.221d)] thermodynamic variable which is additive (2.231b) ≡ *(2.104f) [equal*
(2.232b) ≡ *(2.105e)] in the **adiabatic** equilibrium of two isolated basic thermo-*
dynamic systems which: (i) do not exchange heat (2.230a, b); (ii) have constant total
volume (2.231a).

2.3.5 ISOCHORIC PROCESS AND THERMAL EQUILIBRIUM

Suppose now that instead of not exchanging heat (Figure 2.3a) the two thermodynamic systems are separated by a rigid wall (Figure 2.3b) and cannot change volume
(2.233a, b):

$$dV_1 = 0 = dV_2: \quad dU_1 = T_1 dS_1 = dQ_1, \quad dU_2 = T_2 dS_2 = dQ_2. \quad (2.233a\text{--}f)$$

In the **isochoric process** at constant volume there is no work and the internal energies (2.233c–f) coincide with the heat exchanged. There is no heat exchange with the
outside (2.234a, b) but the two systems can exchange heat with each other through
the wall thus varying the total internal energy (2.234c–e):

$$0 = dS = dS_1 + dS_2: \quad dU = dU_1 + dU_2 = (T_1 - T_2)dS_1 = (T_2 - T_1)dS_2. \quad (2.234a\text{--}e)$$

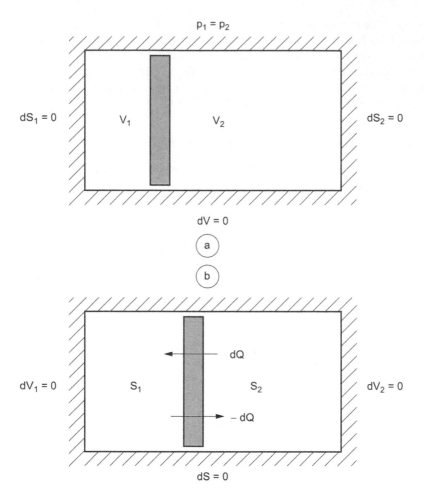

FIGURE 2.3 Two cases of equilibrium (Figure 2.2) for a basic thermodynamic system with two pairs of extensive – intensive parameters: (a) equal pressure with volume variations in the adiabatic case, excluding heat exchanges; (b) equal temperature for heat exchanges at constant volume.

The condition of equilibrium that the total internal energy be stationary (2.235a) implies that the temperature be the same (2.235b):

$$dU = 0 \Leftrightarrow T_1 = T_2. \tag{2.235a, b}$$

This result confirms that *the entropy (temperature) is an extensive (2.221a) [intensive (2.221c)] thermodynamic variable, which is additive (2.234b) ≡ (2.104c) [equal (2.235b) ≡ (2.105b)] in the* **isochoric** *equilibrium of two isolated basic* *thermodynamic systems which have: (i) constant volume (2.233a, b); (ii) do not exchange heat with the exterior (2.234a, b).*

2.3.6 WORK AND HEAT IN AN ISOBARIC PROCESS

In an **isobaric** thermodynamic process the work (2.70c) is the product of the constant pressure (2.236a) by minus the volume change (2.236b, c):

$$p = const \equiv p_0: \quad W_{12} = -\int_1^2 pdV = p_0\left(V_1 - V_2\right). \qquad (2.236\text{a--c})$$

In isobaric conditions (2.228a) $\equiv$ (2.237a) the heat exchanged is (2.228b,c) $\equiv$ (2.237b,c) the difference of enthalpies (2.237d, e):

$$dp = 0: \quad dQ = TdS = dH, \quad Q_{12} = \int_1^2 TdS = \int_1^2 dH = H_2 - H_1. \qquad (2.237\text{a--e})$$

The change in internal energy in an isobaric (2.228a) $\equiv$ (2.236a) process:

$$U_2 - U_1 = \int_1^2 dU = Q_{12} + W_{12} = H_2 - H_1 + p_0\left(V_2 - V_1\right), \qquad (2.238\text{a--c})$$

is (2.238a–c) the sum of: (i) the heat exchanged (2.237e); (ii) the work performed (2.236c).

2.3.7 HEAT AND WORK IN AN ISOTHERMAL PROCESS

In an **isothermal** thermodynamic process (2.239a) the heat exchanged is the constant temperature times the difference in entropy (2.239b, c):

$$T = const \equiv T_0: \quad Q_{12} = \int_1^2 TdS = T_0\left(S_2 - S_1\right). \qquad (2.239\text{a--c})$$

In isothermal conditions (2.240a) the work performed is (2.229b, c) $\equiv$ (2.240b, c) the difference of free energies (2.229a, b) $\equiv$ (2.240d, e):

$$dT = 0: \quad dF = -pdV = dW, \quad W_{12} = \int_1^2 dF = F_2 - F_1. \qquad (2.240\text{a--e})$$

The change in internal energy in an isothermal (2.239a) $\equiv$ (2.240a) process:

$$U_2 - U_1 = \int_1^2 dU = W_{12} + Q_{12} = T_0\left(S_2 - S_1\right) + F_2 - F_1, \qquad (2.241\text{a--c})$$

is the sum (2.241a–c) of: (i) the heat exchanged (2.239c); (ii) the work performed (2.240e).

2.3.8 FOUR THERMODYNAMIC VARIABLES AND FOUR RELATIONS BETWEEN DERIVATIVES (MAXWELL 1867)

From the four functions of state for a basic thermodynamic system, namely the internal energy (2.221d), free energy (2.222b), enthalpy (2.223b) and free enthalpy (2.224d) it follows that: (i) the exact differentials specify, through their first-order partial derivatives, two expressions for each of the four thermodynamic variables, namely the temperature (2.242a, b), the entropy (2.242c, d), the volume (2.242e, f) and the pressure (2.242g, h):

$$\left(\frac{\partial U}{\partial S}\right)_V = T = \left(\frac{\partial H}{\partial S}\right)_p, \quad -\left(\frac{\partial F}{\partial T}\right)_V = S = -\left(\frac{\partial G}{\partial T}\right)_p, \qquad (2.242\text{a--d})$$

$$\left(\frac{\partial H}{\partial p}\right)_S = V = \left(\frac{\partial G}{\partial p}\right)_T, \quad -\left(\frac{\partial U}{\partial V}\right)_S = p = -\left(\frac{\partial F}{\partial V}\right)_T; \qquad (2.242\text{e--h})$$

(ii) *the equality of continuous second-order derivatives leads to the* **four Maxwell relations** *(2.243a–d) between* **thermodynamic derivatives:**

$$\left(\frac{\partial T}{\partial V}\right)_S = \frac{\partial^2 U}{\partial V\,\partial S} = \frac{\partial^2 U}{\partial S\,\partial V} = -\left(\frac{\partial p}{\partial S}\right)_V, \quad \left(\frac{\partial S}{\partial V}\right)_T = -\frac{\partial^2 F}{\partial V\,\partial T} = -\frac{\partial^2 F}{\partial T\,\partial V} = \left(\frac{\partial p}{\partial T}\right)_V, \quad (2.243\text{a, b})$$

$$\left(\frac{\partial T}{\partial p}\right)_S = \frac{\partial^2 H}{\partial p\,\partial S} = \frac{\partial^2 H}{\partial S\,\partial p} = \left(\frac{\partial V}{\partial S}\right)_p, \quad -\left(\frac{\partial S}{\partial p}\right)_T = \frac{\partial^2 G}{\partial p\,\partial T} = \frac{\partial^2 G}{\partial T\,\partial p} = \left(\frac{\partial V}{\partial T}\right)_p. \quad (2.243\text{c, d})$$

The **Mnemonic diagonal** *in Figure 2.4 illustrates: (i) the first-order derivatives (2.242a–h) for thermodynamic variables through individual triangles; (ii) the equalities of thermodynamic derivatives (2.243a–d) as Maxwell relations between opposing triangles.* A basic thermodynamic system (subsections 2.3.11–2.3.13) has 12 distinct thermodynamic derivatives (subsection 2.3.14), of which only 3 are independent (subsection 2.3.15) and their relations (subsections 2.3.10–2.3.13) can be established using the properties of two-dimensional Jacobians (subsection 2.3.9).

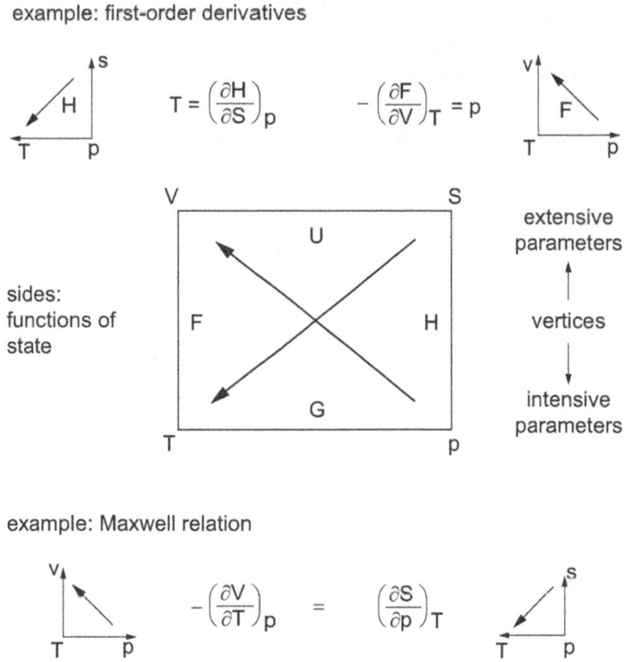

example: first-order derivatives

$$T = \left(\frac{\partial H}{\partial S}\right)_p \qquad -\left(\frac{\partial F}{\partial V}\right)_T = p$$

sides:
functions of
state

extensive
parameters

↑

vertices

↓

intensive
parameters

example: Maxwell relation

$$-\left(\frac{\partial V}{\partial T}\right)_p = \left(\frac{\partial S}{\partial p}\right)_T$$

FIGURE 2.4 Mnemonic diagram for a basic thermodynamic system (Figure 2.3) showing: (i) on opposite corners of a square the pairs of extensive – intensive variables, namely (a) mechanical (volume – pressure) and (b) thermal (entropy – temperature); (ii) on the sides the four functions of state, starting with the internal energy $U(V,S)$, then enthalpy $H(p,S)$ and free energy $F(V,T)$ and free enthalpy $G(p,T)$ on the opposite side. The thermodynamic derivatives of first-order correspond to triangles; the Maxwell relations between thermodynamic derivatives relate opposing triangles.

2.3.9   SKEW-SYMMETRY, INVERSION AND PRODUCT PROPERTIES OF JACOBIANS

A two-dimensional **Jacobian (1891)** is defined by the determinant (2.244c, d) of the partial derivatives of two functions (2.244a, b):

$$A = A(C,D), \quad B = B(C,D): \quad \frac{\partial(A,B)}{\partial(C,D)} \equiv \begin{vmatrix} \dfrac{\partial A}{\partial C} & \dfrac{\partial B}{\partial C} \\ \dfrac{\partial A}{\partial D} & \dfrac{\partial B}{\partial D} \end{vmatrix} \equiv \frac{\partial A}{\partial C}\frac{\partial B}{\partial D} - \frac{\partial A}{\partial D}\frac{\partial B}{\partial C}. \quad (2.244\text{a–d})$$

The partial derivative (2.245d) is a particular case (2.245c) of Jacobian with one coincident upper and lower variable, bearing in mind that the upper (lower) variables are independent (2.245a, b).

$$\frac{\partial A}{\partial B} = 0 = \frac{\partial C}{\partial D}: \quad \frac{\partial(A,B)}{\partial(C,B)} = \frac{\partial A}{\partial C}\frac{\partial B}{\partial B} - \frac{\partial A}{\partial B}\frac{\partial B}{\partial C} = \left(\frac{\partial A}{\partial C}\right)_B. \quad (2.245\text{a–d})$$

The Jacobian (2.244a–d) has the properties: (i) of **skew-symmetry**, changing sign when either the upper (2.246a) [or the lower (2.246b)] pair of variables is inter-changed, and returning to the same sign (2.246c) if both pairs are interchanged:

$$\frac{\partial(A,B)}{\partial(C,D)} = -\frac{\partial(B,A)}{\partial(C,D)} = -\frac{\partial(A,B)}{\partial(D,C)} = \frac{\partial(B,A)}{\partial(D,C)}; \qquad (2.246\text{a–c})$$

(ii) of **inversion**, that the inverse partial derivative (2.247a) of (2.246d) exchanges the numerator and denominator (2.247b, c):

$$\left(\frac{\partial C}{\partial A}\right)_B = \frac{\partial(C,B)}{\partial(A,B)} = \left[\frac{\partial(A,B)}{\partial(C,B)}\right]^{-1} = \left[\left(\frac{\partial A}{\partial C}\right)_B\right]^{-1}; \qquad (2.247\text{a–c})$$

(iii) of the **product** of the Jacobians (2.248c) for two successive transformations (2.244a, b; 2.248a, b):

$$B = B(E,F), \quad D = D(E,F): \quad \frac{\partial(A,B)}{\partial(E,F)} = \frac{\partial(A,B)}{\partial(C,D)}\frac{\partial(C,D)}{\partial(E,F)}. \qquad (2.248\text{a–c})$$

To prove (2.248c) both sides are expanded: (a) the r.h.s. leads to (2.249a) with four factors (2.249b):

$$\frac{\partial(A,B)}{\partial(C,D)} \times \frac{\partial(C,D)}{\partial(E,F)} = \left(\frac{\partial A}{\partial C}\frac{\partial B}{\partial D} - \frac{\partial A}{\partial D}\frac{\partial B}{\partial C}\right)\left(\frac{\partial C}{\partial E}\frac{\partial D}{\partial F} - \frac{\partial C}{\partial F}\frac{\partial D}{\partial E}\right)$$

$$= \frac{\partial A}{\partial C}\frac{\partial B}{\partial D}\frac{\partial C}{\partial E}\frac{\partial D}{\partial F} + \frac{\partial A}{\partial D}\frac{\partial B}{\partial C}\frac{\partial C}{\partial F}\frac{\partial D}{\partial E} - \frac{\partial A}{\partial C}\frac{\partial B}{\partial D}\frac{\partial C}{\partial F}\frac{\partial D}{\partial E} - \frac{\partial A}{\partial D}\frac{\partial B}{\partial C}\frac{\partial C}{\partial E}\frac{\partial D}{\partial F}; \qquad (2.249\text{a, b})$$

(b) the r.h.s. of (2.248c) leads (2.250a) by implicit differentiation of (2.244a, b; 2.248a, b) to (2.250b):

$$\frac{\partial(A,B)}{\partial(E,F)} = \frac{\partial A}{\partial E}\frac{\partial B}{\partial F} - \frac{\partial A}{\partial F}\frac{\partial B}{\partial E}$$

$$= \left(\frac{\partial A}{\partial C}\frac{\partial C}{\partial E} + \frac{\partial A}{\partial D}\frac{\partial D}{\partial E}\right)\left(\frac{\partial B}{\partial C}\frac{\partial C}{\partial F} + \frac{\partial B}{\partial D}\frac{\partial D}{\partial F}\right) - \left(\frac{\partial A}{\partial C}\frac{\partial C}{\partial F} + \frac{\partial A}{\partial D}\frac{\partial D}{\partial F}\right)\left(\frac{\partial B}{\partial C}\frac{\partial C}{\partial E} + \frac{\partial B}{\partial D}\frac{\partial D}{\partial E}\right); \qquad (2.250\text{a, b})$$

(c) of the eight products in (2.250b) the four that do not (do) repeat A, B, C, D in the numerator coincide with (2.249b) [cancel by pairs]; (iv) this proves the equality of (2.249a) ≡ (2.250a) and hence (2.248c). Thus *the two-dimensional Jacobian (2.244a–d) includes the partial derivative (2.245a–d), and has the properties of skew-symmetry (2.246a–c), inversion (2.247a–c) and product (2.248a–c).* The prop-erties of Jacobians can be used to relate thermodynamic derivatives (subsections 2.3.10–2.3.15).

2.3.10 SPECIFIC HEATS AT CONSTANT VOLUME AND PRESSURE

In an isochoric (isobaric) thermodynamic process, that is at constant volume (2.226a) ≡ (2.251a) [pressure (2.228a) ≡ (2.251d)] the heat exchanged is specified by the internal energy (2.226c, d) ≡ (2.251b, c) [enthalpy (2.228b, c) ≡ (2.251e, f):

$$dV = 0: \quad dQ = TdS = dU, \quad dp = 0: \quad dQ = TdS = dH, \qquad (2.251\text{a–f})$$

suggesting the: (i) definition of **specific heat at constant volume (pressure)** as the heat dQ needed for (2.252a) [(2.252d)] a temperature change dT at constant volume (pressure); (ii) that coincides with temperature times the rate-of-change of entropy with temperature (2.252b) [(2.252e)] at constant volume (pressure); (iii) that also coincides with the derivative of the internal energy (2.252c) [enthalpy (2.252f)] with regard to temperature, at constant volume (pressure):

$$C_V \equiv \left(\frac{dQ}{dT} \right)_V = T \left(\frac{\partial S}{\partial T} \right)_V = \left(\frac{\partial U}{\partial T} \right)_V, \quad C_p \equiv \left(\frac{dQ}{dT} \right)_p = T \left(\frac{\partial S}{\partial T} \right)_p = \left(\frac{\partial H}{\partial T} \right)_p \cdot (2.252\text{a–f})$$

The third thermodynamic derivative for a basic thermodynamic system is chosen to be the **coefficient of thermal expansion** as the rate of volume change with temperature at constant pressure (2.253a) that coincides with (2.253b) by (2.243d):

$$K_p \equiv \frac{1}{V} \left(\frac{\partial V}{\partial T} \right)_p = -\frac{1}{V} \left(\frac{\partial S}{\partial p} \right)_T ; \quad K_T \equiv -\frac{1}{V} \left(\frac{\partial V}{\partial p} \right)_T = \frac{1}{\beta V}, \qquad (2.253\text{a–d})$$

the **coefficient of isothermal compression** (2.253c) is minus the rate of volume change with the pressure at constant temperature, and equals (2.253d) in terms of the coefficient of isothermal expansion (2.164c).

*For a basic thermodynamic system (subsection 2.2.14) there are only 3 independent thermodynamic derivatives: (i) the constitutive parameters (2.254c) specific heat at constant volume (2.134a–d) ≡ (2.158a, b) ≡ (2.252a–c), the **thermal pressure coefficient** (2.160c, d) ≡ (2.254a, b) and the coefficient of isothermal compression (2.164c) ≡ (2.253d):*

$$\alpha \equiv \left(\frac{\partial p}{\partial T} \right)_V = \left(\frac{\partial S}{\partial V} \right)_T : \quad \{C_V, \alpha, \beta\} \leftrightarrow \{C_V, C_p, K_p\}; \qquad (2.254\text{a–d})$$

(ii) as an alternative in the sequel are chosen as the three independent thermodynamic derivatives (2.254d) the specific heats at constant volume (2.252a–c) and pressure (2.252d–f) and the coefficient of thermal expansion (2.253a, b). It follows that K_T, α, β are each expressible (Section 2.3.11) as functions of $\{C_V, C_p, K_p\}$ in (2.254d).

2.3.11   COEFFICIENTS OF THERMAL EXPANSION AND ISOTHERMAL COMPRESSION

The coefficient of isothermal compression (2.253d) is related to the three independent thermodynamic derivatives (2.254d) by starting with (2.252b) ≡ (2.255a):

$$
\begin{aligned}
C_V &\equiv T\left(\frac{\partial S}{\partial T}\right)_V = T\frac{\partial(S,V)}{\partial(T,V)} = T\frac{\partial(S,V)}{\partial(T,p)}\frac{\partial(T,p)}{\partial(T,V)} \\
&= T\left(\frac{\partial p}{\partial V}\right)_T\left[\left(\frac{\partial S}{\partial T}\right)_p\left(\frac{\partial V}{\partial p}\right)_T - \left(\frac{\partial S}{\partial p}\right)_T\left(\frac{\partial V}{\partial T}\right)_p\right] \\
&= T\left(\frac{\partial S}{\partial T}\right)_p + T\left[\left(\frac{\partial V}{\partial T}\right)_p\right]^2\left[\left(\frac{\partial V}{\partial p}\right)_T\right]^{-1} = C_p + T\frac{(K_p V)^2}{(-K_T V)} = C_p - \frac{TVK_p^2}{K_T} \quad \text{(2.255a–h)} \\
&= C_p - T\beta V^2 K_p^2,
\end{aligned}
$$

where were used: (i) first (2.245d) in (2.255b); (ii) second (2.248c) in (2.255c); (iii) third (2.244d) [(2.245d)] in the second (first) factor in (2.255d); (iv) fourth (2.247c) [(2.243d)] in the first (second) term of (2.255e); (v) fifth (2.252e) [(2.253a, c)] in the first (second) term of (2.255f), that simplifies to (2.255g); (vi) sixth substitution of (2.253d) leads to (2.255h).

The coefficient of isothermal expansion (2.253d) [compression (2.253c)] have already been related by (2.255h) [(2.255g)] to the three independent thermodynamic derivatives (2.254d). It remains to consider the coefficient of isothermal compression (2.254a, b) that is related to (2.254d) by:

$$
\begin{aligned}
\alpha &\equiv \frac{\partial(p,V)}{\partial(T,V)} = \frac{\partial(p,V)}{\partial(p,T)}\frac{\partial(p,T)}{\partial(T,V)} = -\left(\frac{\partial V}{\partial T}\right)_p\frac{\partial(p,T)}{\partial(V,T)} \\
&= -K_p V\left(\frac{\partial p}{\partial V}\right)_T = \frac{-K_p V}{-K_T V} = \frac{K_p}{K_T} = \beta V K_p, \quad \text{(2.256a–h)}
\end{aligned}
$$

where were used: (i) first (2.245d) in (2.256a); (ii) second (2.248c) in (2.256b); (iii) third (2.245d) and (2.246b) in (2.256c); (iv) fourth (2.253a) and (2.245d) in (2.256d); (v) fifth (2.253c) in (2.256f), simplifying to (2.256g); (vi) sixth (2.253d) in (2.256h).

It has been shown that *taking as the 3 independent thermodynamic derivatives (2.254d) the specific heats at constant volume (2.252a–c) and pressure (2.252d–f) and the coefficient of thermal expansion (2.253a) they specify: (i/ii) the coefficient of isothermal compression (2.253c) [expansion (2.253d)] by (2.255g) ≡ (2.257a) [(2.255h) ≡ (2.257b)]:*

$$
K_T = \frac{K_p^2 TV}{C_p - C_V}; \quad \beta = \frac{C_p - C_V}{TV^2 K_p^2}; \quad \alpha = \frac{K_p}{K_T} = \beta V K_p = \frac{C_p - C_V}{K_p TV}; \quad \text{(2.257a–e)}
$$

(iii) the coefficient of isothermal compression (2.254a, b) by (2.256g, h) ≡ (2.257c, d) or (2.257e) using (2.257a) [(2.257b)] in (2.257c) [(2.257d)]. The adiabatic and

isothermal sound speeds are also specified by thermodynamic derivatives (subsection 2.3.12).

2.3.12 ADIABATIC AND ISOTHERMAL SOUND SPEEDS

The **mass density** is the inverse (2.258a) of the **specific volume**, and the sound speed (subsection II.2.1.4) can be calculated in adiabatic (isothermal) conditions [notes III.7.14 and IV.7.16 (subsections IV.5.5.24–IV.5.5.25)]. The square of the **adiabatic (isothermal) sound speed** (2.258b) ≡ (III.7.301b) ≡ (IV.7.388b) [(2.258d) ≡ (III.5.130a)] is the derivative of the pressure with regard to the mass density at constant entropy (temperature):

$$\rho \equiv \frac{dm}{dV} \equiv \frac{1}{\vartheta}: \quad (c_s)^2 \equiv \left(\frac{\partial p}{\partial \rho}\right)_S = -\vartheta^2 \left(\frac{\partial p}{\partial \vartheta}\right)_S, \quad (c_T)^2 \equiv \left(\frac{\partial p}{\partial \rho}\right)_T = -\vartheta^2 \left(\frac{\partial p}{\partial \vartheta}\right)_T; \quad (2.258\text{a–e})$$

the derivatives with regard to the mass density (2.258b) [(2.258d)] in the adiabatic (isothermal) sound speed can be replaced by (2.258c) [2.258e)] derivatives with regard to the specific volume using (2.258a) in (2.259a–d):

$$\frac{\partial}{\partial \rho} = \frac{\partial \vartheta}{\partial \rho} \frac{\partial}{\partial \vartheta} = \left(\frac{\partial \rho}{\partial \vartheta}\right)^{-1} \frac{\partial}{\partial \vartheta} = \left[\frac{\partial \left(\frac{1}{\vartheta}\right)}{\partial \vartheta}\right]^{-1} \frac{\partial}{\partial \vartheta} = -\vartheta^2 \frac{\partial}{\partial \vartheta} = -\vartheta V \frac{\partial}{\partial V}. \qquad (2.259\text{a–e})$$

In the adiabatic sound speed (2.258c) appears the thermodynamic derivative:

$$\left(\frac{\partial p}{\partial V}\right)_S = \frac{\partial(p,S)}{\partial(V,S)} = \frac{\partial(p,S)}{\partial(p,T)} \frac{\partial(p,T)}{\partial(V,T)} \frac{\partial(V,T)}{\partial(V,S)} = \left(\frac{\partial S}{\partial T}\right)_p \left(\frac{\partial p}{\partial V}\right)_T \left(\frac{\partial T}{\partial S}\right)_V$$

$$= \frac{C_p}{T} \times \frac{1}{-K_T V} \times \frac{T}{C_V} = -\frac{C_p}{C_V K_T V} = -\frac{C_p}{C_V} \frac{C_p - C_V}{K_p^2 V^2 T}, \qquad (2.260\text{a–f})$$

where were used: (i) first (2.245d) in (2.260a); (ii) second twice (2.248c) in (2.260b); (iii) third (2.245d) three times in (2.260c); (iv) fourth (2.252b, e) and (2.253c) in (2.260d) that simplifies to (2.260e); (v) fifth (2.257a) in (2.260f). In the isothermal sound speed (2.258e) appears the thermodynamic derivative (2.261a, b) ≡ (2.253c, d):

$$\left(\frac{\partial p}{\partial V}\right)_T = \left[\left(\frac{\partial V}{\partial p}\right)_T\right]^{-1} = -\frac{1}{K_T V} = -\frac{C_p - C_V}{K_p^2 V^2 T}, \qquad (2.261\text{a–c})$$

and (2.257a) leads to (2.261c).

Substituting (2.260e–f) [(2.261b, c)] in (2.258b, c) [(2.258d, e)] follows that *the adiabatic (isothermal) sound speed is given by (2.258b, c) ≡ (2.262a, b) [(2.258d, e)*

$\equiv (2.263a, b)]$ *and specified by (2.262c, d) [(2.263c, d)] in terms of the independent thermodynamic derivatives (2.254d; 2.257a):*

$$\left(c_s\right)^2 = \left(\frac{\partial p}{\partial \rho}\right)_S = -\vartheta V\left(\frac{\partial p}{\partial V}\right)_S = \frac{C_p}{C_V}\frac{\vartheta}{K_T} = \frac{C_p}{C_V}\frac{C_p - C_V}{K_p^2 T}\frac{\vartheta}{V}, \qquad (2.262a\text{--}d)$$

$$\left(c_T\right)^2 = \left(\frac{\partial p}{\partial \rho}\right)_T = -\vartheta V\left(\frac{\partial p}{\partial V}\right)_T = \frac{\vartheta}{K_T} = \frac{C_p - C_V}{K_p^2 T}\frac{\vartheta}{V}, \qquad (2.263a\text{--}d)$$

and their ratio (2.264a) equals the ratio of specific heats at constant pressure and volume (2.264b) that defines the **adiabatic exponent***:*

$$\left(\frac{c_s}{c_T}\right)^2 = \frac{C_p}{C_V} = \gamma; \quad R \equiv C_p - C_V = C_V\left(\gamma - 1\right) = \left(1 - \frac{1}{\gamma}\right)C_p, \qquad (2.264a\text{--}e)$$

the adiabatic exponent together with the **difference of specific heats** *(2.264c) provide an alternative representation of the specific heats at constant volume (2.264d) and pressure (2.264e).* Two more thermodynamic derivatives are considered (subsection 2.3.13) to complete the set of 12 distinct thermodynamic derivatives of a basic thermodynamic system (subsections 2.3.14–2.3.15).

2.3.13 NON-ADIABATIC PRESSURE AND VOLUME COEFFICIENTS

Two more examples of thermodynamic derivatives are the **non-adiabatic pressure (volume) coefficient** defined as the derivative of the entropy with regard to the pressure at constant volume (2.265a) [vice-versa (2.266a)]. The non-adiabatic pressure coefficient is given by (2.265a–f),

$$S_p \equiv \left(\frac{\partial S}{\partial p}\right)_V = \frac{\partial(S,V)}{\partial(p,V)} = \frac{\partial(S,V)}{\partial(T,V)}\frac{\partial(T,V)}{\partial(p,V)} = \frac{C_V}{T}\left(\frac{\partial T}{\partial p}\right)_V = \frac{C_V}{\alpha T} = \frac{C_V K_p V}{C_p - C_V}, (2.265a\text{--}f)$$

where were used: (i) first (2.245d) in (2.265b); (ii) second (2.248c) in (2.265c); (iii) third (2.252b) and (2.245d) in (2.265d); (iv) fourth (2.254a) in (2.265e); (v) fifth (2.257e) in (2.265f). The derivative of the entropy with regard to the volume at constant pressure is given by (2.266a–e):

$$S_V \equiv \left(\frac{\partial S}{\partial V}\right)_p = \frac{\partial(S,p)}{\partial(V,p)} = \frac{\partial(S,p)}{\partial(T,p)}\frac{\partial(T,p)}{\partial(V,p)} = \left(\frac{\partial S}{\partial T}\right)_p\left(\frac{\partial T}{\partial V}\right)_p = \frac{C_p}{K_p T V}, \qquad (2.266a\text{--}e)$$

where were used: (i) first (2.245d) in (2.266b); (ii) second (2.248c) in (2.266c); (iii) third (2.245d) in (2.266d); (iv) fourth (2.252e) and (2.253a) in (2.266e). Thus *the non-adiabatic pressure (volume) coefficient defined as derivative of the entropy with*

regard to the pressure (volume) at constant volume (pressure) is given by (2.265a–f)
[(2.266a–e)] in terms of the independent thermodynamic derivatives (2.254d). The
preceding thermodynamic coefficients (Table 2.4) are sufficient to (subsections
2.3.7–2.3.12) determine all 12 distinct thermodynamic derivatives of a basic thermodynamic system (subsection 2.3.14) in terms of the set of 3 independent thermodynamic derivatives (2.254d).

2.3.14 Twelve Non-Inverse Thermodynamic Derivatives

A basic thermodynamic system has **12 non-inverse thermodynamic derivatives**
because: (i) each of the four thermodynamic variables (S, V, T, p) can be differentiated with regard to 3 others leading to $4 \times 3 = 12$; (ii) each derivative can be taken
with one of the two remaining variables as constant, leading to $2 \times 12 = 24$; (iii) the
inverse derivatives are not considered as distinct on account of (2.247a–c) leading
back to $24 \div 2 = 12$. The method of identification of the 12 thermodynamic derivatives (2.267a–l):

$$d_{1-12} \equiv \left\{ \begin{array}{l} \left(\dfrac{\partial S}{\partial T}\right)_V, \left(\dfrac{\partial S}{\partial T}\right)_p, \left(\dfrac{\partial S}{\partial p}\right)_T, \left(\dfrac{\partial S}{\partial p}\right)_V, \left(\dfrac{\partial S}{\partial V}\right)_T, \left(\dfrac{\partial S}{\partial V}\right)_p, \\[4mm] \left(\dfrac{\partial T}{\partial p}\right)_S, \left(\dfrac{\partial T}{\partial p}\right)_V, \left(\dfrac{\partial T}{\partial V}\right)_S, \left(\dfrac{\partial T}{\partial V}\right)_p, \left(\dfrac{\partial p}{\partial V}\right)_S, \left(\dfrac{\partial p}{\partial V}\right)_T, \end{array} \right\} \qquad (2.267a–l)$$

can be explained (Table 2.5) as follows: (i) consider the four thermodynamic
variables in some order, for example (S, T, p, V); (ii) the entropy can be differentiated with regard to temperature at constant volume (2.267a) or pressure
(2.267b); (iii) the entropy can be differentiated with regard to pressure at constant temperature (2.267c) or volume (2.267d); (iv) the entropy may be differentiated with regard to volume at constant temperature (2.267e) or pressure (2.267f);
(v) there are no more derivatives of the entropy besides (2.267a–f); (vi) the derivatives with regard to entropy are the inverses of (2.267a–f) and thus excluded;
(vii) the temperature can be differentiated with regard to pressure at constant
entropy (2.267g) or volume (2.267h); (viii) the temperature can be differentiated
with regard to volume at constant entropy (2.267i) or pressure (2.267j); (ix) the
derivatives with regard to temperature are inverses of the preceding and thus
excluded; (x) the pressure can be differentiated with regard to volume at constant
entropy (2.267k) or temperature (2.267l); (xi) other derivatives of the pressure
are inverses of the preceding and thus excluded; (xii) all derivatives of the volume are inverses of the preceding and thus there is none to add to the list of
twelve (2.267a–l). It has been shown that *a basic thermodynamic system has 12*
distinct non-inverse thermodynamic derivatives, for example, the set (2.267a–l),
all of which can be expressed in terms of the independent set of three (2.254d),
as shown next (subsection 2.3.15).

TABLE 2.4
List Of Thermodynamic Derivatives

Definition	Inequalities	Name	General matter	Ideal gas*
$C_v = T\left(\dfrac{\partial S}{\partial T}\right)_v = \left(\dfrac{dQ}{dT}\right)_v = \left(\dfrac{\partial U}{\partial T}\right)_v$	$C_v > 0$	Specific heat at constant volume	$C_v = \dfrac{R}{\gamma-1}$	$C_v = \dfrac{fR}{2}$
$C_p = T\left(\dfrac{\partial S}{\partial T}\right)_p = \left(\dfrac{dQ}{dT}\right)_p = \left(\dfrac{\partial H}{\partial T}\right)_p$	$C_p > C_v > 0$	Specific heat at constant volume pressure	$C_p = \dfrac{\gamma R}{\gamma-1} = \gamma C_v = C_v + R$	$C_p = \dfrac{f+2}{2}R$
$K_p = \dfrac{1}{V}\left(\dfrac{\partial V}{\partial T}\right)_p = -\dfrac{1}{V}\left(\dfrac{\partial S}{\partial p}\right)_v$	$K_p\alpha < 0$	Coefficient of thermal expansion	K_p	$K_p = \dfrac{1}{T}$
$K_T = -\dfrac{1}{V}\left(\dfrac{\partial V}{\partial p}\right)_T = \dfrac{1}{\beta V}$	$K_T > 0$	Coefficient of isothermal compression	$K_T = \dfrac{K_p^2 TV}{C_p - C_v}$	$K_T = \dfrac{1}{p}$
$\alpha = \left(\dfrac{\partial p}{\partial T}\right)_v = \left(\dfrac{\partial S}{\partial V}\right)_T$	$K_p\alpha < 0$	Thermal pressure coefficient	$\alpha = \dfrac{C_p - C_v}{K_p TV}$	$\alpha = \dfrac{R}{V}$
$\beta = -\left(\dfrac{\partial p}{\partial V}\right)_T = \dfrac{E}{3(1-2\sigma)}$	$\beta > 0$	Coefficient of isothermal expansion	$\beta = \dfrac{C_p - C_v}{K_p^2 V^2 T}$	$\beta = \dfrac{p}{V}$
$(c_s)^2 = -\left(\dfrac{\partial p}{\partial \rho}\right)_s = -9V\left(\dfrac{\partial p}{\partial V}\right)_s$	$(c_s)^2 > (c_T)^2$	Adiabatic sound speed	$(c_s)^2 = \dfrac{C_p}{C_v}\dfrac{C_p - C_v}{K_p^2 T}$	$(c_s)^2 = \gamma pV = \gamma RT$
$(c_T)^2 = -\left(\dfrac{\partial p}{\partial \rho}\right)_T = -9V\left(\dfrac{\partial p}{\partial V}\right)_T$	$(c_T)^2 > 0$	Isothermal sound speed	$(c_T)^2 = \dfrac{C_p - C_v}{K_p^2 T}$	$(c_T)^2 = pV = RT$

$S_p = \left(\dfrac{\partial S}{\partial p}\right)_v$	Non-adiabatic pressure coefficient	$S_v S_p > 0$	$S_p = \dfrac{C_v K_p V}{C_p - C_v}$	$S_p = \dfrac{C_v}{p}$
$S_v = \left(\dfrac{\partial S}{\partial V}\right)_p$	Non-adiabatic volume coefficient	$S_p S_v > 0$	$S_v = \dfrac{C_p}{K_p TV}$	$S_v = \dfrac{C_p}{V}$
$\gamma = \dfrac{C_p}{C_v}$	Adiabatic exponent	$\gamma > 1$	$\gamma = 1 + \dfrac{R}{C_v}$	$\gamma = 1 + \dfrac{2}{f}$
$R = C_p - C_v$	Gas constant	$R > 0$	$R = (\gamma - 1)C_v = \left(1 - \dfrac{1}{\gamma}\right)C_p$	$R = \dfrac{8.316}{M}$ JK^{-1}mole^{-1}

* M molecular mass; f-number of degrees-of-freedom of molecules

Note: A basic thermodynamic system is specified by two pairs of extensive – intensive parameters, namely mechanical (volume – pressure) and thermal (entropy – temperature); this allows the definition of twelve thermodynamic derivatives, that are expressible in terms of only three of them that are independent, and simplify for an ideal gas; the thermodynamic coefficients arise from the first principle of thermodynamics, and the inequalities they satisfy can be proved from the second principle of thermodynamics.

TABLE 2.5
Twelve Non-Inverse Thermodynamic Derivatives

Number	Thermodynamic Derivative	Sign	General Matter	Ideal Gas*
d_1	$\left(\dfrac{\partial S}{\partial T}\right)_v$	>0	$\dfrac{C_v}{T}$	$\dfrac{fR}{2T}$
d_2	$\left(\dfrac{\partial S}{\partial T}\right)_p$	>0	$\dfrac{C_p}{T}$	$\dfrac{f+2}{2}\dfrac{R}{T}$
$d_3 = -\dfrac{1}{d_{10}}$	$\left(\dfrac{\partial S}{\partial p}\right)_T = -\left(\dfrac{\partial V}{\partial T}\right)_p$	(1)	$-K_p V$	$-\dfrac{R}{p}$
$d_4 = -\dfrac{1}{d_9}$	$\left(\dfrac{\partial S}{\partial p}\right)_v = -\left(\dfrac{\partial V}{\partial T}\right)_s$	(2)	$\dfrac{C_v K_p V}{C_p - C_v}$	$\dfrac{C_v}{p} = \dfrac{fR}{2p}$
$d_5 = \dfrac{1}{d_8}$	$\left(\dfrac{\partial S}{\partial V}\right)_T = \left(\dfrac{\partial p}{\partial T}\right)_v = \alpha$	(3)	$\dfrac{C_p - C_v}{K_p TV}$	$\dfrac{R}{V}$
$d_6 = \dfrac{1}{d_7}$	$\left(\dfrac{\partial S}{\partial V}\right)_p = \left(\dfrac{\partial p}{\partial T}\right)_s$	(4)	$\dfrac{C_p}{K_p TV}$	$\dfrac{C_p}{V} = \dfrac{f+2}{2}\dfrac{R}{V}$

d_{11} $\left(\dfrac{\partial p}{\partial V}\right)_s = -\dfrac{(c_s)^2}{9V}$ >0 $-\dfrac{C_p}{C_v}\dfrac{C_p - C_v}{K_p^2 V^2 T}$ $-\gamma\dfrac{p}{V}$

d_{12} $\left(\dfrac{\partial p}{\partial V}\right)_T = -\dfrac{(c_T)^2}{9T}$ >0 $-\dfrac{C_p - C_v}{K_p^2 V^2 T}$ $-\dfrac{p}{V}$

* f-number of degrees of freedom of a molecule

f=3 monatomic

f=5 diatomic (or polyatomic with atoms in a row)

f=6 polyatomic (with atomic not in a row)

(1-4) all have the same sign

Note: For a basic thermodynamic system (Table 2.4) there are twenty-four thermodynamic derivatives, consisting of twelve inverse pairs; taking as the variables entropy, temperature, pressure and volume, a thermodynamic derivative differentiates a first variable with regard to a second variable with a third variable constant; the twelve thermodynamic derivatives (Table 2.5) relate to the twelve thermodynamic parameters (Table 2.4), and thus are expressible in terms of three of them, simplify for ideal gases, and satisfy inequalities arising from the second principle of thermodynamics.

2.3.15   THREE INDEPENDENT THERMODYNAMIC DERIVATIVES

Of the twelve non-inverse thermodynamic derivatives (2.267a–l): (i/ii) the first two
(2.267a) [(2.267b)] are specified by the specific heat at constant volume (2.252b) ≡
(2.268a) [pressure (2.252e) ≡ (2.268b)]:

$$d_1:\left(\frac{\partial S}{\partial T}\right)_V = \frac{C_V}{T}, \quad d_2:\left(\frac{\partial S}{\partial T}\right)_p = \frac{C_p}{T}, \quad d_3 = -\frac{1}{d_{10}}:\left(\frac{\partial S}{\partial p}\right)_T = -\left(\frac{\partial V}{\partial T}\right)_p = -K_p V; \quad (2.268\text{a–d})$$

(iii) the third thermodynamic derivative (2.267c) equals another thermodynamic
derivative (2.243d) ≡ (2.268c) and both are specified (2.253a) ≡ (2.268d) by the
coefficient of thermal expansion, completing the set (2.254d) of 3 independent thermodynamic derivatives, and accounting for four derivatives (d_1, d_2, d_3, d_{10}) in (2.267a,
b, c, j); (iv/v) the fourth (2.267d) [fifth (2.267e)] derivative equals another derivative
(2.267d) ≡ (2.243a) ≡ (2.269a, b) [(2.267e) ≡ (2.243b) ≡ (2.269e, f)] leading to
(2.265f) ≡ (2.269c, d) [(2.254a) ≡ (2.257e) ≡ (2.269g, h)]:

$$d_4 = -\frac{1}{d_9}: \quad \left(\frac{\partial S}{\partial p}\right)_V = -\left(\frac{\partial V}{\partial T}\right)_S = S_p = \frac{C_V K_p V}{C_p - C_V}, \qquad (2.269\text{a–d})$$

$$d_5 = \frac{1}{d_8}: \quad \left(\frac{\partial S}{\partial V}\right)_T = \left(\frac{\partial p}{\partial T}\right)_V = \alpha = \frac{C_p - C_V}{K_p T V}, \qquad (2.269\text{e–h})$$

thus accounting for four additional thermodynamic derivatives (d_4, d_5, d_8, d_9) in
(2.267d, e, h, i); (vi/vii) the sixth (2.267f) [seventh (2.267g)] thermodynamic derivative equals another thermodynamic derivative (2.243c) ≡ (2.270a, b) [(2.243c) ≡
(2.270e)] leading to (2.266e) ≡ (2.270c, d) [(2.266e) ≡ (2.270f, g)]:

$$d_6 = \frac{1}{d_7}: \quad \left(\frac{\partial S}{\partial V}\right)_p = \left(\frac{\partial p}{\partial T}\right)_S = S_V = \frac{C_p}{K_p T V}, \quad \left(\frac{\partial T}{\partial p}\right)_S = \left(\frac{\partial V}{\partial S}\right)_p = \frac{1}{S_V} = \frac{K_p T V}{C_p}, \quad (2.270\text{a–g})$$

and since (2.270b, c, d) and (2.270e, f, g) are inverses of each other, they specify
(2.270a) two new derivatives (d_6, d_7) in (2.267f, g); (viii/ix) the remaining two thermodynamic derivatives d_{11} (d_{12}) in (2.267k) [(2.267l)] relate to the adiabatic (2.262a–d)
≡ (2.271a, b) [isothermal (2.263a–d) ≡ (2.271e, f) and (2.253d) ≡ (2.271d)] sound
speeds:

$$d_{11}:\left(\frac{\partial p}{\partial V}\right)_S = -\frac{(c_s)^2}{9V} = -\frac{C_p}{C_V}\frac{C_p - C_V}{K_p^2 V^2 T},$$

$$d_{12}:\left(\frac{\partial p}{\partial V}\right)_T = -\beta = -\frac{(c_T)^2}{9V} = -\frac{C_p - C_V}{K_p^2 V^2 T}. \qquad (2.271\text{a–e})$$

Thus *the 12 non-inverse thermodynamic derivatives (2.267a–l) for a basic thermodynamic system are expressed in terms of the 3 independent thermodynamic*

derivatives (2.254d): (i/ii) by (2.267a) ≡ (2.268a) [(2.267b) ≡ (2.268b)] defining the specific heat at constant volume (2.252a–c) [pressure (2.252d–f)]; (iii–iv) by (2.267c) ≡ (2.267j) ≡ (2.268c, d) the coefficient of thermal expansion (2.253a, b); (v–viii) by (2.267d) ≡ (2.267i) ≡ (2.269a, b) [(2.267e) ≡ (2.267h) ≡ (2.269e, f)] involving in (2.269c, d) [2.269g, h)] the non-adiabatic pressure coefficient (2.265a–f) [the coefficient of isothermal compression (2.254a, b)]; (ix/x) by (2.267f) ≡ (2.267g) ≡ (2.270a–g) that are inverses apart from sign; (xi/xii) by (2.267k) ≡ (2.271a, b) [(2.267l) ≡ (2.271c–e)] involving the adiabatic (isother-mal) sound speed. The detailed consideration of the thermodynamic derivatives (subsections 2.3.8–2.3.15) of a basic thermodynamic system (2.3.1–2.3.7) leads to the second (third) principles of thermodynamics [subsections 2.3.16–2.3.23 (2.3.28–2.3.29)].

2.3.16 THE SECOND PRINCIPLE OF THERMODYNAMICS AND ENTROPY GROWTH

The first principle of thermodynamics (2.221d) implies that the internal energy is a function of entropy and volume U (S, V) and is stationary at equilibrium, that is has zero first-order differential (2.272a); *the **principle of stability** states that for the equilibrium to be stable the internal energy must be minimum, and thus at lowest order the second-order differential must be positive (2.272b). The **principle of minimum internal energy** (2.272a, b) will be shown (subsections 2.3.17–2.3.24) to imply a **principle of maximum entropy** (2.272c, d).*

$$\text{stability:} \quad dU = 0 < d^2U \Rightarrow dS = 0 > d^2S \qquad (2.272a\text{–}d)$$

*The principle of maximum entropy (2.272c, d) relates to the **second principle of thermodynamics** that can be stated: for an isolated system the entropy cannot decrease, leading to two possibilities: (i) for a **reversible process**, consisting of equilibrium states, there is no change in entropy (2.273a); (ii) for an **irreversible process**, including at least some non-equilibrium states, the entropy must grow (2.273b):*

$$\text{isolated system:} \quad dS \geq 0 \begin{cases} dS = 0 \text{ for reversible process,} \\ dS > 0 \text{ for irreversible process.} \end{cases} \qquad (2.273a, b)$$

An **isolated system** is not in contact with any other system thus excluding exchanges of heat and/or work. If a system is not isolated, then all the systems with which it interacts can be added, until there are neither heat nor work exchanges with the exterior, and the system extended in this way becomes isolated. The second principle of thermodynamics may not hold for a non-isolated system, for example: (i) the internal energy may increase if work is supplied; (ii) the entropy may reduce if heat is extracted. Neither (i) nor (ii) apply to an isolated system.

Since for an isolated system the internal energy has a minimum (2.272a, b) as a function of the entropy and specific volume, its second-order differential (2.274a) is a quadratic form in two variables that must be positive-definite (2.274b), leading to the necessary (2.202a, b) $\equiv$ (2.274c, d) [and sufficient (2.205a, b; 2.200c, f) $\equiv$ (2.274c, e)] conditions:

$$d^2U = \left(\frac{\partial^2 U}{\partial S^2}\right)_V (dS)^2 + 2\frac{\partial^2 U}{\partial S \partial V}dSdV + \left(\frac{\partial^2 U}{\partial V^2}\right)_S (dV)^2 > 0:$$

$$U_{11} \equiv \left(\frac{\partial^2 U}{\partial S^2}\right)_V > 0 < \left(\frac{\partial^2 U}{\partial V^2}\right)_S \equiv U_{22}, \quad U_{12} \equiv \left(\frac{\partial^2 U}{\partial S^2}\right)_V \left(\frac{\partial^2 U}{\partial V^2}\right)_S - \left(\frac{\partial^2 U}{\partial V \partial S}\right)^2 > 0.$$

(2.274a–e)

The necessary and sufficient conditions (2.274c–e) for minimum internal energy (2.274a, b) lead (subsedctions 2.3.18–2.3.19) to thermodynamic inequalities (subsection 2.3.20) that are compared (subsection 2.3.23) with those derived from a principle of maximum entropy (subsections 2.3.21–2.3.22). The principle of maximum entropy states that the second-order differential of the entropy as a function of the internal energy and specific volume (2.225a) is a negative-definite quadratic form (2.275b) with two variables, with necessary (2.207a, b) $\equiv$ (2.275c, d) [and sufficient (2.208a–c) $\equiv$ (2.275c, e)] conditions:

$$d^2S = \left(\frac{\partial^2 S}{\partial U^2}\right)_V (dU)^2 + 2\frac{\partial^2 S}{\partial V \partial U}dUdV + \left(\frac{\partial^2 S}{\partial V^2}\right)_U (dU)^2 < 0:$$

$$S_{11} \equiv \left(\frac{\partial^2 S}{\partial U^2}\right)_V < 0 > \left(\frac{\partial^2 S}{\partial V^2}\right)_U \equiv S_{22}, \quad S_{12} \equiv \left(\frac{\partial^2 S}{\partial U^2}\right)_V \left(\frac{\partial^2 S}{\partial V^2}\right)_U - \left(\frac{\partial^2 S}{\partial V \partial U}\right)^2 > 0.$$

(2.275a–e)

The condition that the second-order differential is a definite quadratic form, namely positive (2.272b) [negative (2.272d)] for the internal energy (entropy), generally does not apply to other functions of state; for example, the free energy (subsection 2.2.5) has a second-order differential that is an indefinite quadratic form (subsection 2.2.29) and this is also the case for the enthalpy but not for free enthalpy that has a maximum (subsection 2.3.23). The formal proof (subsections 2.3.18–2.3.23) of the implications (2.274a–e; 2.275a–e) is preceded (subsection 2.2.17) by geometric interpretation of (2.272a–d) in terms of the shape of thermodynamic surface and the possible paths on it.

2.3.17   PATHS ON THE CONVEX THERMODYNAMIC SURFACE

In a three-dimensional space with coordinates entropy, specific volume and internal energy the thermodynamic state of a system is represented (Figure 2.5) by a point P. An isolated system has constant internal energy and *the locus of equilibrium states is the **thermodynamic surface** of constant internal energy. The second principle of thermodynamics (2.272a–d) implies that the thermodynamic surface must be **convex*** since (Figure 2.5) a decrease in internal energy implies an increase in entropy along the path c. If the thermodynamic state surface were concave at some point (Figure 2.6) a path g could be found with increasing energy and decreasing entropy, which would contradict the above statement of the second principle of thermodynamics.

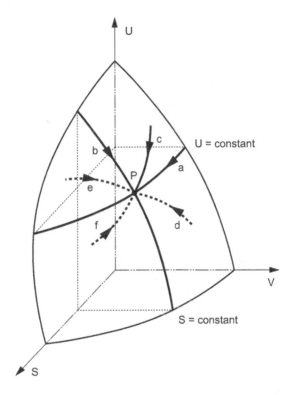

FIGURE 2.5 The second principle of thermodynamics has a geometrical interpretation in the thermodynamic space with variables entropy S, volume V and internal energy U, where the latter is represented by a surface $U(V, S)$. A thermodynamic state is represented by a point P on the internal energy surface, that is convex, so that: (i) the accessible states lie along paths "c" of decreasing internal energy U and increasing entropy S, that is in the sector between $S = const$ and $U = const$; (ii) the states outside this sector are not accessible, because they violate the second principle of thermodynamics, namely the path "d" has increasing internal energy, the path "e" has decreasing entropy and the path "f" both decreasing entropy and increasing internal energy.

Thus *the locus of states of stable thermodynamic equilibrium must be a surface devoid of saddle points, unlike Q in Figure 2.6, that is, must be a convex surface, as shown in Figure 2.5.*

Consider next isolated systems, that is thermodynamic systems not interacting with the exterior, that is exchanging no heat and performing no work. If a thermodynamic system is not isolated, it can be extended to include all other systems with which it interacts, until an isolated system is obtained. The second principle of thermodynamics states that the entropy of an isolated system cannot decrease, and the internal energy cannot increase. If the thermodynamic system undergoes a **reversible process**, defined as consisting of a succession of equilibrium states, the internal energy is constant, and it maximizes entropy along the path a in Figure 2.5; if the thermodynamic system follows an **irreversible process**, involving intermediate

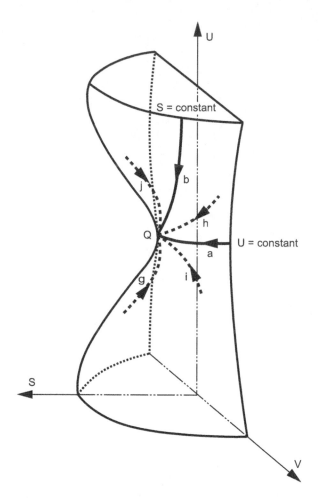

FIGURE 2.6   The second principle of thermodynamics requires that the surface of constant internal energy U in the thermodynamic space of coordinates volume V, entropy S and U, has to be convex (Figure 2.5) and excludes a concave surface (Figure 2.6) that has at least one saddle point Q, because there is no path with increasing entropy and decreasing internal energy: (i) along the path "h" the entropy increases and the internal energy decreases, but the saddle point Q is a minimum of entropy, contradicting the second principle of thermodynamics; (ii–iv) the entropy decreases along paths "j" and "g" and the internal energy increases along path "i". Thus there is no path satisfying the second principle of thermodynamics near the saddle point Q of a concave thermodynamic surface.

non-equilibrium states, for example, in the presence of dissipation, the energy decreases and entropy increases, so that the path c on the thermodynamic surface in Figure 2.5 must lie between those of constant entropy b and constant energy a. Along the path b of constant entropy the internal energy decreases and the volume increases in agreement with the negative sign in (2.242g). Other paths are excluded for an isolated thermodynamic system, because: (d, f) involve increasing energy, which is not possible for an isolated system, i.e. energy is conserved (decreased) in the absence (presence) of dissipation; (e) involves decreasing entropy for an isolated system, which again is not permissible because the entropy must be constant (increase) in a reversible (irreversible) process.

Thus non-isolated systems can follow paths anywhere on the thermodynamic surface, and it is only isolated systems that are confined to the sector bounded by the paths of constant energy and entropy. In the case of a concave thermodynamic surface (Figure 2.6) near a **saddle point** Q the internal energy would decrease, and the entropy would increase along a path h in the sector between the paths a of constant internal energy and b of constant entropy, but Q would be a minimum of the entropy, contradicting the second principle of thermodynamics. Along the path h both the volume and internal energy would decrease contradicting (2.242g). The paths g, i have increasing energy and j has decreasing entropy, so none is possible for an isolated system. The convex thermodynamic surface (Figure 2.6) cannot extend to infinity, so the internal energy, specific volume and entropy must all three be finite. An alternative formulation of the second principle of thermodynamics states that **(Caratheodory 1909)** *in the neighbourhood of a thermodynamic state there are inaccessible states. The geometrical representation using the convex surface of constant internal energy (Figure 2.5) identifies precisely those neighbourhoods: (i) accessible between the lines of constant entropy b and constant internal energy a, that is the path c in the sector (a, b); (ii) inaccessible outside this sector.* The geometrical interpretation of the principle of stability (2.272a–d) is followed by the implications (2.274c–e) of minimum internal energy (2.274a, b) as regards the sign of thermodynamic derivatives (Sections 2.3.18–2.3.19).

2.3.18 STABLE EQUILIBRIUM AND MINIMUM INTERNAL ENERGY

The condition of stable equilibrium is that the internal energy has a minimum (2.272a, b), and hence at lowest-order its second-order differential must be positive definite (2.274a, b); the necessary (2.274c, d) [and sufficient (2.274c, e)] conditions involve 3 thermodynamic derivatives evaluated next. The first necessary and sufficient condition of stability (2.274c) $\equiv$ (2.276a, b) concerns the second-order derivative of the internal energy with regard to the entropy at constant volume (2.276b), that is evaluated by (2.242a) $\equiv$ (2.276c) by specific heat at constant volume (2.252b) $\equiv$ (2.276d) and must be positive (2.276d):

$$0 < U_{11} \equiv \left(\frac{\partial^2 U}{\partial S^2} \right)_V = \left(\frac{\partial T}{\partial S} \right)_V = \frac{T}{C_V}, \qquad (2.276a\text{–}d)$$

and implies that the temperature and specific heat at constant volume must have the same sign, that is, both positive or negative.

In order to eventually exclude the second possibility, is considered next the second necessary condition of stability (2.274d) ≡ (2.277a, b) specified by the second-order derivative of the internal energy with regard to the volume at constant entropy:

$$
\begin{aligned}
U_{22} &\equiv \left(\frac{\partial^2 U}{\partial V^2}\right)_S = -\left(\frac{\partial p}{\partial V}\right)_S = -\frac{\partial(p,S)}{\partial(V,S)} = -\frac{\partial(p,S)}{\partial(p,V)}\frac{\partial(p,V)}{\partial(V,S)} \\
&= -\frac{\partial(p,S)}{\partial(p,T)}\frac{\partial(p,T)}{\partial(p,V)}\frac{\partial(p,V)}{\partial(V,T)}\frac{\partial(V,T)}{\partial(V,S)} \\
&= \left(\frac{\partial S}{\partial T}\right)_p \left(\frac{\partial T}{\partial V}\right)_p \left(\frac{\partial p}{\partial T}\right)_V \left(\frac{\partial T}{\partial S}\right)_V = \frac{C_p}{T}\times\frac{1}{K_p V}\times\alpha\times\frac{T}{C_V} \\
&= \frac{C_p\alpha}{C_V K_p V} = \frac{C_p}{C_V K_T V},
\end{aligned}
\qquad (2.277\text{a–h})
$$

that is evaluated by: (i) first using (2.242g) in (2.277b); (ii) second using (2.245d) in (2.277c); (iii) third using (2.248c) three times in (2.277d); (iv) fourth using (2.245d) four times in (2.277e); (v) fifth times (2.252c, e; 2.253a; 2.254a) in (2.277f) that simplifies to (2.277g); (vi) sixth using (2.257c) leading to (2.277h).

The second sufficient condition of stability (2.274e) involves the second-order mixed derivative of the internal energy with regard to entropy and volume (2.278a):

$$
\begin{aligned}
\frac{\partial^2 U}{\partial V\,\partial S} &= \left(\frac{\partial T}{\partial V}\right)_S = \frac{\partial(T,S)}{\partial(V,S)} = \frac{\partial(T,S)}{\partial(V,T)}\frac{\partial(V,T)}{\partial(V,S)} \\
&= -\left(\frac{\partial S}{\partial V}\right)_T \left(\frac{\partial T}{\partial S}\right)_V = -\frac{T}{C_V}\left(\frac{\partial p}{\partial T}\right)_V = -\frac{\alpha T}{C_V},
\end{aligned}
\qquad (2.278\text{a–f})
$$

that is evaluated by: (i) first using (2.242a) in (2.278); (ii) second using (2.245d) in (2.278b); (iii) third using (2.248c) in (2.278c); (iv) fourth using (2.245d) twice in (2.278d); (v) fifth using (2.252e; 2.243b) in (2.278e); (vi) sixth using (2.254a) in (2.278f). Substituting the three second-order derivatives of the internal energy (2.276d; 2.277g; 2.278f) in the second sufficient stability condition (2.274e) leads to (2.279a–c):

$$
\begin{aligned}
0 < U_{12} &\equiv \frac{T}{C_V}\times\frac{C_p\alpha}{C_V K_p V} - \frac{\alpha^2 T^2}{C_V^2} = \frac{\alpha T}{C_V^2 K_p V}\left(C_p - \alpha K_p VT\right) \\
&= \frac{T}{C_V^2 K_T V}\left(C_p - \frac{K_p^2 VT}{K_T}\right) = \frac{T}{C_V K_T V},
\end{aligned}
\qquad (2.279\text{a–e})
$$

that is evaluated using: (i) first (2.257c) in (2.279d); (ii) second (2.255g) in (2.279e).

The three stability conditions (2.276d) ≡ (2.280a), (2.277h) ≡ (2.280b) and (2.279e) ≡ (2.280c):

$$
\frac{T}{C_V} > 0, \quad \frac{C_p}{C_V K_T V} > 0, \quad \frac{T}{C_V K_T V} > 0: \quad K_T = \frac{K_p}{\alpha} > 0, \quad \frac{C_p - C_V}{\alpha^2 TV} > 0, \quad (2.280\text{a–f})
$$

are taken together to prove: (i) that (2.280a, c) imply (2.280d) and hence (2.257c) also (2.280e); (ii) that (2.280e) and (2.257e) imply (2.280f) leading to (2.281a):

$$\frac{C_p - C_V}{T} > 0; \quad \frac{C_p - C_V}{C_V} > 0 \Rightarrow C_p > C_V, \quad T > 0, \quad C_V > 0, \quad (2.281a\text{-}e)$$

the ratio of (2.281a) and (2.280a) gives (2.281b) that implies (2.281c); using (2.281c) in (2.281a) gives (2.281d) and from (2.280a) follows (2.281e). Thus the stability principle of minimum internal energy (2.274a–e) leads to (2.280d, e; 2.281c, d, e) that specify or relate the signs of all thermodynamic derivatives (subsection 2.3.19).

2.3.19 INEQUALITIES FOR THERMODYNAMIC DERIVATIVES

*The stability principle of minimum internal energy (2.272a, b) ≡ (2.282a, b) implies that: (i) the temperature T, that appears in the first principle of thermodynamics (2.221d), and specifies thermal equilibrium (2.235b), is always positive (2.281d) ≡ (2.282c) and thus is the **absolute temperature**; (ii) the specific heat at constant volume (2.252a–c) is positive (2.281e) ≡ (2.282e) implying that in order to increase the temperature at constant volume it is necessary to supply heat; (iii) the specific heat at constant pressure (2.252d–f) is larger than the specific heat at constant volume (2.281c) ≡ (2.282d), implying that for the same increase in temperature, more heat is needed at constant pressure than at constant volume, because in the case of constant pressure some heat is used to increase the volume, and only the remaining heat is available to raise the temperature; (iv) the coefficient of isothermal compression (2.253c) is positive (2.280d) ≡ (2.282f) implying that at constant temperature the volume increases when pressure decreases and vice-versa:*

$$dU = 0 < d^2 U: <=> T > 0, \quad C_p > C_V > 0, \quad K_T > 0. \qquad (2.282a\text{-}f)$$

*The **thermodynamic inequalities** (2.282c–f) are valid for any substance. The set of four thermodynamic inequalities (2.282c–f) is equivalent to (2.282a, b) = (2.272a, b). The stability principle of minimum internal energy; it follows that if the thermodynamic inequalities (2.282e–f) are satisfied then the internal energy is minimum as a function of the two variables volume and entropy (2.274a–e).*

From (2.280e) follows that the coefficient of thermal expansion (2.253a, b) and the thermal pressure coefficient (2.254a, b) have the same sign (2.283a–e)

$$sgn(K_p) = sgn(\alpha) = sgn\left(\frac{\partial V}{\partial T}\right)_p = -sgn\left(\frac{\partial S}{\partial p}\right)_T = sgn\left(\frac{\partial S}{\partial V}\right)_T = sgn\left(\frac{\partial p}{\partial T}\right)_V, \quad (2.283a\text{-}g)$$

$$\equiv \theta = \pm 1$$

but does not indicate whether the sign or **indicator** is positive or negative; for most substances it is positive $\theta = 1$. Thus *the principle of minimum internal energy (2.282a, b) implies that the coefficient of thermal expansion (2.253a, b) and thermal pressure coefficient (2.254a, b) have the same sign (2.283a–e), that is positive $\theta = 1$ in (2.283f, g) for most substances, implying that: (i) the volume increases with temperature at* *constant pressure (2.283b); (ii–iii) the entropy decreases with pressure (2.283c) and increases with volume (2.283d) at constant temperature; (iv) the pressure increases with temperature at constant volume (2.283e). Substitution of (2.281c–e) in (2.262d; 2.263d) shows that for all substances the adiabatic sound speed (2.262a–d) ≡ (2.284a) exceeds (2.284b) the isothermal sound speed (2.284c) and both are real (2.284d):*

$$(c_S)^2 \equiv \left(\frac{\partial p}{\partial \rho}\right)_S = \frac{C_p}{C_V}\left(\frac{\partial p}{\partial \rho}\right)_T = \gamma (c_T)^2 > 0. \qquad (2.284a\text{–}d)$$

Substitution of (2.281c–e) and (2.283a, b) in (2.265f) [(2.266e)] shows that *the non-adiabatic pressure (2.265a–f) ≡ (2.285a, b) [volume (2.266a–e) ≡ (2.285c, d)] coefficients are positive for most substances*:

$$S_p \equiv \left(\frac{\partial S}{\partial p}\right)_V > 0 < \left(\frac{\partial S}{\partial V}\right)_p \equiv S_V, \qquad (2.285a\text{–}d)$$

for which the entropy increases with volume at constant pressure (2.285a, b) and vice-versa (2.285c, d). Using again (2.280e, f; 2,281d–f) in (2.267a–f) follows that among the **twelve non-inverse thermodynamic derivatives** (2.267a–f) the principle of minimum internal energy (2.272a, b) specifies: (i) the sign of four (2.286a–d) for all substances:

$$\left(\frac{\partial S}{\partial T}\right)_p > \left(\frac{\partial S}{\partial T}\right)_V > 0 > \left(\frac{\partial p}{\partial V}\right)_T > \left(\frac{\partial p}{\partial V}\right)_S ; \qquad (2.286a\text{–}d)$$

(ii) the same sign for four equal pairs (2.287a–h):

$$-sgn\left(\frac{\partial S}{\partial p}\right)_T = sgn\left(\frac{\partial S}{\partial p}\right)_V = sgn\left(\frac{\partial S}{\partial V}\right)_T = sgn\left(\frac{\partial S}{\partial V}\right)_p$$

$$= sgn\left(\frac{\partial T}{\partial p}\right)_S = sgn\left(\frac{\partial T}{\partial p}\right)_V = -sgn\left(\frac{\partial T}{\partial V}\right)_S = -sgn\left(\frac{\partial T}{\partial V}\right)_T = \theta, \qquad (2.287a\text{–}h)$$

with the sign being positive for most substances. Whereas the principle of minimum internal energy (2.282a, b) specifies relations for all thermodynamic derivatives (2.282c–f; 2.283a–g; 2.284a–d; 2.285a–d; 2.286a–d; 2.287a–h), the principle of maximum entropy (2.272c, d) will be shown (subsections 2.3.20–2.3.22) to prove only (2.282d, e), and hence is a weaker statement of the second principle of thermodynamics, justifying one-sided implication ⇒ in (2.272a–d).

2.3.20   FIRST AND SECOND-ORDER DERIVATIVES OF ENTROPY

The necessary (2.275c, d) [sufficient (2.275c, e)] conditions for maximum entropy (2.272c, d) or negative-definite quadratic form for the second-order differential (2.275a, b) involve 3 second-order derivatives that are evaluated next. The entropy (2.225a) is a function of the internal energy and volume with first-order derivatives (2.288a, b):

$$\left(\frac{\partial S}{\partial U}\right)_V = \frac{1}{T}, \quad \left(\frac{\partial S}{\partial V}\right)_U = \frac{p}{T}, \quad \left[\frac{\partial}{\partial V}\left(\frac{1}{T}\right)\right]_U = \frac{\partial^2 S}{\partial V \partial U} = \frac{\partial^2 S}{\partial U \partial V} = \left[\frac{\partial}{\partial U}\left(\frac{p}{T}\right)\right]_V \quad (2.288\text{a–e})$$

and the equality of second-order derivatives leads to (2.288c–e). The first necessary and sufficient condition (2.275c) for maximum entropy is specified by the second-order derivative of the entropy with regard to the internal energy at constant volume (2.275a) ≡ (2.289a):

$$S_{11} \equiv \left(\frac{\partial^2 S}{\partial U^2}\right)_V = \left[\frac{\partial}{\partial U}\left(\frac{1}{T}\right)\right]_V = -\frac{1}{T^2}\left(\frac{\partial T}{\partial U}\right)_V = -\frac{1}{C_V T^2} < 0, \quad C_V > 0, \quad (2.289\text{a–e})$$

that is evaluated by: (i) first using (2.288a) in (2.289a, b); (ii) second using (2.252b) in (2.289c). *The first necessary and sufficient condition for maximum entropy (2.289d) implies that the specific heat at constant volume is positive (2.289e) with no assumption on the temperature.*

The second necessary (2.275d) and sufficient (2.275e) conditions for maximum entropy involve: (i) the mixed second-order derivative of the entropy with regard to the internal energy and volume, given by (2.288c) ≡ (2.290a, b):

$$\frac{\partial^2 S}{\partial V \partial U} = \left[\frac{\partial}{\partial V}\left(\frac{1}{T}\right)\right]_U = -\frac{1}{T^2}\left(\frac{\partial T}{\partial V}\right)_U; \quad (2.290\text{a, b})$$

$$\left(\frac{\partial^2 S}{\partial V^2}\right) = \left[\frac{\partial}{\partial V}\left(\frac{p}{T}\right)\right]_U = \frac{1}{T}\left(\frac{\partial p}{\partial V}\right)_U - \frac{p}{T^2}\left(\frac{\partial T}{\partial V}\right)_U; \quad (2.290\text{c, d})$$

(ii) the second-order derivative of the entropy with regard to the volume at constant internal energy is given by (2.288h) ≡ (2.290c, d). In order to apply the conditions (2.275c–e) of maximum entropy (2.275a, b) it is necessary to evaluate two distinct derivatives in (2.290b, d) at constant internal energy (subsection 2.3.21).

2.3.21 THERMODYNAMIC DERIVATIVES AT CONSTANT INTERNAL ENERGY

The derivative of the pressure with regard to volume at constant internal energy appears in (2.290d) ≡ (2.291a) where are used (2.245d; 2.248c) leading to (2.291b, c):

$$\left(\frac{\partial p}{\partial V}\right)_U = \frac{\partial(p,U)}{\partial(V,U)} = \frac{\partial(p,U)}{\partial(p,T)}\frac{\partial(p,T)}{\partial(V,T)}\frac{\partial(V,T)}{\partial(V,U)} = \left(\frac{\partial U}{\partial T}\right)_p\left(\frac{\partial p}{\partial V}\right)_T\left(\frac{\partial T}{\partial U}\right)_V; \quad (2.291\text{a–c})$$

the first derivative in (2.291c) is:

$$\left(\frac{\partial U}{\partial T}\right)_p = \left[\frac{\partial(H - pV)}{\partial T}\right]_p = \left(\frac{\partial H}{\partial T}\right)_p - p\left(\frac{\partial V}{\partial T}\right)_p = C_p - pVK_p, \quad (2.292\text{a–c})$$

evaluated by: (i) first using (2.223a) in (2.292a, b); (ii) second using (2.252e) and (2.253a) in (2.192c). Substituting (2.292c; 2.253c; 2.252b) in (2.291c) leads to (2.293a), and using (2.257a) follows (2.293b, c):

$$
\begin{aligned}
\left(\frac{\partial p}{\partial V}\right)_U &= \left(C_p - pVK_p\right) \times \frac{1}{\left(-K_T V\right)} \times \frac{1}{C_V} = \left(-\frac{C_p}{VC_V} + \frac{pK_p}{C_V}\right)\frac{C_p - C_V}{K_p^2 TV} \\
&= -\frac{C_p}{C_V}\frac{C_p - C_V}{K_p^2 V^2 T} + \frac{C_p - C_V}{C_V}\frac{p}{K_p TV},
\end{aligned}
\tag{2.293a–c}
$$

completing the evaluation of (2.291a) appearing in (2.290d).

Both in (2.290b, d) appear the derivative of the temperature with regard to the volume at constant internal energy (2.294a):

$$
\left(\frac{\partial T}{\partial V}\right)_U = \frac{\partial(T,U)}{\partial(V,U)} = \frac{\partial(T,U)}{\partial(V,S)}\frac{\partial(V,S)}{\partial(V,U)} = \left(\frac{\partial S}{\partial U}\right)_V
\begin{vmatrix}
\left(\dfrac{\partial T}{\partial V}\right)_S & \left(\dfrac{\partial U}{\partial V}\right)_S \\[2mm]
\left(\dfrac{\partial T}{\partial S}\right)_V & \left(\dfrac{\partial U}{\partial S}\right)_V
\end{vmatrix}
\tag{2.294a–e}
$$

$$
= \frac{1}{T}
\begin{vmatrix}
-\dfrac{C_p - C_V}{C_V K_p V} & -p \\[2mm]
\dfrac{T}{C_V} & T
\end{vmatrix}
= -\frac{C_p - C_V}{C_V K_p V} + \frac{p}{C_V},
$$

that is evaluated by: (i) first using (2.245d) in (2.294a); (ii) second using (2.248c) in (2.294b); (iii) third using (2.244c; 2.245d) in (2.294c); (iv) fourth using (2.288a; 2.269b; 2.242a, g; 2.252a) in (2.294d) that simplifies to (2.294e). The thermodynamic derivatives at constant internal energy (2.293c) and (2.294e) can be used in (2.290b, d) together with (2.289d) to prove (subsection 2.3.22) the stability conditions (2.275c–e) for maximum entropy (2.275a, b).

2.3.22 THERMODYNAMIC STABILITY AND MAXIMUM ENTROPY

Substitution of (2.294e) in (2.290b) specifies the cross-derivative of entropy with regard to internal energy and specific volume (2.295a):

$$
\frac{\partial^2 S}{\partial V \partial U} = \frac{C_p - C_V}{C_V K_p T^2 V} - \frac{p}{T^2 C_V},
\tag{2.295a}
$$

$$
\left(\frac{\partial^2 S}{\partial V^2}\right)_U = -\frac{C_p}{C_V}\frac{C_p - C_V}{K_p^2 V^2 T^2} + 2\frac{C_p - C_V}{C_V}\frac{p}{K_p T^2 V} - \frac{p^2}{C_V T^2},
\tag{2.295b}
$$

the second-order derivative of entropy with regard to specific volume at constant internal energy (2.295b) follows form substitution of (2.293c; 2.294e) in (2.290d). Substituting (2.289d; 2.295a, b) in (2.275e) gives (2.296b, c):

$$0 < S_{12} = \frac{C_p}{C_V^2} \frac{C_p - C_V}{K_p^2 V^2 T^4} - 2 \frac{C_p - C_V}{C_V^2} \frac{p}{K_p T^4 V} + \frac{p^2}{C_V^2 T^4} - \left(\frac{C_p - C_V}{C_V K_p T^2 V} - \frac{p}{T^2 C_V} \right)^2$$

$$= \frac{C_p - C_V}{C_V^2 K_p^2 T^4 V^2} \left[C_p - \left(C_p - C_v \right) \right] = \frac{\dfrac{C_p}{C_V} - 1}{K_p^2 T^4 V^2}, \tag{2.296a–d}$$

that is positive (2.296a). The second necessary condition (2.275d) is met by (2.295b) in the form (2.297a–d):

$$\left(\frac{\partial^2 S}{\partial V^2} \right)_U = -\frac{1}{C_V T^2} \left[\frac{C_p \left(C_p - C_V \right)}{K_p^2 V^2} - 2 \frac{\left(C_p - C_V \right) p}{K_p V} + p^2 \right]$$

$$= -\frac{1}{C_V T^2} \left[\frac{C_p \left(C_p - C_V \right)}{K_p^2 V^2} + \left(\frac{C_p - C_V}{K_p V} - p \right)^2 - \left(\frac{C_p - C_V}{K_p V} \right)^2 \right] \tag{2.297a–d}$$

$$= -\frac{1}{C_V T^2} \left[\left(\frac{C_p - C_V}{K_p V} - p \right)^2 + \frac{C_V \left(C_p - C_V \right)}{K_p^2 V^2} \right] < 0.$$

Thus *the principle of maximum entropy (2.272c, d) ≡ (2.298a, b), that is most often associated with the second principle of thermodynamics, proves only that the specific heat at constant pressure is larger than the specific heat at constant volume (2.296d) ≡ (2.298c) and that the latter is positive (2.289e) ≡ (2.298d):*

$$dS = 0 < d^2 S: \iff C_p > C_V > 0. \tag{2.298a–d}$$

The set of two thermodynamic inequalities (2.298c–d) is equivalent to (2.298a, b) = (2.272c, d) the principle of maximum entropy; it follows that if the thermodynamic inequalities (2.298c, d) are satisfied then the entropy is maximum as a function of the two variables volume and entropy (2.275a–e). The principle of stability or minimum internal energy (2.282a, b): (i) proves (2.298c, d) ≡ (2.282d, e) and thus implies (2.272a–d) the principle of maximum entropy; (ii) proves in addition other thermodynamic inequalities (2.282c, f; 2.283a–g) with important implications. In particular the condition of positive absolute temperature (2.282c) follows from the principle of minimum internal energy (2.282a, b) but not from the principle of maximum entropy. For all these reasons the stability principle of minimum internal energy (2.272a, b) is a more restrictive or less general statement of the second principle of thermodynamics than the principle of maximum entropy. The principle of minimum internal energy, as both a principle of stability and second principle of thermodynamics is closer to the first principle of thermodynamics, that requires stationary internal energy. The principle of maximum entropy is more general than the principle of minimum internal energy because it imposes less restrictions namely: (i) the same conditions (2.298c, d) ≡ (2.282d, e); (ii) but not the extra conditions (2.282c, f).

For example, the principle of maximum entropy allows negative absolute temperature (2.299a) implying: (i) from (2.276a; 2.277h; 2.279e), respectively (2.299b–d); (ii) it follows (2.207a, b; 2.208a–c) that the internal energy is maximum (2.229e, f):

$$T < 0 \Rightarrow U_{11} < 0 < U_{22}, \quad U_{12} < 0 \Rightarrow dU = 0 > d^2U. \qquad (2.299a\text{–}f)$$

Therefore, maximum entropy implies minimum (2.299g) [maximum (2.299b)] internal energy for positive (negative) absolute temperature:

$$S\,maximum \begin{cases} U\,minimum \; if\,T > 0, \\ U\,maximum \; if\,T < 0. \end{cases} \qquad (2.99g,h)$$

The principle of minimum internal energy yields more results (subsections 2.3.17–2.3.19) with much less cumbersome calculations (subsections 2.3.20–2.3.22) than the principle of maximum entropy. It is shown next (subsection 2.3.23) that in comparison with the minimum internal energy and maximum entropy, the other three functions of state, namely, the enthalpy and free energy (free enthalpy) have no extremals (is maximum).

2.3.23   INEXISTENCE OF EXTREMALS FOR ENTHALPY AND FREE ENERGY AND MAXIMUM FOR FREE ENTHALPY

The equilibrium of a thermodynamic system is equivalently specified by five thermodynamic functions of state being stationary (2.299a–e), namely the entropy (2.225a, b), internal energy (2.221d), free energy (2.222a, b), enthalpy (2.223a, b) and free enthalpy (2.224a–d):

$$\text{equilibrium:} \quad 0 = dS = dU = dF = dH = dG; \qquad (2.300a\text{–}e)$$

$$\text{stability:} \quad d^2S < 0 < d^2U, \quad d^2F \gtrless 0, \qquad (2.300f\text{–}h)$$

$$d^2H \gtrless 0, \quad d^2G < 0. \qquad (2.300i, j)$$

The stability requires by maximum entropy (2.300a) [minimum internal energy (2.300b)] that have negative (2.275a–e) [positive (2.274a–e)] definite second-order differentials (2.300f) [(2.300g)]; in comparison with the free energy (2.300c), and enthalpy (2.300d) [free enthalpy (2.300e)] have indefinite (2.300h, i) [negative-definite (2.300j)] second-order differentials, and thus do not have extremals (does have a maximum). It follows that there is no maximum or minimum principle for the free energy and enthalpy (there is a maximum principle for the free enthalpy) and hence thermodynamic inequalities cannot (can) be derived from them.

In order to prove that a second-order differential is indefinite it is sufficient to show that the diagonal second-order derivatives have opposite signs. For example, for the free energy (2.301a, b):

$$\left(\frac{\partial^2 F}{\partial T^2} \right)_V \times \left(\frac{\partial^2 F}{\partial V^2} \right)_T < 0 \Rightarrow d^2F \gtrless 0. \qquad (2.301a, b)$$

The second-order diagonal derivatives are evaluated for: (i) the free energy (2.222a, b) by (2.242c; 2.252b) [(2.242h; 2.253c)] leading to (2.302a, b) [(2.302d, e)]:

$$F_{11} \equiv \left(\frac{\partial^2 F}{\partial T^2}\right)_V = -\left(\frac{\partial S}{\partial T}\right)_V = -\frac{C_V}{T} < 0 < \frac{1}{K_T V} = -\left(\frac{\partial p}{\partial V}\right)_T = \left(\frac{\partial^2 F}{\partial V^2}\right) \equiv F_{22}; \quad (2.302a\text{–}e)$$

(ii) the enthalpy (2.223a, b) by (2.242b; 2.252e) [(2.242e; 2.262b)] leading to (2.303a, b) [(2.303d, e)]:

$$H_{11} \equiv \left(\frac{\partial^2 H}{\partial S^2}\right)_p = \left(\frac{\partial T}{\partial S}\right)_p = \frac{T}{C_p} > 0 > -\frac{9V}{\left(c_S\right)^2} = \left(\frac{\partial V}{\partial p}\right)_S = \left(\frac{\partial^2 H}{\partial p^2}\right)_S \equiv H_{22}. \quad (2.303a\text{–}e)$$

Since the second-order diagonal derivatives of the free energy (2.302a–e) [enthalpy (2.303a–e)] have opposite signs, their second-order differentials are indefinite (2.300c) [(2.300d)], and there is no maximum or minimum principle from which thermodynamic inequalities can be obtained.

For example, in the case of the free energy the second-order differential is given by (2.304a) involving the diagonal (2.302a, b) $\equiv$ (2.304b, c) and (2.302d, e) $\equiv$ (2.304d, e) derivatives and non-diagonal derivative (2.304f):

$$d^2 F = F_{11}\left(dT\right)^2 + F_{22}\left(dV\right)^2 + 2F_{12}dTdV:$$

$$F_{11} \equiv \left(\frac{\partial^2 F}{\partial T^2}\right)_V > 0, \quad F_{22} \equiv \left(\frac{\partial^2 F}{\partial V^2}\right)_T < 0, \quad F_{12} \equiv \frac{\partial^2 F}{\partial T \partial V}. \quad (2.304a\text{–}f)$$

For the present purpose the sign of (2.304f) is not needed, since for an isochoric (2.305a) [isothermal (2.305c)] process the second-order differential of the free enthalpy is positive (2.305b) [negative (2.305d)]:

$$dV = 0: \quad d^2 F = F_{11}\left(dT\right)^2 > 0; \quad dT = 0: \quad d^2 F = F_{22}\left(dV\right)^2 < 0. \quad (2.305a\text{–}d)$$

implying that it is indefinite (2.301a, b).

In the case of the free enthalpy (2.224a–d) the second-order diagonal derivatives are evaluated by (2.242c; 2.252e) [(2.242f; 2.253c)] and are both negative (2.306a, b) [(2.306d, e)]:

$$G_{11} \equiv \left(\frac{\partial^2 G}{\partial T^2}\right)_p = -\left(\frac{\partial S}{\partial T}\right)_p = -\frac{C_p}{T} < 0 > -K_T V = \left(\frac{\partial p}{\partial V}\right)_T = \left(\frac{\partial^2 G}{\partial p^2}\right)_T \equiv G_{22}, \quad (2.306a\text{–}e)$$

meeting the two necessary conditions (2.207a, b) for negative definiteness and also the first sufficient condition (2.208b); the second sufficient condition (2.208c) involves the second-order cross-derivative (2.307a, b) that is evaluated by (2.242f; 2.253a):

$$\frac{\partial^2 G}{\partial T \partial p} = \left(\frac{\partial V}{\partial T}\right)_p = K_p V:$$

$$G_{12} \equiv \left(\frac{\partial^2 G}{\partial T^2}\right)_p \left(\frac{\partial^2 G}{\partial p^2}\right)_T - \left(\frac{\partial^2 G}{\partial T \partial p}\right)^2 = \frac{C_p K_T V}{T} - K_p^2 V^2 = \frac{C_V K_T V}{T} > 0,$$

$$(2.307a\text{–}f)$$

implying (2.307c, d) leading by (2.255g) to (2.307e) that is positive (2.307f).

    Thus the sufficient conditions for negative definiteness (2.208b, c) are met and *the second-order differential for the free enthalpy is negative-definite (2.300j) ≡ (2.308b) implying that the free enthalpy is maximum (2.300e) ≡ (2.308a), and leading (2.306b, c; 2.307f) to the thermodynamic inequalities (2.308c–e) implying (2.308f):*

$$dG = 0 < d^2 G: \iff \frac{C_p}{T} > 0, \quad K_T > 0, \quad \frac{C_V}{T} > 0 \Rightarrow \frac{C_p}{C_V} > 0. \quad (2.308a\text{–}f)$$

The thermodynamic inequalities (2.308c–e) are equivalent to the condition of maximum free enthalpy (2.300j), and thus if they are satisfied the enthalpy is maximum as a function of the two variables pressure and temperature. Next examples are given examples of thermodynamic stability for interacting bodies, namely [subsections 2.3.24–2.3.25 (2.3.26–2.3.27)] two communicating reservoirs (a small body immersed in a large environment).

2.3.24   FLUID TRANSFER BETWEEN FULL AND EMPTY RESERVOIRS

Consider two reservoirs (Figure 2.7a) with volumes V_1 and V_2, initially at time $t = 0$ the first with fluid at pressure $p_1 \neq 0$ and the second empty $p_2 = 0$. A hole is opened in the separating wall, leading to an unsteady flow from the full to the empty reservoir over a time scale T; for much longer time $t \gg T$ the pressure has stabilized at p_3 for the whole volume. Assuming that the two reservoirs are isolated from the outside, and there are no heat exchanges (2.309a, b), the internal energy is conserved, and is the same at times $t \gg T$ and $t = 0$ so the final pressure satisfies (2.309c)

$$dS_1 = 0 = dS_2: \quad U(p_1, V_1) = U(p_3, V_1 + V_2). \quad (2.309a\text{–}c)$$

The rate of change of temperature with volume is calculated at constant internal energy (2.310a):

$$\left(\frac{\partial T}{\partial V}\right)_U = \frac{\partial(T, U)}{\partial(V, U)} = \frac{\partial(V, T)}{\partial(V, U)} \frac{\partial(T, U)}{\partial(V, T)} = -\left(\frac{\partial T}{\partial U}\right)_V \left(\frac{\partial U}{\partial V}\right)_T$$

$$= -\frac{1}{C_V}\left[-p + T\left(\frac{\partial S}{\partial V}\right)_T\right] = \frac{1}{C_V}\left[p - T\left(\frac{\partial p}{\partial T}\right)_V\right],$$

$$(2.310a\text{–}e)$$

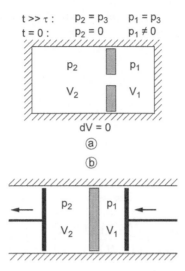

$$t \gg \tau: \quad p_2 = p_3 \quad p_1 = p_3$$
$$t = 0: \quad p_2 = 0 \quad p_1 \neq 0$$

FIGURE 2.7 Two examples of irreversible thermodynamic processes are gas transfer: (a) from a full to an empty reservoir through a hole in the wall separating them; (b) across a porous wall between two reservoirs kept at constant pressure by sliding pistons moving in the direction of lower pressure.

by: (i) first using (2.245d) in (2.310b); (ii) second using (2.248c) in (2.310c); (iii) third using (2.252c) and (2.221d) in (2.310d); (v) fourth using (2.243b) in (2.310e). The variation of entropy with volume at constant internal energy is given by (2.282b) = (2.311a) that is positive (2.311b) showing that the increase in volume (2.311c) is associated with an increase in entropy (2.311d):

$$\left(\frac{\partial S}{\partial V}\right)_U = \frac{p}{T} > 0: \quad dV > 0 \Rightarrow dS > 0. \tag{2.311a–d}$$

It has been shown that *in the expansion of a fluid from a full to an empty reservoir, both thermally insulated (2.309a, b) and with rigid walls (Figure 2.7a): (i) the final equilibrium pressure is determined by the conservation of internal energy (2.309c); (ii) the rate of change of temperature with volume is given by (2.310e); (iii) the expansion (2.311c) leads to an entropy increase (2.311d) at a rate (2.311a, b) in agreement with the second principle of thermodynamics for an irreversible process (2.273b).* The entropy always increases in irreversible processes, and this is also the case (subsection 2.3.25) for a fluid exchange between reservoirs at constant pressure through a porous wall.

2.3.25 TRANSFER AT CONSTANT PRESSURE THROUGH A POROUS WALL (JOULE–THOMSON 1882)

Consider next the fluid transfer through a porous wall (Figure 2.7b) between two reservoirs kept at constant pressure by a sliding pistons moving in the direction of lower pressure. The fluid at initial pressure p_1 and volume V_1 eventually stabilizes at pressure p_2 and volume V_2, performing a work (2.312c) equal to the difference of internal energies (2.312d), assuming that both reservoirs are thermally insulated (2.312a, b):

$$dS_1 = 0 = dS_2: \quad p_1V_1 - p_2V_2 = W_{\hat{12}} = U_2 - U_1. \tag{2.312a–d}$$

From $(2.312d) \equiv (2.313a)$ follows (2.223a) that the enthalpy is conserved (2.313b) specifying the final volume:

$$U_1 + p_1V_1 = U_2 + p_2V_2: \quad H(p_1,V_1) = H(p_2,V_2). \tag{2.313a, b}$$

The rate of change of the temperature with the pressure at constant enthalpy (2.414a):

$$\begin{aligned}
\left(\frac{\partial T}{\partial p}\right)_H &= \frac{\partial(T,H)}{\partial(p,H)} = \frac{\partial(p,T)}{\partial(p,H)}\frac{\partial(T,H)}{\partial(p,T)} = -\left(\frac{\partial T}{\partial H}\right)_p\left(\frac{\partial H}{\partial p}\right)_T \\
&= -\frac{1}{C_p}\left[V + T\left(\frac{\partial S}{\partial p}\right)_T\right] = \frac{1}{C_p}\left[T\left(\frac{\partial V}{\partial T}\right)_p - V\right],
\end{aligned} \tag{2.314a–e}$$

is obtained by: (i) first using (2.245d) in (2.314a); (ii) second using (2.248c) in (2.314b); (iii) third using (2.245c) in (2.314c); (iv) fourth using (2.252e) and (2.223b) in (2.314d); (v) fifth using (2.243d) in (2.314e). The variation of entropy with pressure at constant enthalpy is given (2.225b) by (2.315a), and is negative (2.315b) showing that the reduction in pressure (2.315c) leads to an increase in entropy (2.315d):

$$\left(\frac{\partial S}{\partial p}\right)_H = -\frac{V}{T} < 0: \quad dp < 0 \Rightarrow dS > 0. \tag{2.315a–d}$$

It has been shown that *in the transfer of fluid through a porous wall between thermally insulated (2.312a, b) reservoirs at constant pressure (Figure 2.7b): (i) the final volume is specified by the conservation of enthalpy (2.313b); (ii) the rate-of-change of temperature with pressure at constant enthalpy is given by (2.314e); (iii) the decrease in pressure (2.315c) leads to an increase in entropy (2.315d) at a rate (2.315a, b) in agreement with the second principle of thermodynamics for an irreversible process (2.273b).* Instead of two comparable (Figure 2.7a, b) reservoirs (subsections 2.3.24–2.3.25) next (subsection 2.3.26) is considered a "small" body in a "large" environment, that tends to reduce the external disturbances.

2.3.26 BODY REDUCING ENVIRONMENTAL DISTURBANCES (LE CHATELIER 1898)

Consider a small **body** whose thermodynamic state is specified by an extensive X and intensive Y parameter. The body is immersed in a large medium or **environment**, that the body cannot change, though the body is affected by the environment; the environment has extensive x and intensive y parameters. The combined thermodynamic system is isolated, has extensive (X,x) and intensive (Y,y) parameters, and is in equilibrium if the entropy is stationary (2.316a, b):

$$Y = -\frac{\partial S}{\partial X} = 0 = -\frac{\partial S}{\partial x} = y; \quad S_1 \equiv \frac{\partial^2 S}{\partial X^2} = -\left(\frac{\partial Y}{\partial X}\right)_x < 0 > -\left(\frac{\partial y}{\partial x}\right)_X = \frac{\partial^2 S}{\partial x^2} \equiv S_{22}, \quad \text{(2.316a–d)}$$

$$S_{12} \equiv \frac{\partial^2 S}{\partial X^2}\frac{\partial^2 S}{\partial x^2} - \left(\frac{\partial^2 S}{\partial x \partial X}\right) = \left(\frac{\partial Y}{\partial X}\right)_x \left(\frac{\partial y}{\partial x}\right)_X - \left[\left(\frac{\partial Y}{\partial x}\right)_X\right]^2 > 0, \quad \text{(2.316e)}$$

for a stable equilibrium are necessary (sufficient) the conditions of maximum entropy (2.316c, d) [(2.316c, e)].

A perturbation in the medium leads to an immediate perturbation in the body (2.317a) before equilibrium is re-established:

$$(\Delta Y)_x = \left(\frac{\partial Y}{\partial X}\right)_x \Delta x; \quad (\Delta Y)_y = \left(\frac{\partial Y}{\partial X}\right)_y \Delta x, \quad \text{(2.317a, b)}$$

after equilibrium is re-established the variation (2.317b) is evaluated at constant intensive parameter of the environment. The equilibrium (2.317b) $\equiv$ (2.318a):

$$\left(\frac{\partial Y}{\partial X}\right)_y = \frac{\partial(Y,y)}{\partial(X,y)} = \frac{\partial(Y,y)}{\partial(X,x)}\frac{\partial(X,x)}{\partial(X,y)} = \left(\frac{\partial x}{\partial y}\right)_X \left[\left(\frac{\partial Y}{\partial X}\right)_x\left(\frac{\partial y}{\partial x}\right)_X - \left(\frac{\partial Y}{\partial x}\right)_X\left(\frac{\partial y}{\partial X}\right)_x\right]$$
$$\text{(2.318a–c)}$$

is evaluated by: (i) first using (2.245d) in (2.318a); (ii) second using (2.248c) in (2.318b); (iii) third using (2.244d; 2.245d) in (2.318c). In (2.318c) appear (2.319a) and (2.319b–d):

$$\left(\frac{\partial x}{\partial y}\right)_X \left(\frac{\partial y}{\partial x}\right)_X = 1, \quad \left(\frac{\partial Y}{\partial x}\right)_X = -\frac{\partial^2 S}{\partial x \partial X} = -\frac{\partial^2 S}{\partial X \partial x} = \left(\frac{\partial y}{\partial X}\right)_x. \quad \text{(2.319a–d)}$$

Substituting (2.319a, d) in (2.318c) leads to (2.320a) and (2.316d) implies that is negative (2.320b)

$$\left(\frac{\partial Y}{\partial X}\right)_y - \left(\frac{\partial Y}{\partial X}\right)_x = -\left(\frac{\partial x}{\partial y}\right)_X \left[\left(\frac{\partial Y}{\partial x}\right)_X\right]^2 < 0. \quad \text{(2.320a, b)}$$

This proves the **principle of Le Chatelier (1898)**: *if a body is inserted in a large environment it tends to reduce (2.320b) $\equiv$ (2.321a, b) the perturbations due to the environment (2.317a, b) $\equiv$ (2.321c, d)*

$$\left(\frac{\partial Y}{\partial X}\right)_x > \left(\frac{\partial Y}{\partial X}\right)_y > 0 \Leftrightarrow (\Delta Y)_x > (\Delta Y)_y > 0. \quad \text{(2.321a–d)}$$

Two examples are given next of a body in pressure or thermal equilibrium with the environment (subsection 2.3.27).

2.3.27 BODY IN PRESSURE/THERMAL EQUILIBRIUM WITH ENVIRONMENT

For a body in pressure equilibrium with the environment the extensive variable is minus volume (2.322a) and the intensive variable is pressure (2.322b). A pressure variation in the environment is felt by the body as an immediate adiabatic (2.322c) volume perturbation (2.322e) that becomes isothermal (2.322d) at equilibrium (2.322f) so that (2.321a) implies (2.322e):

$$X = -V, Y \equiv p, x \equiv S, y \equiv T: \quad -\left(\frac{\partial p}{\partial V}\right)_S > -\left(\frac{\partial p}{\partial V}\right)_T. \qquad (2.322\text{a--e})$$

The relation (2.322e) implies (2.323a) from (2.262b; 2.263b) that the adiabatic sound speed (2.262a–d) exceeds the isothermal sound speed (2.263a–d) in agreement with (2.323b) ≡ (2.284d):

$$\left(\frac{c_s}{V}\right)^2 > \left(\frac{c_T}{V}\right)^2 \Rightarrow c_s > c_T \Rightarrow 1 < \left(\frac{c_s}{c_T}\right)^2 < \gamma = \frac{C_p}{C_V} \Rightarrow C_p > C_V, \quad (2.323\text{a--e})$$

from (2.323b) ≡ (2.323c) also follow (2.264a, b) ≡ (2.323d) [(2.323e)] in agreement with (2.282c) ≡ [(2.298c].

For a body in temperature equilibrium with the environment the extensive parameter is entropy (2.324a) and the intensive parameter is temperature (2.324b). A temperature variation in the environment is felt by the body as an immediate isochoric (2.324c) perturbation (2.324e) that becomes isobaric (2.324d) at equilibrium (2.324f) so that (2.321a) implies (2.324e):

$$X \equiv S, Y \equiv T, x \equiv V, y \equiv p: \quad \left(\frac{\partial T}{\partial S}\right)_V > \left(\frac{\partial T}{\partial S}\right)_p; \qquad (2.324\text{a--e})$$

from (2.252b, e) follows that (2.324f) is equivalent to (2.325a)

$$\frac{T}{C_V} > \frac{T}{C_p} \Rightarrow C_p > C_V, \qquad (2.325\text{a, b})$$

implying (2.325b) in agreement with (2.325b) ≡ (2.282d) ≡ (2.298b) ≡ (2.323e). A consequence of the second principle of thermodynamics is that the absolute temperature is positive (2.282c) leading to the third principle of thermodynamics on properties at zero temperature (subsection 2.3.28).

2.3.28 THERMODYNAMIC PROPERTIES AT ZERO ABSOLUTE TEMPERATURE

It is assumed that the internal energy (2.221a–d) [enthalpy (2.223a, b)] is a twice differentiable function of temperature near zero temperature (2.326a) [(2.326b)]

$$U(T,V) = U_0 + a(V)T + \frac{1}{2}b(V)T^2, \quad H(T,p) = H_0 + A(p)T + \frac{1}{2}B(p)T^2; \qquad \text{(2.326a, b)}$$

the coefficients in (2.326a) [(2.326b)] may be functions of the volume (pressure):

$$\{a(V), b(V)\} \equiv \lim_{T \to 0}\left\{\left(\frac{\partial U}{\partial T}\right)_V, \left(\frac{\partial^2 U}{\partial T^2}\right)_V\right\} = \lim_{T \to 0}\left\{C_V(T,V), \left(\frac{\partial}{\partial T}[C_V(T,V)]\right)_V\right\}, \qquad \text{(2.327a, b)}$$

$$\{A(p), b(p)\} \equiv \lim_{T \to 0}\left\{\left(\frac{\partial H}{\partial T}\right)_p, \left(\frac{\partial^2 H}{\partial T^2}\right)_p\right\} = \lim_{T \to 0}\left\{C_p(T,p), \left(\frac{\partial}{\partial T}[C_p(T,p)]\right)_p\right\}, \qquad \text{(2.327c, d)}$$

namely: (i) the first-order coefficient is the specific heat at constant volume (2.252c) $\equiv$ (2.327a) [pressure (2.252f) $\equiv$ (2.327c)]; (ii) the second-order coefficient is the derivative of the first with regard to temperature (2.327b) [(2.327d)].

In a process at constant volume (2.328a) [pressure (2.329a)] the entropy (2.225a) [(2.225b)] is given by (2.328b) [(2.329b)] implying (2.328c, d) [(2.329c)]:

$$dV = 0: \quad dS = \frac{dU}{T} = a(V)\frac{dT}{T} + b(V)dT, \quad S(T,V) = a(V)\log T + b(V)T, \quad \text{(2.328a–c)}$$

$$dp = 0: \quad dS = \frac{dH}{dT} = A(p)\frac{dT}{T} + B(p)dT, \quad S(p,T) = A(p)\log T + B(p)T. \quad \text{(2.329a–c)}$$

Thus a finite entropy at zero temperature (2.330b) [(2.331b)] in an isochoric (2.330a) [isobaric (2.331a)] process implies (2.330c) [(2.331c)]

$$dV = 0: \quad \lim_{T \to 0} S(T,V) < \infty \Rightarrow a(V) = 0, \qquad \text{(2.330a–c)}$$

$$dp = 0: \quad \lim_{T \to 0} S(T,p) < \infty \Rightarrow A(p) = 0. \qquad \text{(2.331a–c)}$$

These results lead to the third principle of thermodynamics (subsection 2.3.29).

2.3.29 THIRD PRINCIPLE OF THERMODYNAMICS (NERST 1907)

*The **third principle of thermodynamics** (Nerst 1907) states that an internal energy (2.221d) [enthalpy (2.223a, b)] that is a twice differentiable function of temperature (2.326a) [(2.326b)] and a finite entropy at zero temperature (2.330b) [(2.331b)] in an isochoric (2.330a) $\equiv$ (2.332a) [isobaric (2.331a) $\equiv$ (2.333a)] process:*

$$dV = 0: \quad \lim_{T \to 0}\frac{S(T,V)}{T} = b(V) = \lim_{T \to 0}\frac{C_V(T,V)}{T}, \quad \lim_{T \to 0}\frac{U(T,V) - U_0(V)}{T^2} = \frac{b(V)}{2}, \quad \text{(2.332a–d)}$$

$$dp = 0: \quad \lim_{T \to 0}\frac{S(T,p)}{T} = B(p) = \lim_{T \to 0}\frac{C_p(T,p)}{T}, \quad \lim_{T \to 0}\frac{H(T,p) - H_0(p)}{T^2} = \frac{B(V)}{2}, \quad \text{(2.333a–d)}$$

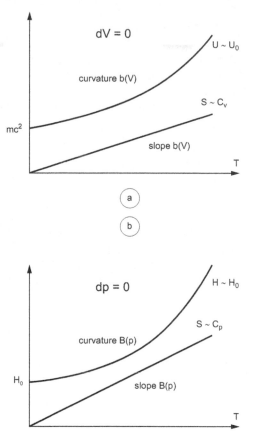

FIGURE 2.8 The third principle of thermodynamics specifies the properties of matter neat zero absolute temperature, and states that in an (a) isochoric [(b) isobaric] process at lowest order in powers of temperature: (i,ii) the entropy and specific heat at constant volume (pressure) are linear functions of temperature; (iii) the internal energy (enthalpy) is a quadratic function of temperature.

imply [Figure 2.8a (b)] that: (i/ii) the entropy (2.332b) [(2.333b)] and specific heat at constant volume (2.332c) [pressure (2.333c)] vanish in proportion to the temperature; (iii) the internal energy (2.332d) [enthalpy (2.333d)] vanish as the square of temperature, apart from a possible constant value at zero temperature.

The proof of (2.332c) [(2.333c)] follows from (2.252b) ≡ (2.334a) [(2.252d)] ≡ (2.335a)] applied to (2.326a; 2.330c) [(2.326b; 2.331c)] leading to (2.334b) [(2.335b)] that simplifies to (2.334c) ≡ (2.332c) [(2.335c) ≡ (2.331c)],

$$\lim_{T \to 0} \frac{C_V(T,V)}{T} \equiv \lim_{T \to 0} \frac{1}{T}\left(\frac{\partial S}{\partial T}\right)_p = \lim_{T \to 0} \frac{1}{T}\left\{\frac{\partial}{\partial T}\left[\frac{1}{2}b(V)T^2\right]\right\}_V = b(V), \quad (2.334\text{a-c})$$

$$\lim_{T \to 0} \frac{C_p(T,p)}{T} \equiv \lim_{T \to 0} \frac{1}{T} \left(\frac{\partial S}{\partial T} \right)_V = \lim_{T \to 0} \frac{1}{T} \left\{ \frac{\partial}{\partial T} \left[\frac{1}{2} B(p) T^2 \right] \right\}_p = B(p). \qquad (2.335a\text{--}c)$$

From (2.282d, e) ≡ (2.298c, d) follows the limit (2.336b)

$$B(p) = \lim_{T \to 0} \frac{C_p(T,p)}{T} \geq \lim_{T \to 0} \frac{C_v(T,V)}{T} = b(V) > 0 \qquad (2.336a\text{--}c)$$

so that a finite (2.333c) ≡ (2.336a) implies (2.336b) a finite (2.336c) ≡ (2.332c). Thus combining the second and third principles of thermodynamics only the condition (2.326b; 2.331b) need be imposed to obtain (2.326a; 2.330b), (2.332a–d) and (2.333a–d).

The conditions that, in the limit of zero temperature, in an isobaric process (2.331a), the entropy is finite (2.331b) and the enthalpy twice differentiable with regard to temperature (2.326b), imply that: (i) the entropy (2.333b) and specific heat at constant pressure (2.333c) decay like the temperature, and the enthalpy (2.333d) decays like the square of the temperature to within a constant value at zero temperature; (ii) in an isochoric process (2.330a) the entropy is finite (2.330b) and the internal energy is a twice differentiable function of temperature (2.326a), and more precisely the entropy (2.332b) and specific heat at constant volume (2.332c) decay like the temperature, and the internal energy decays like the square of the temperature (2.332d) to within a constant value at zero temperature.

The macroscopic temperature corresponds at microscopic level to the average kinetic energy of molecules. Thus zero temperature (2.337a) corresponds to matter at rest at molecular level, that has only one state of equilibrium (2.337b). The entropy is proportional to the logarithm of the number of states through the Boltzmann constant (2.337c) and thus is zero at zero temperature (2.337d) in agreement with (2.332b; 2.333b):

$$T = 0: \quad n = 1, \quad S = k \log n = 0; \quad E = mc^2, \qquad (2.337a\text{--}e)$$

the **rest energy** is shown in relativity to equal (2.337e) the product of the mass by the square of the speed of light in vacuo. The absolute value of the energy cannot be determined in non-relativistic thermodynamics, that applies only to energy changes. Since an increase in the number of accessible states increase entropy, the latter may be interpreted as a measure of "disorganization". The thermodynamic surface being convex, it cannot extend to infinity, and thus: (i) energy is always finite; (ii) the entropy of an isolated system cannot grow without bound, that is there is no "total disorganization" corresponding to an infinite number of accessible states. The second (third) principle of thermodynamics has been stated [subsections 2.3.1–2.3.27 (2.3.28–2.3.29)] for a basic thermodynamic system (2.221a–d) for conciseness of exposition, and also apply more generally to a general thermodynamic system with variable number of particles (subsection 2.3.30) and subject to force or stress fields (subsection 2.3.31).

2.3.30   System with a Variable Number of Particles

Extending the basic thermodynamic system to allow for a variable number of parti-cles N adds to the differentials of internal energy (2.86c)/free energy (2.222b)/enthalpy (2.223b)/free enthalpy (2.224d) an additional term in, respectively (2.338a/b/c, d) involving the chemical potential:

$$dU = TdS - pdV + \mu dN, \quad dF = -SdT - pdV + \mu dN, \qquad (2.338a, b)$$

$$dH = TdS + Vdp + \mu dN, \quad dG = -SdT + Vdp + \mu dN; \qquad (2.338c, d)$$

also (2.338a) [(2.338c)] imply (2.338e) [(2.338f)] for the entropy,

$$\frac{dU}{T} + \frac{p}{T}dV - \frac{\mu}{T}dN = dS = \frac{dH}{T} - \frac{V}{T}dp - \frac{\mu}{T}dN, \qquad (2.338e, f)$$

generalizing (2.225a) [(2.225b)] to allow for a variable number of particles. From (2.338a–f) it follows *that the **chemical potential** is given by (2.339a–f):*

$$\mu = \left(\frac{\partial U}{\partial N}\right)_{S,V} = \left(\frac{\partial F}{\partial N}\right)_{T,V} = \left(\frac{\partial H}{\partial N}\right)_{S,p} = \left(\frac{\partial G}{\partial N}\right)_{T,p} = -T\left(\frac{\partial S}{\partial N}\right)_{U,V} = -T\left(\frac{\partial S}{\partial N}\right)_{H,p}. \quad (2.339a–f)$$

The functions of state and extensive (intensive) variables must be proportional to (independent of) the number of particles, for example, the internal energy, free energy, enthalpy, free enthalpy, entropy and volume (pressure and temperature). Thus *for a basic thermodynamic system with a variable number of particles the inter-nal energy (2.340a)/free energy (2.340b)/enthalpy (2.340c)/free enthalpy (2.340d)/entropy (2.340e, f):*

$$U(S,V,N) = N\bar{u}\left(\frac{S}{N}, \frac{V}{N}\right), \quad F(T,V,N) = N\bar{f}\left(T, \frac{V}{N}\right),$$

$$H(S,p,N) = N\bar{h}\left(\frac{S}{N}, p\right), \qquad (2.340a–c)$$

$$G(T,p,N) = N\bar{g}(T,p), \quad S(U,V,N) = N\bar{s}_u\left(\frac{U}{N}, \frac{V}{N}\right),$$

$$S(H,p,N) = N\bar{s}_h\left(\frac{H}{N}, p\right), \qquad (2.340d–f)$$

involve homogeneous functions of the first degree $\bar{u}/\bar{f}/\bar{h}/\bar{g}/\bar{s}_u/\bar{s}_h$, with first-order dif-ferentials, respectively (2.338a)/(2.338b)/(2.338c)/(2.338d)/(2.338e, f) involving the chemical potential (2.339a–f). The free enthalpy (2.340d) is singled out in the sequel (Section 2.3.31) because it leads to an expression for the differential of the chemical potential.

2.3.31   DIFFERENTIALS OF CHEMICAL POTENTIALS (GIBBS 1876–1878, DUHEM 1886)

Differentiation of the free enthalpy (2.340d) with regard to the particle number leads to (2.341a, b):

$$\frac{G}{N} = \bar{g}(T,p) = \frac{dG}{dN} = \mu, \quad G = \mu N, \qquad (2.341a\text{–}d)$$

showing that *the chemical potential (2.341c) ≡ (2.338d) is the free enthalpy per particle (2.341c) ≡ (2.341d)*. Differentiating (2.341d) leads to (2.342a):

$$dG - \mu dN = N d\mu, \quad N d\mu = -S dT + V dp, \qquad (2.342a,\, b)$$

and substitution of (2.338d) in (2.342a) yields the **Gibbs (1876 - 1878) – Duhem (1886) relation** (2.342b) *for the differential of the chemical potential of a single chemical species.*

In the case of L chemical species the free enthalpy (2.341d) is additive (2.343a), (2.86c; 2.88f; 2.87j; 2.89f) ≡ (2.343a–d):

$$G = \sum_{\ell=1}^{L} \mu_\ell N_\ell = \sum_{\ell=1}^{L-1} \bar{\mu}_\ell N_\ell = \sum_{\ell=1}^{L} v_\ell \xi_\ell = \sum_{\ell=1}^{L-1} \bar{v}_\ell \xi_\ell, \qquad (2.343a\text{–}d)$$

and can be expressed in terms of: (i) mole numbers (2.343a, b) or mass fractions (2.343c, d); (ii) chemical potential (2.343a) [affinities (2.343c)] for all L substances; (iii) relative chemical potentials (2.343b) [relative affinities (2.343d)] for L − 1 substances. Differentiating (2.343b) leads to (2.343e) that may be compared with (2.120) ≡ (2.343f):

$$\sum_{\ell=1}^{L-1} \left(\bar{\mu}_\ell dN_\ell + N_\ell d\bar{\mu}_\ell \right) = dG = -S dT - \vec{D}\cdot d\vec{E} - \vec{B}\cdot d\vec{H} - S_{ij} dT_{ij} + \sum_{\ell=1}^{L-1} \bar{\mu}_\ell dN_\ell, \quad (2.343e,\, f)$$

thus proving the **generalized Gibbs – Duhem relation** (2.344a–d) *in terms of mole numbers (mass fractions) and absolute or relative chemical potentials (affinities), including besides chemical also thermal, electric, magnetic and elastic effects:*

$$\sum_{\ell=1}^{L} N_\ell d\mu_\ell = \sum_{\ell=1}^{L-1} N_\ell d\bar{\mu}_\ell = \sum_{\ell=1}^{L} \xi_\ell dv_\ell = \sum_{\ell=1}^{L-1} \xi_\ell d\bar{v}_\ell \qquad (2.344a\text{–}d)$$
$$= -S dT - \vec{D}\cdot d\vec{E} - \vec{B}\cdot d\vec{H} - S_{ij} dT_{ij}.$$

The first (second) principle of thermodynamics [Section 2.1 (2.3)] concerns equilibrium (stability) and leads to the constitutive (dissipative) properties of matter [Section 2.2 (2.4)].

2.4   ENTROPY PRODUCTION AND DIFFUSIVE PROPERTIES

There are analogies in the way the first (second) principle of thermodynamics [Section 2.1 (2.3)] lead to the constitutive (diffusive) tensors [Section 2.2 (2.4)] that specify the properties of matter in an equilibrium state (non-equilibrium evolution between two equilibrium states). A first case is the heat flux in a thermally conducting medium with non-uniform temperature (subsections 2.4.1–2.4.2), hence outside thermal equilibrium (subsections 2.1.24 and 2.3.5). The Ohmic electric current in an electrically conducting medium subject to an electric field (subsections 2.4.4) is another case of a dissipative process, in this instance electromagnetic (subsection 2.4.3). The combination of thermal and electrical conduction (subsections 2.4.5–2.4.6) is an example of the coexistence of more than one dissipative process, relating gradients to fluxes through diffusive tensors (subsections 2.4.5–2.4.8), that include couplings and symmetries, as for the constitutive tensors (subsections 2.2.5–2.2.16). The relation between viscous stresses and rates of strain in a viscous fluid (subsections 2.4.9–2.4.11) is a third dissipative process, with formal similarities with the stress-strain constitutive relations in elasticity (subsections 2.1.15–2.1.19 and 2.2.12–2.2.14).

The quadratic forms (subsections 2.2.20–2.2.22 and 2.2.25–2.2.27) play an important role in establishing conditions for: (i) constitutive coefficients to ensure positive energy (subsections 2.2.17–2.2.19, 2.2.23–2.2.24, 2.2.28); (ii) thermodynamic derivatives to ensure (subsections 2.3.16–2.3.17) minimum internal energy (maximum entropy) at stable equilibrium [subsections 2.3.18–2.3.19 (2.3.20–2.3.22)] and definiteness or indefiniteness of other functions of state (subsections 2.2.29 and 2.3.23); (iii) diffusive tensors to ensure entropy production in irreversible processes (subsections 2.4.1–2.4.11). The condition of positive-definite entropy production is if anything still more important because it shows that: (i) although thermal (electrical) conduction [subsections 2.4.1–2.4.2 (2.4.3–2.4.4)] can be coupled (subsections 2.4.5–2.4.8) the viscous stresses (subsections 2.4.9–2.4.11) are decoupled (subsection 2.4.12) as well as the pressure gradient (subsection 2.4.13); (ii) the electric current can (cannot) depend on the electric (magnetic) field (subsection 2.4.14) and can depend on their outer vector product (subsection 2.4.15).

The dissipative effects include: (i) the Fourier law relating the heat flux to the temperature gradient through the thermal conductivity (subsections 2.4.1–2.4.2); (ii) the Ohm law relating the electric current to the electric field through the electrical conductivity (subsections 2.4.3–2.4.4); (iii) the relation between viscous stresses and rates of strain for a Newtonian viscous fluid through the shear and bulk viscosities (subsections 2.4.9–2.4.11); (iv) the Hall electric current involving the outer product of the electric and magnetic fields (subsection 2.4.15); (v/vi) the Thomson (Peltier) effect of heat generation or absorption at the junction of two thermoelectric materials [subsection 2.4.16 (2.4.17)]; (vii) the Fick law relating the mass flux to the gradient of the concentration in a two-phase medium (subsections 2.4.18–2.4.19). The diffusion relations lead to heat type equations that can be generalized in the case of coupled diffusion, such as thermoelectric (subsection 2.4.20). The constitutive (diffusion) tensors and relations for anisotropic matter [Section 2.2 (2.4)] simplify considerably in the isotropic case. This is illustrated by considering the general equations for unsteady anisothermal piezoelectromagnetism in a linear anisotropic medium (subsection 2.4.21) and their simplifications for isotropic media and steady conditions (subsection 2.4.22). The latter equations are solved explicitly in the one-dimensional case of a parallel-sided slab with constant electric charges and currents and heat conduction specifying the electric and magnetic fields and the displacement and temperature (subsection 2.4.23).

2.4.1 HEAT CONDUCTION AND FLUX IN A DOMAIN

The total entropy in a three-dimensional domain D_3 of volume element dV is given by (2.345a) and the second principle of thermodynamics states that for an irreversible process of an isolated system it must be an increasing function (2.273b) of time (2.345b):

$$\bar{S} = \int_{D_3} S dV: \quad 0 < \dot{\bar{S}} \equiv \frac{d\bar{S}}{dt} = \int_{D_3} \frac{\partial S}{\partial t} dV = \int_{D_3} \frac{1}{T} \frac{\partial Q}{\partial t} dV, \qquad (2.345a\text{--}c)$$

where in (2.345c) was used the relation (2.6b) with the heat. The heat balance (1.307b) $\equiv$ (2.346b) without heat sources (1.307a) $\equiv$ (1.346a) specifies the entropy production (2.345c) in terms of the energy flux in (2.346c):

$$w = 0: \quad \frac{\partial Q}{\partial t} + \nabla \cdot \vec{G} = 0, \quad \dot{\bar{S}}_q = -\int_{D_3} \frac{1}{T} \left(\partial_i G_i \right) dV > 0. \qquad (2.346a\text{--}c)$$

Performing an integration by parts in the integrand of (2.346c) leads to (2.347):

$$\dot{\bar{S}}_q = -\int_{D_3} \left[\partial_i \left(\frac{G_i}{T} \right) - G_i \partial_i \left(\frac{1}{T} \right) \right] dV. \qquad (2.347)$$

The first term on the r.h.s. of (2.347) is transformed to a surface integral (2.348b) by the divergence theorem (III.5.163a–c):

$$\vec{G} \cdot \vec{N} \Big|_{\partial D} = 0: \quad \int_{D_3} \partial_i \left(\frac{G_i}{T} \right) dV = \int_{\partial D_3} \frac{G_i}{T} dS_i = 0; \qquad (2.348a\text{--}c)$$

since the system is isolated, the boundary must extend sufficiently far that it does not interact with the exterior, and thus the energy flux through the boundary in the normal direction is zero (2.348a) and the surface integral (2.348b) vanishes (2.348c). Thus only the second term on the r.h.s. of (2.347) remains

$$\dot{\bar{S}}_q \equiv \frac{dS_q}{dt} = -\int_{D_3} \frac{1}{T^2} G_i \left(\partial_i T \right) dV = -\int_{D_3} \frac{\vec{G} \cdot \nabla T}{T^2} dV, \qquad (2.349a\text{--}c)$$

showing that *the entropy production by heat conduction in a domain D_3 with volume element dV is given by (2.349a–c) in terms of the heat flux (2.346b) and temperature.* Applying to (2.349a–c) the Fourier (1818) law of heat conduction specifies the sign of the thermal conductivity (Section 2.4.2).

2.4.2   Thermal Conductivity (Fourier 1818) Scalar and Tensor

The **Fourier (1818) law** of heat conduction for a linear isotropic material states that (1.308b) ≡ (2.350a) the heat flux is proportional to the temperature gradient through minus the thermal conductivity scalar:

$$\vec{G} = -k\nabla T: \quad \dot{S}_q = \int_{D_3} k\frac{\nabla T \cdot \nabla T}{T^2} dV > 0 \Rightarrow k > 0, \qquad (2.350\text{a–d})$$

substitution of (2.350a) in (2.349c) leads to (2.350b); thus entropy production (2.350c) requires a positive thermal conductivity (2.350d) implying (2.350a) that heat flows from the higher to the lower temperatures.

In the case of **linear anisotropic heat conduction** the heat flux is still proportional to but not necessarily anti-parallel to the temperature gradient, and thus the properties of matter appear through the **thermal conductivity tensor** in (2.351a):

$$G_i = -k_{ij}\partial_j T: \quad \dot{S}_q = \int_{D_3} \frac{1}{T^2} k_{ij} (\partial_i T)(\partial_j T) dV > 0, \qquad (2.351\text{a–c})$$

substitution of (2.351a) in (2.349b) leads to (2.351b), showing that entropy production requires that the thermal conductivity tensor is positive-definite, that is satisfies the relations (2.199a–e; 2.205a, b) ≡ (2.352a–c):

$$k_{11} > 0, \quad \begin{vmatrix} k_{11} & k_{12} \\ k_{12} & k_{22} \end{vmatrix} > 0, \quad \begin{vmatrix} k_{11} & k_{12} & k_{13} \\ k_{12} & k_{22} & k_{23} \\ k_{13} & k_{23} & k_{33} \end{vmatrix} > 0, \qquad (2.352\text{a–c})$$

similar to those for the dielectric permittivity (2.215a–e) [magnetic permeability (2.216a–e)].

Thus *the condition of entropy production (2.350b) [(2.351c)] for the Fourier law of heat conduction in a linear isotropic (2.350a) [anisotropic (2.351a)] medium implies that the thermal conductivity scalar (tensor) is positive (2.350d) [positive-definite (2.351c) ≡ (2.352a–c)]*. Similar results apply to thermal (electric) conduction [subsection 2.4.2 (2.4.4)] based on the heat (electromagnetic energy) balance [subsection 2.4.1 (2.4.3)].

2.4.3   Electromagnetic Energy Density, Flux and Dissipation

In order to consider the electric conductivity/resistivity (subsection 2.4.4) it is necessary to start with the energy balance for the electromagnetic field. The rate of change with time of the electric (2.25b) [magnetic (2.50b)] energy is given by (2.353a) [(2.353b)]:

$$\frac{\partial E_e}{\partial t} = \vec{E} \cdot \frac{\partial \vec{D}}{\partial t}, \quad \frac{\partial E_h}{\partial t} = \vec{H} \cdot \frac{\partial \vec{B}}{\partial t} : \quad E_{eh} = E_e + E_h, \quad \frac{\partial E_{eh}}{\partial t} = \vec{E} \cdot \frac{\partial \vec{D}}{\partial t} + \vec{H} \cdot \frac{\partial \vec{B}}{\partial t}, \quad (2.353\text{a--d})$$

leading to (2.353d) for the rate of change with time of the total **electromagnetic energy** (2.353c). The **Maxwell (1863) equations** for the unsteady electromagnetic field (2.354a, b):

$$\nabla \wedge \vec{E} = -\frac{1}{c} \frac{\partial \vec{B}}{\partial t}, \quad c \nabla \wedge \vec{H} = \vec{J} + \frac{\partial \vec{D}}{\partial t}, \quad\quad (2.354\text{a, b})$$

where $\vec{J}$ is the **electric current**, substituted in (2.353d) lead to (2.355a):

$$\frac{\partial E_{eh}}{\partial t} = c \left[\vec{E} \cdot \left(\nabla \wedge \vec{H} \right) - \vec{H} \cdot \left(\nabla \wedge \vec{E} \right) \right] - \vec{J} \cdot \vec{E} = -c \nabla \cdot \left(\vec{E} \wedge \vec{H} \right) - \vec{J} \cdot \vec{E} \quad (2.355\text{a, b})$$

where in (2.355b) was used the identity (2.46b).

Thus (2.355b) ≡ (2.356a–c) can be rewritten as *the electromagnetic energy balance (2.356a)*:

$$\frac{\partial E_{eh}}{\partial t} + \nabla \cdot \vec{G}^{eh} = -\dot{Q}_e : \quad \vec{G}^{eh} = c \vec{E} \wedge \vec{H}, \quad \dot{Q}_e = \vec{J} \cdot \vec{E}, \quad\quad (2.356\text{a--c})$$

*involving: (i) the time rate-of-change of the electromagnetic energy (2.353c) that is the sum of the electric (2.353a) and magnetic (2.353b) energies; (ii) the divergence of the **electromagnetic energy flux** (2.356b) or **Poynting (1884) vector** that is the outer vector product of the electric and magnetic fields multiplied by the speed of light in vacuo; (iii) the **Joule (1847) effect** of electrical dissipation (2.356c) specified by the inner vector product of the electric field and current. The electrical dissipation (iii) is considered next for the Ohmic electric current (subsection 2.4.4).*

2.4.4   Joule (1847) Effect, Ohm (1827) Law and Electrical Conductivity/Resistivity

The Ohm (1827) law in a linear isotropic medium states that the electric current is parallel to the electric field (2.315a) through the scalar **electrical conductivity** σ_e, whose inverse is the **electrical resistivity** $\dfrac{1}{\sigma_e}$:

$$\vec{J} = \sigma_e \vec{E} : \quad T \frac{dS_e}{dt} = \dot{Q}_e = \vec{J} \cdot \vec{E} = \frac{J^2}{\sigma_e} = E^2 \sigma_e > 0, \quad \sigma_e > 0. \quad\quad (2.357\text{a--g})$$

*A positive rate of dissipation (2.357f) by the Joule effect implies that: (i) the electrical conductivity (and resistivity) are positive (2.357g); (ii) the **conduction electric current** (2.357a) is parallel to the electric field; (iii/iv) the heat dissipation equals* *the square of the electric current (field) multiplied (2.357d) [(2.357e)] by the electric resistivity (conductivity). In the case of a linear anisotropic medium the Ohm's law states that the electric current is proportional (2.358a) but not necessarily parallel to the electric field, through the* **electric conductivity tensor**, *whose inverse (2.358b) is the* **electric resistivity tensor**:

$$J_i = \sigma_{ij} E_j, \quad E_i = \sigma_{ij}^{-1} J_j: \quad \dot{Q}_e = \vec{J} \cdot \vec{E} = \sigma_{ij} E_i E_j = \sigma_{ij}^{-1} J_i J_j > 0 \quad (2.358\text{a–f})$$

both positive-definite (2.358f) so that the dissipation by Joule effect is positive (2.358c–e). The coupling of thermal (electrical) conduction [subsections 2.4.1–2.4.2 (2.4.3–2.4.4)] is considered next (subsection 2.4.5).

2.4.5 COUPLING OF HEAT FLUX AND ELECTRIC CURRENT

As an example of coupling of two dissipative effects, in the presence of both a temperature gradient (an electric field), the Fourier (2.350a) [Ohm (2.357a)] laws gain cross terms (2.359a) [(2.359b)] involving the coupling diffusion parameters D_1 (D_2):

$$\vec{G} = -k\nabla T + D_1 \vec{E}, \quad \vec{J} = \sigma_e \vec{E} + D_2 \nabla T. \qquad (2.359\text{a, b})$$

The total entropy production per unit volume (2.360a) is the sum (2.360b) of the contribution of thermal (2.349c) [electrical (2.358c)] conduction

$$T \dot{S}_{qe} = T \dot{S}_q + \dot{Q}_e = -\frac{\vec{G} \cdot \nabla T}{T} + \vec{J} \cdot \vec{E}. \qquad (2.360\text{a, b})$$

Substituting in (2.360b) the coupled Fourier (2.359a) [Ohm (2.359b)] laws leads to the total **thermoelectric conductive entropy production** (2.361a):

$$\dot{S}_{qe} = \frac{k(\nabla T \cdot \nabla T)}{T^2} + \sigma_e \frac{(\vec{E} \cdot \vec{E})}{T} + \frac{1}{T}\left(D_2 - \frac{D_1}{T}\right)\left(\vec{E} \cdot \nabla T\right) > 0 \qquad (2.361\text{a, b})$$

that must be positive (2.361b).

The electric field $\vec{E}$ and temperature gradient ∇T can be chosen so large that their inner vector product is negative and violates (2.361b); this can be prevented iff the diffusion coefficients are related by (2.362a) so that the entropy production (2.361b; 2.362a) ≡ (2.362b) is always positive (2.362c):

$$D \equiv D_2 = \frac{D_1}{T}: \quad \dot{S}_{qe} = \frac{k}{T^2}|\nabla T|^2 + \frac{\sigma_e}{T} E^2 > 0. \qquad (2.362\text{a–c})$$

Thus *in order to have positive entropy production (2.361b) ≡ (2.362b) the* **coupling thermoelectric diffusion coefficient** *(2.362a) appears in both coupling terms of the combined Fourier (2.359a) ≡ (2.363a) [Ohm (2.359b) ≡ (2.363b)] laws of thermal (electrical) conduction:*

$$\vec{G} = -\nabla T + TD\vec{E}, \quad \vec{J} = \sigma_e \vec{E} + D\nabla T \qquad (2.363a, b)$$

*In the case of a linear anisotropic medium the combined **Fourier-Ohm law** of thermoelectric conduction specify the heat flux (2.364a) [electric current (2.364b)] involving besides the thermal (2.351a) [electrical (2.358a)] conductivity tensors a **coupling thermoelectric conductivity tensor**:*

$$G_i = -k_{ij}\partial_j T + TD_{ij}E_j, \quad J_i = \sigma_{ij}E_j + D_{ij}\partial_j T, \qquad (2.364a, b)$$

that cancels out of the entropy production (2.360b) ≡ (2.365a, b) that is always positive (2.365c):

$$\dot{S}_{qe} = -\frac{1}{T^2} G_i \partial_i T + \frac{1}{T} J_i E_i = \frac{1}{T^2} k_{ij}(\partial_i T)(\partial_j T) + \frac{1}{T}\sigma_{ij}E_i E_j > 0, \quad (2.365a\text{–}c)$$

for positive-definite thermal (2.352a–c) and electrical conductivity tensors. There is an analogy between the symmetry of the constitutive (diffusive) coefficients [subsection 2.2.5 (2.4.6)].

2.4.6   GRADIENTS, FLUXES AND RECIPROCITY (ONSAGER 1931)

*The analogues of the extensive (2.103b) [intensive (2.103d)] thermodynamic variables for the augmented internal energy (2.103a) are the **gradients** (2.366a) [**fluxes** (2.366b)] in the **entropy production** (2.365c) ≡ (2.366c):*

$$n = 1,\ldots,\tilde{N}: \quad \tilde{X}_n = \left\{-\frac{\nabla T}{T^2}, \frac{\vec{E}}{T}\right\}, \quad \tilde{Y}_n = \left\{\vec{G},\vec{J}\right\}; \quad \tilde{S} = \sum_{n=1}^{\tilde{N}} \tilde{Y}_n \tilde{X}_n \equiv \tilde{Y}_n \tilde{X}_n. \quad (2.366a\text{–}c)$$

For a linear matter the fluxes (2.366b) are proportional to the gradients (2.366a) through the **diffusion tensor** (2.367a):

$$\tilde{Y}_n = \tilde{Z}_{nm}\tilde{X}_m, \quad \dot{S} = \tilde{Z}_{nm}\tilde{X}_n\tilde{X}_m > 0, \quad \tilde{Z}_{nm} = \tilde{Z}_{mn}. \qquad (2.367a\text{–}d)$$

*The **diffusion relations** (2.367a) lead to an entropy production (2.367b) that is: (i) a quadratic form of the gradients; (ii) the coefficients are the diffusion tensors that form symmetric set (2.367d) leading to the **reciprocity principle (Onsager 1931)**; (iii) the requirement of entropy production (2.367c) implies that the quadratic form (2.367b) [diffusion tensor (2.367d)] is positive-definite.* The combined Fourier-Ohm law of thermoelectric conduction for linear isotropic (2.363a, b) [anisotropic (2.364a, b)] media is reconsidered [subsection 2.4.7 (2.4.8)] as an application of the irreversible thermodynamics of fluxes and gradients (subsection 2.4.6).

2.4.7   ISOTROPIC THERMOELECTRIC DIFFUSION SCALARS

In the case of the thermoelectric effect the fluxes (2.368a) [gradients (2.368b)] are: (i) for thermal conduction (2.349c; 2.366c) the heat flux $\vec{G}$ (minus T^{-2} times the temperature gradient); (ii) for electrical conduction (2.357b) ≡ (2.366c) the electric current $\bar{J}$ (T^{-1} times the electric field $\bar{E}$):

$$n = 1, ..., 6 \equiv \tilde{N}: \quad \bar{Y}_n = \{\bar{G}, \bar{J}\}, \quad \bar{X}_n = \{-T^{-2}\nabla T, T^{-1}\bar{E}\}. \qquad (2.368a, b)$$

The diffusion relations (2.363a, b) ≡ (2.367a) are (2.369):

$$\begin{bmatrix} \bar{G} \\ \bar{J} \end{bmatrix} = \begin{bmatrix} -kT^2 & DT^2 \\ DT^2 & \sigma_e T \end{bmatrix} \begin{bmatrix} T^{-2}\nabla T \\ T^{-1}\bar{E} \end{bmatrix}, \qquad (2.369)$$

using a symmetric set (2.367d) of diffusion coefficients in the matrix in (2.369) with the factors involving the temperature.

The thermoelectric diffusion relations (2.369) ≡ (2.367a): (i) coincide with the combined generalized Fourier-Ohm laws (2.363a, b); (ii) coincide with (2.367d) with the symmetry relation (2.362a) between thermoelectric diffusion coefficients in (2.359a, b); (iii) the latter is needed to comply with the second principle of thermodynamics ensuring positive entropy production (2.360b) ≡ (2.370a, b):

$$\dot{S}_{qe} = -\frac{\nabla T}{T^2} \cdot \left(-k\nabla T + TD\bar{E}\right) + \frac{\bar{E}}{T} \cdot \left(\sigma_e \bar{E} + D\nabla T\right) = k\frac{(\nabla T \cdot \nabla T)}{T^2} + \sigma_e \frac{(\bar{E} \cdot \bar{E})}{T} > 0: \text{(2.370a–d)}$$

$$k > 0 < \sigma_e$$

(iv) the condition of positive entropy production (2.370c) ≡ (2.362b) implies that the thermal (2.350d) ≡ (2.370d) [electrical (2.357g) ≡ (2.370e)] conductivities are both positive; (v) the thermoelectric coupling scalar D that appears in the diffusion relations (2.363a, b) ≡ (2.369) does not appear in the entropy production (2.362b) ≡ (2.370b) and thus has unrestricted sign. The combined Fourier-Ohm law (Section 2.4.5) for linear isotropic (anisotropic) media [subsection 2.4.7 (2.4.8)] both comply with positive entropy production leading to constraints on the diffusion scalars (tensors).

2.4.8 ANISOTROPIC THERMOELECTRIC DIFFUSION TENSORS

The combined thermoelectric entropy production (2.365a) ≡ (2.366c) involves the same fluxes (2.368a) and gradients (2.368b) in linear isotropic (anisotropic) media, and scalar (tensor) diffusion coefficients in (2.369) [(2.371)]:

$$\begin{bmatrix} G_i \\ J_i \end{bmatrix} = \begin{bmatrix} -k_{ij}T^2 & D_{ij}T^2 \\ D_{ij}T & \sigma_{ij}T \end{bmatrix} \begin{bmatrix} T^{-2}\partial_j T \\ T^{-1}E_j \end{bmatrix}, \qquad (2.371)$$

in agreement with the isotropic (anisotropic) combined Fourier-Ohm law (2.363a, b) ≡ (2.369) [(2.364a, b) ≡ (2.371)]. The entropy production (2.360b) is given for linear anisotropic thermoelectric conduction (2.371) by (2.372a):

$$\dot{S}_{qe} = -\frac{1}{T^2}\left(\partial_i T\right)\left(-k_{ij}\partial_j T + TD_{ij}E_j\right) + \frac{1}{T}E_i\left(\frac{1}{T}D_{ij}\partial_j T + \sigma_{ij}E_j\right)$$

$$= \frac{1}{T^2}k_{ij}\left(\partial_i T\right)\left(\partial_j T\right) + \frac{1}{T}\sigma_{ij}E_i E_j > 0, \qquad (2.372\text{a–c})$$

and is positive iff the thermal (2.352a–c) [electrical (2.373a–c)] conduction tensors are positive-definite:

$$\sigma_{11} > 0, \quad \begin{vmatrix} \sigma_{11} & \sigma_{12} \\ \sigma_{12} & \sigma_{22} \end{vmatrix} > 0, \quad \begin{vmatrix} \sigma_{11} & \sigma_{12} & \sigma_{13} \\ \sigma_{12} & \sigma_{22} & \sigma_{23} \\ \sigma_{13} & \sigma_{23} & \sigma_{33} \end{vmatrix} > 0. \qquad (2.373\text{a–c})$$

Besides thermal (electric) conduction [subsections 2.4.1–2.4.2 (2.4.3–2.4.4)] and their combination (subsections 2.4.5–2.4.8), another diffusion process is the viscous stresses associated with the rates of strain in a viscous fluid (subsection 2.4.9).

2.4.9 STRESSES AND RATES-OF-STRAIN FOR A VISCOUS FLUID

There is an analogy between the constitutive (diffusion) relation of the elastic (viscous) stresses with the strains (rates-of-strain) in an elastic solid (2.137a) [viscous fluid (2.374a)]:

$$\tau_{ij} = D_{ijk\ell}\dot{S}_{k\ell}: \quad D_{ijk\ell} = D_{jik\ell} = D_{ij\ell k} = D_{k\ell ij}, \qquad (2.374\text{a–d})$$

with the elastic stiffness (2.137e–g) [**viscosity tensor** (2.374b–d)] having the same symmetries for a linear anisotropic medium. The strain (2.82d) (rate of strain (2.375b,c)] tensor is a similar symmetric part of the matrix of partial spatial derivatives of the displacement [velocity (2.375a)] vector:

$$v_i = \frac{\partial u_i}{\partial t} = \dot{u}_i: \quad \dot{S}_{ij} = \frac{1}{2}\left(\partial_i \dot{u}_j + \partial_j \dot{u}_i\right) = \frac{1}{2}\left(\partial_i v_j + \partial_j v_i\right). \qquad (2.375\text{a–c})$$

The trace or sum of diagonal elements of the strain (rate of strain) tensor is the volume change (2.85f) [**dilatation** or divergence of the velocity (2.376a)]:

$$\dot{S}_{ii} = \partial_i v_i = \nabla \cdot \vec{v}; \quad \tilde{\dot{S}}_{ij} \equiv \dot{S}_{ij} - \frac{1}{3}\dot{S}_{kk}\delta_{ij}, \quad \tilde{\dot{S}}_{ii} = 0, \qquad (2.376\text{a–c})$$

the **sliding rate of shear tensor** (2.376b) subtracts from the diagonal elements of the rate of strain tensor one-third of the dilatation, so that the trace (2.376c) becomes zero, and thus it represents a pure rate-of-shear without dilatation. In the case of an isotropic fluid the viscosity tensor reduces to two scalars, namely the shear and bulk viscosities for a Newtonian fluid (subsection 2.4.10).

2.4.10   SHEAR AND BULK VISCOSITIES FOR A NEWTONIAN FLUID

For a linear isotropic or **Newtonian fluid** *the viscous stresses (2.376a) can depend on the rates-of-strain (2.375a–c) only as a linear combination of the sliding rate of shear tensor (2.376b) [dilatation (2.376a)] with scalar coefficients, namely the* **shear (bulk) viscosity** ζ (η):

$$\tau_{ij} = 2\zeta \dot{S}_{ij} + \eta \dot{S}_{kk} \delta_{ij} = \zeta \left(\partial_i v_j + \partial_j v_i \right) + \left(\eta - \frac{2}{3}\zeta \right) \dot{S}_{kk} \delta_{ij}. \qquad (2.377a, b)$$

The momentum equation balances the inertia force against the external forces per unit volume and divergence of the viscous stress tensor (2.378a) $\equiv$ (IV.6.538):

$$\rho \dot{v}_i = f_i + \partial_j \tau_{ij}: \quad \dot{W} = f_i v_i = \rho v_i \dot{v}_i - v_i \partial_j \tau_{ij}, \qquad (2.378a\text{–}c)$$

and the power or activity or work per unit time of the external forces (2.378b) is given by (2.378c). The first term on the r.h.s. of (2.378c) is the rate of change with time of the kinetic energy (2.11b) $\equiv$ (2.379a):

$$\dot{W} - \dot{E}_k = -v_i \partial_j \tau_{ij} = -\partial_j \left(v_i \tau_{ij} \right) + \tau_{ij} \partial_j v_i. \qquad (2.379a\text{–}c)$$

   The difference between the power of the external forces and the rate of change of the kinetic energy is written in the form (2.379c) $\equiv$ (2.380a):

$$\frac{\partial}{\partial t}\left(W - E_k \right) + \nabla \cdot \vec{G}_v = \dot{Q}_v, \quad G_{vi} = v_j \tau_{ij}, \quad \dot{Q}_v = \tau_{ij} \dot{S}_{ij}, \qquad (2.380a\text{–}c)$$

of the energy balance involving: (i) the time rate-of-change of the power of the external forces (2.378b) minus the time rate-of-change of the kinetic energy (2.11b); (ii) the divergence of the **viscous energy flux** *(2.380b) that is the inner product of the velocity (2.375a) by the viscous stress tensor; (iii) the* **viscous dissipation** *(2.380c) that is the double inner product of the viscous stress tensor by the rate of strain tensor (2.375a–c) because* τ_{ij} *is symmetric, and thus* $\partial_j v_i$ *on the r.h.s. of (2.379c) can be replaced by* $\dot{S}_{ij}$ *on the r.h.s. of (2.380c), as follows from (2.381a–c):*

$$\tau_{ij} \partial_j v_i = \frac{1}{2}\left(\tau_{ij} + \tau_{ji} \right)\left(\partial_j v_i \right) = \frac{1}{2}\tau_{ij}\left(\partial_j v_i + \partial_i v_j \right) = \tau_{ij}\dot{S}_{ij}. \qquad (2.381a\text{–}c)$$

It is shown next that a positive dissipation implies that the shear and bulk viscosities are positive for a Newtonian fluid (2.377a, b) for which the momentum equation becomes the Navier-Stokes equation (subsection 2.4.11).

2.4.11 POSITIVE VISCOSITIES IN THE NAVIER (1822) – STOKES (1845) EQUATION

The viscous dissipation (2.380c) for a Newtonian fluid (2.377a) is given by (2.382a) where are used (2.376a–c) leading to (2.382b, c):

$$
\dot{Q}_v = \left(2\zeta \dot{\bar{S}}_{ij} + \eta \dot{S}_{kk}\delta_{ij}\right)\left(\dot{\bar{S}}_{ij} + \frac{1}{3}\dot{S}_{kk}\delta_{ij}\right)
$$

$$
= 2\zeta \dot{\bar{S}}_{ij}\dot{\bar{S}}_{ij} + \left(\eta + \frac{2}{3}\zeta\right)\dot{S}_{kk}\dot{S}_{ii} + \frac{\eta}{3}\left(\dot{S}_{kk}\right)^2 \delta_{ii} \qquad (2.382a\text{–}c)
$$

$$
= \frac{\zeta}{2}\left(\partial_i v_j + \partial_j v_i - \frac{2}{3}\partial_k v_k\right)^2 + \eta \left(\nabla \cdot \vec{v}\right)^2.
$$

The condition of positive (2.383a) viscous dissipation (2.382c) implies that both the shear (2.383b) and bulk (2.383c) viscosities are positive:

$$
Q_v > 0 \Rightarrow \zeta > 0 < \eta. \qquad (2.383a\text{–}c)
$$

In the momentum equation (2.378a) the isotropic pressure is subtracted (2.384a) from the viscous stresses (2.384b):

$$
f_i = -\partial_i p: \quad \partial_j \left(\tau_{ij} - p\delta_{ij}\right) = \rho \dot{v}_i = \rho \frac{dv_i}{dt} = \rho \left(\frac{\partial v_i}{\partial t} + \frac{\partial v_i}{\partial x_j}\frac{dx_j}{dt}\right)
$$

$$
= \rho \left(\frac{\partial v_i}{\partial t} + v_j \frac{\partial v_j}{\partial x_j}\right), \qquad (2.384a\text{–}e)
$$

and the **acceleration** is understood as the total time derivative of the velocity (2.384c) leading to the **material derivative** (2.384d, e).

The divergence (2.385c) of the viscous stresses for a Newtonian fluid (2.377b) with constant shear (2.385a) and bulk (2.385b) viscosities:

$$
\zeta, \eta = const: \quad \partial_j \tau_{ij} = \zeta \left(\partial_{jj} v_i + \partial_i \partial_j v_j\right) + \left(\eta - \frac{2}{3}\zeta\right)\partial_i \partial_j v_j
$$

$$
= \zeta \partial_{jj} v_i + \left(\eta + \frac{\zeta}{3}\right)\partial_i \partial_j v_j, \qquad (2.385a\text{–}c)
$$

can be put in vector form (2.385c) for substitution in the momentum equation (2.384d) leading to (2.386):

$$
\rho\left[\frac{\partial \vec{v}}{\partial t} + \left(\vec{v}\cdot\nabla\right)\vec{v}\right] + \nabla p = \vec{f} + \zeta \nabla^2 \vec{v} + \left(\eta + \frac{\zeta}{3}\right)\nabla\left(\nabla\cdot\vec{v}\right), \qquad (2.386)
$$

that is (2.386) ≡ (II.4.339) the **Navier (1822) - Stokes (1845)** equation *balancing: (i) the inertia force, equal to the mass density (2.7a) multiplied by the acceleration (2.384c–e); (ii) the pressure gradient; (iii) the external forces per unit volume; (iv) the divergence of the viscous stresses (2.385c) for a Newtonian fluid (2.377b) with constant shear (2.385a) and bulk (2.385b) viscosities.* It is shown next (subsection 2.4.12) that unlike thermal (electric) conduction [subsections 2.4.1–2.4.2 (2.4.3–2.4.4)] that are coupled (subsections 2.4.5–2.4.8) viscous dissipation (subsections 2.4.9–2.4.11) is decoupled.

2.4.12 DECOUPLING OF VISCOSITY FROM THERMAL AND ELECTRICAL CONDUCTION

In hydrodynamics (magnetohydrodynamics) the presence of thermal (and electrical) conduction in a viscous (and electrically conducting) fluid could lead to a coupling of both (interaction of all three). It is shown next that this possibility is excluded by the second principle of thermodynamics requiring positive entropy production in the case of linear isotropic media. Taking as gradients (2.387a) the temperature gradient, electric field and rates of strain and as fluxes (2.387b) the heat flux, electric current and viscous stresses the entropy production is given (2.360b; 2.380c) by (2.387c–e):

$$n = 1,...,12: \quad \tilde{X}_n = \left\{ -T^{-2}\nabla T, T^{-1}\vec{E}, T^{-1}\dot{S}_{ij} \right\}, \quad \tilde{Y}_n = \left\{ \vec{G}, \vec{J}, \tau_{ij} \right\} \quad (2.387a, b)$$

$$\dot{S}_{qev} = \dot{S}_{qe} + T^{-1}\dot{Q}_v = -\frac{G_i \partial_i T}{T^2} + \frac{J_i E_i}{T} + \frac{\tau_{ij}\dot{S}_{ij}}{T} = \sum_{n=1}^{12} \tilde{X}_n \tilde{Y}_n. \quad (2.387c–e)$$

Starting with the most general case of all fluxes depending on all gradients in a linear anisotropic medium leads to the diffusion relations:

$$G_i = -k_{ij}\partial_j T + TD_{ij}E_j + TD_{3ijk}\dot{S}_{jk}, \quad (2.388a)$$

$$J_i = D_{ij}\partial_j T + \sigma_{ij}E_j + D_{4ijk}\dot{S}_{jk}, \quad (2.388b)$$

$$\tau_{ij} = D_{3ijk}\partial_k T - D_{4ijk}E_k + D_{ijk\ell}\dot{S}_{k\ell}, \quad (2.388c)$$

where in addition to the thermal (electrical) conductivity tensors (2.351a) [(2.358a)] and viscosity double tensor (2.374a), and thermoelectric coupling tensor (2.364a, b) appear two new tensors with three indices accounting for **thermoviscous (electroviscous) coupling** if it exists. Substituting (2.388a–c) in (2.387c), the cross-terms involving the tensor D_{3ijk} and D_{4ijk} cancel, leading to the combined thermal, electric and viscous entropy production (2.389a):

$$\dot{S}_{qev} = \frac{1}{T^2}k_{ij}(\partial_i T)(\partial_j T) + \frac{1}{T}\sigma_{ij}E_i E_j + \frac{1}{T}D_{ijk\ell}\dot{S}_{ij}\dot{S}_{k\ell} > 0, \quad (2.389a, b)$$

that is positive (2.389b). In an isotropic medium the thermoviscous (2.390a) [thermo-electric (2.390b)] coupling diffusion tensors would have to be proportional to the permutation symbol (2.54a–d) that is skew-symmetric:

$$\left\{D_{3ijk}, D_{4ijk}\right\} = \left\{D_3, D_4\right\} e_{ijk} = \left\{D_{3ikj}, D_{4ikj}\right\} \Rightarrow D_3 = 0 = D_4, \qquad (2.390a–f)$$

since both diffusion tensors must be symmetric in the last two indices (2.390c) [(2.390d)] both must vanish (2.390e) [(2.390f)]. Thus *in a linear anisotropic medium heat conduction (2.388a), electric currents (2.388b) and viscous stresses (2.388c) can interact leading to positive entropy production (2.389a, b) in the presence of thermoviscous D_{3ijk} and thermoelectric D_{4ijk} coupling. The latter vanish in the isotropic case (2.390a–f) so that the thermal and electric conduction remain coupled (2.364a, b) and the viscous stresses (2.377a, b) decouple from both.* Another possibility excluded by the second principle of thermodynamics is the coupling of the pressure gradient to thermoelectric conduction (Section 2.4.13).

2.4.13 DECOUPLING OF THERMOELECTRIC CONDUCTION FROM THE PRESSURE GRADIENT

If in the case of thermoelectric fluxes (2.368a) ≡ (2.391a) the pressure gradient was added (2.391a) to the gradients (2.368b) the system would no longer be square with:

$$\tilde{Y}_n = \left\{\vec{G}, \vec{J}\right\}, \quad \tilde{X}_n = \left\{-T^{-2}\nabla T, T^{-1}\vec{E}, T^{-1}\nabla p\right\}, \qquad (2.391a, b)$$

and the combined Fourier–Ohm law (2.364a, b) would gain two new terms proportional to the pressure gradient (2.392a, b):

$$G_i = -k_{ij}\partial_j T + TD_{ij}E_j + D_{5ij}\partial_j p, \qquad (2.392a)$$

$$J_i = D_{ij}\partial_j T + \sigma_j E_j + D_{6ij}\partial_j p, \qquad (2.392b)$$

involving two new diffusion tensors. The entropy production (2.360b) would be given by (2.393):

$$\dot{S}_{qep} = \frac{1}{T^2}k_{ij}\left(\partial_i T\right)\left(\partial_j T\right) + \frac{1}{T}\sigma_{ij}E_i E_j + \frac{1}{T}\left(D_{6ij}E_i - \frac{1}{T}D_{5ij}\partial_i T\right)\partial_j p. \qquad (2.393)$$

The last two terms on the r.h.s. of (2.392) have no fixed sign, and could lead to a negative value for the overall sum. Thus the requirement of positive entropy production (2.394a) implies (2.394b, c):

$$\dot{S}_{qep} > 0 \Rightarrow D_{5ij} = 0 = D_{6ij}, \qquad (2.394a–c)$$

that the pressure gradient cannot appear in the thermoelectric conduction law (2.392a, b) ≡ (2.364a, b) for a linear anisotropic medium, and does not affect entropy production (2.393) ≡ (2.365c). Next the dependence of the electric current is considered (subsection 2.4.14) both for: (i) the electric and magnetic fields; (ii) for isotropic and anisotropic media; (iii) for linear and non-linear quadratic media.

2.4.14 ELECTRIC CURRENT IN LINEAR AND NON-LINEAR MEDIA

Considering a linear anisotropic medium the electric current could possibly depend on both the electric and magnetic fields, adding to the Ohm's law (2.358a) another term (2.394a) involving an electromagnetic diffusion tensor, if it exists:

$$J_i = \sigma_{ij} E_j + D_{7ij} H_j; \quad \dot{Q}_{eh}^{(1)} = \sigma_{ij} E_i E_j + D_{7ij} E_i H_j > 0 \Rightarrow D_{7ij} = 0, \quad (2.395a\text{–}d)$$

the dissipation (2.358c) gains an extra term (2.395b) that has no fixed sign, and thus the requirement that it be positive (2.395c) excludes this term (2.395d) from Ohm's law (2.395a) ≡ (2.358a). It has been shown that *for a linear anisotropic medium the Ohm's law specifies an electric current that is (cannot be) proportional to the electric (magnetic) field, and thus involves only the electrical conductivity.*

For a non-linear quadratic anisotropic medium to Ohm's law (2.358a) could be added (2.396) quadratic terms in the electric and magnetic fields and in their cross–product:

$$J_i - \sigma_{ij} E_j = D_{8ijk} E_j E_k + D_{9ijk} H_j H_k + D_{10ijk} E_j H_k. \tag{2.396}$$

The Joule dissipation (2.357c) would be given by (2.397a) and would have to be positive (2.397b)

$$\dot{Q}_{eh}^{(2)} = J_i E_i = \sigma_{ij} E_i E_j + D_{8ijk} E_i E_j E_k + D_{9ijk} E_i H_j H_k + D_{10ijk} E_i E_j H_k > 0. \quad (2.397a, b)$$

In an isotropic medium the three new constitutive tensors have to be proportional (2.398a–c) to the permutation symbol (2.54a–d):

$$\{D_{8ijk}, D_{9ijk}, D_{10ijk}\} = \{D_8, D_9, \sigma_h\} e_{ijk} \Rightarrow D_8 = D_9 = 0 \neq \sigma_h, \tag{2.398a–f}$$

and: (i) since D_{8ijk} and D_{9ijk} are symmetric in jk in (2.396) they must both vanish (2.398d, e); (ii) there no symmetry in jk in D_{10jk} so that it need not vanish (2.398f). Substituting (2.398a–f) in (2.396):

$$J_i = \sigma_{ij} E_j + \sigma_h e_{ijk} E_j H_k, \tag{2.398g}$$

shows that only one non-linear term remains in the electric current (2.398d). In vector form *the electric current in an isotropic medium (2.398g) ≡ (2.399a) consists of: (i) the Ohm effect (2.357a) that is linear in the electric field through the Ohmic electrical conductivity; (ii) the **Hall (1879) effect** that is quadratic and proportional to the outer vector product of the electric and magnetic field through the **Hall electrical conductivity**:*

$$\vec{J} = \sigma_e \vec{E} + \sigma_h \vec{E} \wedge \vec{H}: \quad \dot{Q}_{eh} = \vec{J} \cdot \vec{E} = \sigma_e E^2 = \dot{Q}_e, \tag{2.399a–c}$$

the Hall effect causes no dissipation (2.357b) ≡ (2.398b) that is due entirely to the Ohm effect (2.398c). The Ohm and Hall effects are compared next (subsection 2.4.15) including their relevance for ionic propulsion.

2.4.15 Ohm (1827) and Hall (1879) Effects and Ionic Propulsion

*An electric charge q (current $\vec{J}$) in an electric $\vec{E}$ (magnetic $\vec{H}$) field is associated with an **electric (magnetic) force** (2.19a) ≡ (2.400a) [(2.39a) ≡ (2.400b)]:*

$$\vec{F}_e = q\vec{E} = \frac{q}{\sigma_e}\vec{J}, \quad \vec{F}_h = \frac{\vec{J}\wedge\vec{B}}{c} = \frac{\mu}{c}\left(\vec{J}\wedge\vec{H}\right) = \frac{\mu\sigma_e}{c}\left(\vec{E}\wedge\vec{H}\right) = \frac{\mu\sigma_e}{c^2}\vec{G}_{eh}, \quad (2.400\text{a–f})$$

in the magnetic force (2.400c) c is the speed of light in vacuo and may be introduced: (i) the magnetic permeability in (2.400d) for an isotropic medium (2.53b); (ii) the electrical conductivity in (2.400e) from Ohm's law (2.357a); (iii) the Poyinting vector or electromagnetic energy flux (2.356b) in (2.400f). The Ohm law (2.357a) may also be introduced in the electric force (2.400b).

The Ohm (Hall) effect in the first (second) term of the electric current (2.399a) corresponds to the linear (spiralling) motion of electric charges in an electric (magnetic) field (2.401a) [(2.401c)]:

$$\vec{J}_e = \sigma_e\vec{E}, \quad \vec{F}_e = \frac{q}{\sigma_e}\vec{J}; \quad (2.401\text{a, b})$$

$$\vec{J}_{eh} = \sigma_h\vec{E}\wedge\vec{H}, \quad \vec{F}_{eh} = \frac{\mu\sigma_h}{c}\left[\left(\vec{E}\wedge\vec{H}\right)\wedge\vec{E}\right] = \frac{\mu\sigma_h}{c}\left[E^2\vec{H} - \left(\vec{E}\cdot\vec{H}\right)\vec{E}\right] \quad (2.401\text{c–e})$$

and leads to the electric (2.401b) [magnetic (2.401d, e)] force. The electric and magnetic forces used to accelerate electric charges lead to **ionic propulsion**. Even if the thrust is small, over a long time in the vacuum of space, in the absence of drag, very high speeds can be reached. Besides the thermoelectric coupling in the Fourier-Ohm law (subsections 2.4.1–2.4.8) the Thomson (1851) [Peltier (1834)] effects are considered next [subsection 2.4.16 (2.4.17)].

2.4.16 Thomson (1851) Thermoelectric Effect Including Convection

The electrostatic energy (2.20b) ≡ (2.402a) associated with electric charges q moving with velocity $\vec{v}$ in an electric potential Φ_e leads to a non-diffusive energy flux (2.402b) involving (2.402c, d); the **convective electric current** (2.402e) is equal to the electric charge density multiplied by the velocity.

$$\vec{E}_e = q\Phi_e: \quad \vec{G}_e = E_e\vec{v} = q\vec{v}\Phi_e = \Phi_e\vec{J}, \quad \vec{J} = q\vec{v}, \quad (2.402\text{a–e})$$

The latter is included together with the diffusive energy flux (2.363a) in the total energy flux (2.403a):

$$\vec{G}_* = \vec{G} + \vec{G}_e = \Phi_e\vec{J} + TD\vec{E} - k\nabla T = \Phi_e\vec{J} + TD\frac{\vec{J} - D\nabla T}{\sigma_e} - k\nabla T, \quad (2.403\text{a–c})$$

where (2.363a) [(2.363b)] is used in (3.403b) [(2.403c)]. Rewriting (2.403c) ≡ (2.404a):

$$k_* = k + \frac{TD^2}{\sigma_e} > 0, \quad \vec{G}_* = \left(\Phi_e + \frac{TD}{\sigma_e} \right) \vec{J} - k_* \nabla T, \quad \text{(2.404a, b)}$$

follows that *the total energy flux (2.404b) consists of: (i) the convective energy flux (2.402b–e) associated with the electric energy (2.402a) due to the electric charge and potential; (ii) the diffusive energy flux proportional to the electric current through* $\dfrac{TD}{\sigma_e}$ *involving thermoelectric coupling diffusion coefficient; (iii) the latter appears in the* **extended thermal conductivity** *in (2.404a) including the thermoelectric effect in the generalized Fourier-Ohm law; (iv) in the absence of thermo-electric coupling then* $k_* = k$ *in (2.404a) is the ordinary thermal conductivity in the Fourier law (2.350a).*

The heat release per unit time associated with the energy flux (2.404b) is given by (2.405c, d):

$$\nabla \cdot \vec{J} = 0, \vec{E} = -\nabla \Phi_e:$$

$$\dot{Q}_* = -\nabla \cdot \vec{G}_* = \nabla \cdot \left(k_* \nabla T \right) - \vec{J} \cdot \nabla \Phi_e - \vec{J} \cdot \nabla \left(\frac{TD}{\sigma_e} \right) = \nabla \cdot \left(k_* \nabla T \right) + \vec{J} \cdot \vec{E} - \vec{J} \cdot \nabla \left(\frac{TD}{\sigma_e} \right)$$

$$= \nabla \cdot \left(k_* \nabla T \right) + \vec{J} \cdot \frac{\vec{J} - D\nabla T}{\sigma_e} - \vec{J} \cdot \nabla \left(\frac{TD}{\sigma_e} \right) \qquad \text{(2.405a–g)}$$

$$= \nabla \cdot \left(k_* \nabla T \right) + \frac{J^2}{\sigma_e} - \vec{J} \cdot \left[\frac{2D}{\sigma_e} \nabla T + T \nabla \left(\frac{D}{\sigma_e} \right) \right],$$

where were used: (i) the conservation of the electric current (2.405a) in (2.405d); (ii) the relation between the electric field and potential (2.405b) in (2.405e); (iii) the Fourier-Ohm law (2.363b) in (2.405f, g). From (2.405g) $\equiv$ (2.406a, 2.46a, 2.46a, b):

$$\nabla \bar{D} = \nabla \left(\frac{D}{\sigma_e} \right) + \frac{2D}{\sigma_e T} \nabla T: \quad \dot{Q}_* = \nabla \cdot \left(k_* \nabla T \right) + \frac{J^2}{\sigma_e} - T \vec{J} \cdot \nabla \bar{D}, \quad \text{(2.406a, b)}$$

it follows that *the heat release per unit time (2.406b) is the sum of three effects: (i) thermal conduction by the* **Fourier effect** *(2.404a); (ii) electrical resistivity* $\dfrac{1}{\sigma_e}$ *by the* **Joule effect** *(2.357d);(iii) the coupling leads to the Thomson effect which differs from the Joule effect in being proportional to the electric current, so that both heat release or absorption is possible, for example, changing the direction of the electric current. The third term (iii) in the thermoelectric heat release rate (2.406b)* $\equiv$ *(2.406a) involves the* **Thomson diffusion coefficient** *given by (2.407b):*

$$\dot{Q}_* = \nabla \cdot \left(k_* \nabla T \right) + \frac{J^2}{\sigma_e} + D_t \left(\vec{J} \cdot \nabla T \right): \quad D_t \equiv -T \frac{\partial \bar{D}}{\partial T}, \quad \text{(2.407a, b)}$$

as follows from (2.407c, d):

$$-T \vec{J} \cdot \nabla \bar{D} = -T \left(\vec{J} \cdot \nabla T \right) \frac{\partial \bar{D}}{\partial T} = D_t \frac{\partial \bar{D}}{\partial T}. \quad \text{(2.407c, d)}$$

The Thomson (Peltier) effect [subsection 2.4.16 (2.4.17)] relates the electric current to temperature gradient inside a body (at the junction of two bodies).

2.4.17 VOLTA (1821) – SEEBACK (1822) – PELTIER (1834) EFFECT AND THERMOELECTROMOTIVE FORCE

Another thermoelectric effect appears at the junction of two metals (Figure 2.9). The continuity of the temperature (2.407a) and normal components of the energy flux (2.408b) and electric current (2.408d, e) shows (2.406b) there is a surface heat release:

$$T^+ = T^-, \quad \vec{N} \cdot \vec{G}_*^+ = \vec{N} \cdot \vec{G}_*^-, \quad \vec{N} \cdot \vec{J}^+ = \vec{N} \cdot \vec{J}^- \equiv J_n :$$
$$G_{tn} = \vec{N} \cdot \left(k^- \nabla T^- - k^+ \nabla T^+ \right) + J_n T \left(\bar{D}^- - \bar{D}^+ \right),$$

(2.408a–e)

namely the **Volta (1821) - Seeback (1822) - Peltier (1834) effect**: *at the junction of two distinct thermoelectric materials there is a heat release or absorption proportional to the difference of thermoelectric diffusion coefficient multiplied by minus the normal component of the electric current; thus the heat release reverses into absorption changing the direction of the current, in proportion (2.409a, b) to the **Peltier diffusion coefficient** (2.409c):*

$$\left[-k \frac{\partial T}{\partial n} \right] = -J_n T \left(\bar{D}^+ - \bar{D}^- \right) = -J_n D_p, \quad D_p \equiv T \left(\bar{D}^+ - \bar{D}^- \right).$$

(2.409a–c)

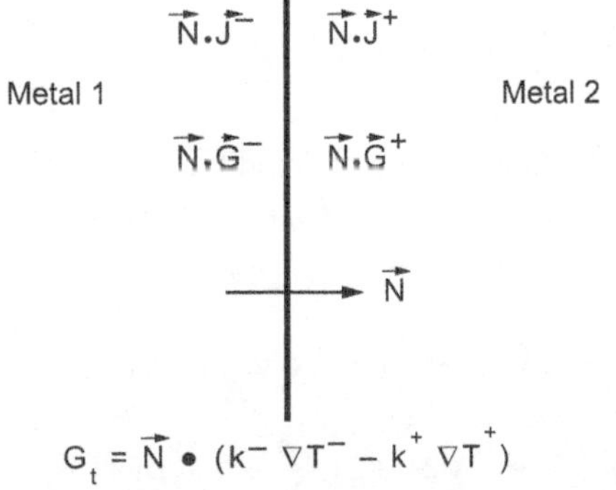

FIGURE 2.9 The Volta-Seebeck-Peltier thermoelectric effect at the junction of two metals at different temperatures: (i) preserves the continuity of the normal components of the electric current and energy flux; (ii) when one is reversed the other is also reversed, so that changing the direction of the electric current changes the energy flux from absorption to generation.

The Thomson (2.407b) and Peltier (2.409c) diffusion coefficients are related by (2.410a, 2.46a, 2.46a, b):

$$D_t = -T \frac{\partial \bar{D}}{\partial T} = -T \frac{\partial}{\partial T}\left(\frac{D_p}{T}\right), \qquad (2.410\text{a, b})$$

showing that both effects coexist in thermoelectric materials. A thermoelectric material in the absence of electric currents (2.411a, b) is subject to a **thermoelectromotive force** *per unit charge (2.411c–e):*

$$0 = \frac{\bar{J}}{\sigma_e} = \bar{E} + \frac{D}{\sigma_e}\nabla T: \quad \bar{F}_t = q\int \bar{E}\cdot d\bar{x} = -q\int \frac{D}{\sigma_e}\left(\nabla T \cdot d\bar{x}\right) = -q\int \frac{D}{\sigma_e}dT$$

$$= -q\int \bar{D}dT = -q\int \frac{D_p}{T}dT, \qquad (2.411\text{a–g})$$

that relates through (2.406a; 2.409c) to the Peltier coefficient (2.411f, g). Besides thermoelectric (viscous) diffusion [subsections 2.4.1–2.4.8 and 2.4.14–2.4.17 (2.4.8–2.4.13)] there is also mass diffusion that is considered for a two-phase medium (subsection 2.4.18).

2.4.18   MASS DIFFUSION (FICK 1855) IN A TWO-PHASE MEDIUM

Consider a medium with two phases with mass densities ρ_1 and ρ_2 moving at the same velocity $\bar{v}$, so that the conservation of mass is specified by (2.412a,2.46a, 2.46a, b):

$$\frac{\partial \rho_1}{\partial t} + \nabla\cdot\left(\rho_1\bar{v}\right) = \nabla\cdot\bar{I}, \quad \frac{\partial \rho_2}{\partial t} + \nabla\cdot\left(\rho_2\bar{v}\right) = -\nabla\cdot\bar{I}, \qquad (2.412\text{a, b})$$

where the **diffusive mass flux** $\bar{I}$ has opposite signs, so that the total mass (2.413a) is conserved (2.413b, c):

$$\rho \equiv \rho_1 + \rho_2: \quad 0 = \frac{\partial}{\partial t}\left(\rho_1 + \rho_2\right) + \nabla\cdot\left[\left(\rho_1 + \rho_2\right)\bar{v}\right] = \frac{\partial \rho}{\partial t} + \nabla\cdot\left(\rho\bar{v}\right). \quad (2.413\text{a–c})$$

The **mass fraction** of the first phase is defined as the ratio of its mass density to the total mass density (2.413d) and satisfies (2.412a) ≡ (2.413e) where (2.413c) is used in (2.413f):

$$\xi \equiv \frac{\rho_1}{\rho}: \quad \nabla\cdot\bar{I} = \frac{\partial}{\partial t}\left(\rho\xi\right) + \nabla\cdot\left(\rho\xi\bar{v}\right)$$

$$= \rho\left(\frac{\partial \xi}{\partial t} + \bar{v}\cdot\nabla\xi\right) + \xi\left[\frac{\partial \rho}{\partial t} + \nabla\cdot\left(\rho\bar{v}\right)\right] = \rho\frac{d\xi}{dt}, \qquad (2.413\text{d–g})$$

and in (2.413g) appears the material derivative (2.413h):

$$\frac{d}{dt} \equiv \frac{\partial}{\partial t} + \vec{v} \cdot \nabla; \quad \vec{I} = -\rho \chi_m \nabla \xi, \quad \rho \frac{d\xi}{dt} = \nabla \cdot \left(\chi_m \rho \nabla \xi \right), \qquad \text{(2.413h–j)}$$

the Fick (1855) law states that the mass flux is proportional to the gradient of the mass fraction (2.413i) through the mass diffusivity χ_m and leads to a convected diffusion equation (2.413j) for the mass fraction (2.413d). The mass density ρ is inserted as a factor in the Fick Law (2.413b) so that the mass diffusivity χ_m has the dimensions $L^2 T^{-1}$ of square of length L divided by time T, and to comply with the second principle of thermodynamics on entropy production, must be positive, as shown next.

2.4.19 ENTROPY PRODUCTION BY MASS DIFFUSION

The entropy change associated with mass diffusion is specified at constant internal energy given by the heat (2.6b) $\equiv$ (2.414a) equal to the internal energy (2.414b) due to the chemical work (2.91g) $\equiv$ (2.414c) leading to the entropy per unit mass (2.414d) and (2.414e, f) per unit volume:

$$T ds = du = dW_c = \bar{v} \, d\xi, \quad ds = \frac{\bar{v}}{T} d\xi, \quad dS = \rho ds = \frac{\bar{v}\rho}{T} d\xi, \qquad \text{(2.414a–f)}$$

where ξ is the mass fraction (2.413d) and $\bar{v}$ the relative mass chemical potential or relative affinity (2.91h). From (2.414f) follows the entropy production (2.414g) over a domain D_3 of volume:

$$\dot{S}_m = \int_{D_3} \frac{\bar{v}\rho}{T} \frac{\partial \xi}{\partial t} dV = \int_{D_3} \frac{\bar{v}\rho}{T} \frac{d\xi}{dt} dV = \int_{D_3} \frac{\bar{v}}{T} \left(\nabla \cdot \vec{I} \right) dV; \qquad \text{(2.414g–i)}$$

in a convected domain the velocity is zero $\vec{v} = 0$, and the material derivative (2.413h) coincides with the local partial time derivative and substitution of (2.413g) leads to (2.414i). An integration by parts of (2.414i) leads to (2.414j)

$$\dot{S}_m = \int_{D_3} \nabla \cdot \left(\frac{\bar{v}\vec{I}}{T} \right) dV - \int_{D_3} \vec{I} \cdot \nabla \left(\frac{\bar{v}}{T} \right) dV. \qquad \text{(2.414j)}$$

The first term on the r.h.s. of (2.414j) becomes a surface integral by the divergence theorem (III.5.163a–c) $\equiv$ (2.414l) and vanishes (2.414m) for zero mass flux across the boundary (2.414k):

$$\vec{I} \cdot \vec{N} \big|_{\partial D_3} = 0: \quad \int_{D_3} \nabla \cdot \left(\frac{\bar{v}\vec{I}}{T} \right) dV = \int_{\partial D_3} \frac{\bar{v}}{T} \vec{I} \cdot d\vec{S} = 0. \qquad \text{(2.414k–m)}$$

Thus the entropy production per unit volume is given (2.415b, c) by integrand of the second term on the r.h.s. of (2.414j):

$$T = const: \quad \dot{S}_m = \frac{d\dot{\bar{S}}_m}{dV} = -\bar{I} \cdot \nabla\left(\frac{\bar{v}}{T}\right) = -\frac{\bar{I} \cdot \nabla\bar{v}}{T}; \qquad (2.415\text{a–d})$$

for purely mass diffusion, the temperature is constant (2.415a), otherwise there would be coupling (notes 2.12–2.14) to heat conduction (subsections 2.4.1–2.4.2). Thus (2.415c) simplifies to (2.415d), where the relative affinity depends only on the mass fraction (2.415e), implying (2.415f):

$$\bar{v} = \bar{v}(\xi): \quad \nabla\bar{v} = \left(\frac{\partial\bar{v}}{\partial\xi}\right)_T \nabla\xi; \quad \dot{S}_m = \rho\frac{\chi_m}{T}\left(\frac{\partial\bar{v}}{\partial\xi}\right)_T (\nabla\xi)^2; \qquad (2.415\text{e–g})$$

substitution of (2.415f) and (2.413i) in (2.415d) leads to (2.415g), showing that *entropy production (2.415h) with relative affinity increasing with mass fraction (2.415i) implies positive mass diffusivity (2.415j) and mass flux from higher to lower mass fraction (2.415k):*

$$\dot{S}_m > 0 < \left(\frac{\partial\bar{v}}{\partial\xi}\right)_T \Rightarrow \chi_m > 0, \quad \bar{I} \cdot \nabla\xi < 0. \qquad (2.415\text{h–k})$$

The diffusion equations become coupled for combined diffusion, for example, thermoelectric conduction (subsection 2.4.20).

2.4.20 Coupled Thermoelectric Diffusion Equations

The divergence (2.416b, c) of the Maxwell equation (2.354b) together with the Maxwell equation (2.21b) ≡ (2.416a) leads to (2.416d):

$$\nabla \cdot \bar{D} = q: \quad 0 = c\nabla \cdot (\nabla \wedge \bar{H}) = \nabla \cdot \bar{J} + \frac{\partial}{\partial t}(\nabla \cdot \bar{D}) = \frac{\partial q}{\partial t} + \nabla \cdot \bar{J}, \qquad (2.416\text{a–d})$$

the conservation (2.416d) of electric charges and currents, that is analogous to the conservation of heat (2.346b). Substituting in the equations of conservation of heat (2.346b; 1.324c) ≡ (2.417a) [electric charges and currents (2.416d) ≡ (2.418a)] the combined Fourier-Ohm laws for the heat flux (2.364a) [electric current (2.364b)] leads to:

$$\rho C_V \frac{\partial T}{\partial t} = -\partial_i G_i = \partial_i\left(k_{ij}\partial_j T\right) - \partial_i\left(TD_{ij}E_j\right), \qquad (2.417\text{a, b})$$

$$\frac{\partial q}{\partial t} = -\partial_i J_i = -\partial_i\left(\sigma_{ij}E_j\right) - \partial_i\left(D_{ij}\partial_j T\right), \qquad (2.418\text{a, b})$$

that are *the coupled thermoelectric diffusion equations (2.417b; 2.418b) in a linear anisotropic medium, involving the thermal (2.352a–c) [electrical (2.373a–c)] conductivity tensors and the thermoelectric coupling tensor. In the case of an isotropic medium (2.419a–c) the coupled thermoelectric diffusion equation (2.417b; 2.418b) involve (2.419d, e) the scalar thermal (electric) conductivity and thermoelectric coupling:*

$$\left\{k_{ij}, \varepsilon_{ij}, D_{ij}\right\} = \left\{k, \varepsilon, D\right\}\delta_{ij}:$$
$$\rho C_V \frac{\partial T}{\partial t} = \nabla \cdot \left(k\nabla T\right) - \nabla \cdot \left(TD\vec{E}\right), \quad \frac{\partial q}{\partial t} = -\nabla \cdot \left(\sigma_e \vec{E}\right) - \nabla \cdot \left(D\nabla T\right). \qquad (2.419\text{a–e})$$

If the diffusion coefficients are constant (2.420a–c) the thermal (thermoelectric) diffusivities (2.420d) [(2.420e)] may be introduced in (2.419d, e) the coupled thermoelectric diffusion equations (2.420f, g):

$$k, \sigma_e, D = const, \quad \left\{\chi_q, \chi_D\right\} = \frac{\left\{k, TD\right\}}{\rho C_V}:$$
$$\frac{\partial T}{\partial t} = \chi_q \nabla^2 T - \chi_D \nabla \cdot \left(T\vec{E}\right), \quad \frac{\partial q}{\partial t} = -\sigma_e \nabla \cdot \vec{E} - D\nabla^2 T. \qquad (2.420\text{a–g})$$

The electric displacement is proportional to the electric field through the dielectric permittivity (2.27b) ≡ (2.421a), and if the latter is constant (2.421b) the divergence of the electric field (2.421c) is given by (2.421d, e) using (2.416a):

$$\vec{D} = \varepsilon\vec{E}; \quad \varepsilon = const: \quad \nabla \cdot \vec{E} = \nabla \cdot \left(\frac{\vec{D}}{\varepsilon}\right) = \frac{\nabla \cdot \vec{D}}{\varepsilon} = \frac{q}{\varepsilon}. \qquad (2.421\text{a–e})$$

In the absence of coupling (2.422a) the temperature satisfies the heat equation (2.422b) ≡ (1.311b) without heat sources:

$$D = 0: \quad \rho C_V \frac{\partial T}{\partial t} = k\nabla^2 T, \quad \frac{\partial e}{\partial t} = -\frac{\sigma_e}{\varepsilon}q \Rightarrow q(t) = q(0)\exp\left(-\frac{\sigma_e}{s}t\right), \qquad (2.422\text{a–d})$$

and the electric charge density (2.422c) decays exponentially with time (2.422d).

The Table 2.6 compares the linear anisotropic/isotropic thermal/electric/viscous diffusion (subsections 2.4.1–2.4.2/2.4.3–2.4.4/2.4.9–2.4.11) and the Table 2.7 concerns thermoelectric coupling (subsections 2.4.5–2.4.8). The constitutive (diffusion) relations [Section 2.2 (2.4)] are combined next in unsteady anisothermal dissipative piezoelectromagnetism of anisotropic matter (subsection 2.4.20) including: (i) thermal, electric, magnetic and elastic constitutive coupling (2.133a, b); (ii) coupled electrical and thermal diffusion (2.364a, b); (iii) as variables temperature, displacement vector and electric and magnetic fields.

TABLE 2.6
Three Diffusive Properties

Property		Thermal Conduction	Electric Conductivity	Viscous Dissipation
variable	independent	Temperature gradient: $\partial_i T$	Electric field: E_i	Rate-of-strain tensor: $\dot{S}_{ij}$
	dependent	Heat flux: G_i	Electric current: J_k	Viscous stress tensor: τ_{ij}
diffusion tensor	anisotropic	$k_{ij} = \partial^2 S/[(\partial_i T)(\partial_j T)]$	$\sigma_{ij} = \partial^2 S/(\partial E_i \partial E_j)$	$D_{ijkn} = \partial^2 S / \partial \dot{S}_{ij} \partial \dot{S}_{kn}$
	isotropic	$k = \partial^2 S/\partial(\nabla T)^2$	$\sigma = \partial^2 S/\partial(|\vec{E}|)^2$	$D_{ijkn} = \eta \partial_{ik}\delta_{jn} + \left(\zeta - \dfrac{\eta}{3}\right)\delta_{ij}\delta_{kn}$
diffusion relation	anisotropic	$G_i = -k_{ij}\,\partial_j T$	$J_i = \sigma_{ij}\,E_j$	$\tau_{ij} = D_{ijkn}\,\dot{S}_{kn}$
	isotropic	$\vec{G} = -k\nabla T$	$\vec{J} = \sigma\vec{E}$	$\tau_{ij} = \eta \dot{S}_{kn} + \left(\zeta - \dfrac{\eta}{3}\right)\dot{S}_{kk}\delta_{ij}$
one-half entropy production	anisotropic	$k_{ij}\,(\partial_i T)(\partial_j T)$	$\sigma_{ij} E_i E_j$	$D_{ijkn}\,\dot{S}_{ij}\,\dot{S}_{kn}$
	isotropic	$k\,(\nabla T)^2$	$E^2\,\sigma$	$\eta\left(\bar{\bar{S}}_{ij}\right)^2 + \zeta\left(\dot{S}_{kk}\right)^2$

Note: The three diffusive properties, namely (a) heat conduction by the Fourier law, (b) electric current specified by Ohm's law, and (c) viscous stresses for a Newtonian fluid are compared as concerns: (i) the independent and dependent variables; (ii–iii) the diffusion relation between them through a diffusion tensor; (iv) the entropy production since all three processes (a, b, c) are irreversible; in (ii, iii, iv) anisotropic and isotropic forms are considered.

TABLE 2.7
A Diffusive Coupling

	Gradient	Flux	Diffusion Anisotropic	Coefficient Isotropic
thermal	$\partial_e T$	G_j	$-k_{ij}$	$-k\,\delta_{ij}$
electric	E_i	J_j	σ_{ij}	$\sigma\,\delta_{ij}$
thermo-electric	$\partial_i T$	E_j	$T D_{ij}$	$T D\,\delta_{ij}$
electro-thermal	E_i	G_j	D_{ij}	$D\,\delta_{ij}$

Note: The coupling between (a) thermal and (b) electrical diffusion involves: (i, ii) the thermal and electrical conductivities in isolation (Table 2.6); (iii) two coupling coefficients that are related and reduce to one (Table 2.7) In order to satisfy the Onsager reciprocity principle to ensure that there is entropy production, since thermo-electric diffusion is an irreversible process.

2.4.21 ANISOTHERMAL DISSIPATIVE PIEZOELECTROMAGNETISM (CAMPOS, SILVA & MOLEIRO 2020)

It is assumed that all constitutive and diffusion coefficients are constant when deriving the fundamental equations that follow, where dot denotes derivative with regard to time (2.423a). In the Maxwell equation (2.354a) ≡ (2.423b) for the electric field is substituted the constitutive relation for the magnetic induction (2.133c) leading to (2.423c):

$$\dot A \equiv \frac{\partial A}{\partial t}: \quad -ce_{ijk}\partial_j E_k = \dot B_i = h_i \dot T + \vartheta_{ji}\dot E_j + \mu_{ij}\dot H_j + \frac{1}{2}q_{ijk}\left(\partial_j \dot u_k + \partial_k \dot u_j\right), \quad (2.423a\text{–}c)$$

where was used the strain tensor (2.82c). In the Maxwell equation for the magnetic field (2.354b) ≡ (2.424a) are substituted the constitutive (2.133b) [diffusion (2.364b)] relations for the electric displacement (current) leading to (2.424b):

$$ce_{ijk}\partial_j H_k = \dot D_i + J_i$$
$$= f_i \dot T + \varepsilon_{ij}\dot E_j + \vartheta_{ij}\dot H_j + \frac{1}{2}p_{ijk}\left(\partial_j \dot u_k + \partial_k \dot u_j\right) + \sigma_{ij}E_j + D_{ij}\partial_j T. \quad (2.424a,\ b)$$

In the energy equation (2.346b; 2.6b) ≡ (2.425a) are substituted the constitutive (diffusion) relation for the entropy (2.133a) [energy flux (2.364a)] leading to (2.425b):

$$0 = \rho T \dot S + \partial_i G_i$$
$$= \rho C_V \dot T + \rho T\left(f_i \dot E_i + h_i \dot H_i\right) + \frac{1}{2}\rho T\alpha_{ij}\left(\partial_i \dot u_j + \partial_j \dot u_i\right) \quad (2.425a,\ b)$$
$$- k_{ij}\partial_i\partial_j T + D\partial_i\left(TE_i\right).$$

The remaining fundamental equation is for the momentum.

The momentum equation balances the inertia force (2.426) against: (i) the electric force (2.29a); (ii) the magnetic force (2.39a); (iii) the divergence of the stress tensor:

$$\rho \ddot{u}_i = E_i \left(\partial_j D_j \right) + \frac{1}{c} e_{ijk} J_j B_k + \partial_i T_{ij}. \tag{2.426}$$

Substituting the constitutive (diffusion) relations for the electric displacement (2.133b), magnetic induction (2.133c) and stress tensor (2.133d) [electric current (2.364b)] leads to (2.427):

$$\rho \ddot{u}_i = E_i \left[f_j \partial_j T + \varepsilon_{ij} \partial_i E_j + \vartheta_{ij} \partial_i H_j + \frac{1}{2} p_{\ell jk} \partial_\ell \left(\partial_j u_k + \partial_k u_j \right) \right]$$

$$+ \frac{1}{c} e_{ijk} \left(\sigma_{j\ell} E_\ell + D_{j\ell} \partial_\ell T \right)$$

$$\times \left[h_k T + \vartheta_{mk} E_m + \mu_{km} H_m + \frac{1}{2} q_{kmn} \left(\partial_m u_n + \partial_n u_m \right) \right] \tag{2.427}$$

$$- \alpha_{ij} \partial_j T - p_{kij} \partial_j E_k - q_{kij} \partial_j H_k + \frac{1}{2} C_{ijk\ell} \partial_j \left(\partial_k u_\ell + u_\ell u_k \right).$$

Thus *the electric $\bar{E}$ and magnetic $\bar{H}$ fields, displacement $\bar{u}$ and temperature T are specified by the fundamental equations of **unsteady anisothermal anisotropic piezo-electromagnetism** consisting of the equations of electricity (2.423c), magnetism (2.424b), energy (2.425b) and momentum (2.426) for: (i) an unsteady elastic solid subsect to electromagnetic forces; (ii) allowing for heat and electrical diffusion; (iii) assuming linear anisotropic medium with constant constitutive (2.133a–d) and diffusion (2.364a, b) coefficients.* The system of equations simplifies considerably for isotropic media and steady conditions (Section 2.4.22).

2.4.22   Isotropic Media and Steady Fields

The case of isotropic media (2.147a–c; 2.148a–c; 2.172d) simplifies the equations: (i/ii) of electricity (2.423c) [magnetism (2.424b)] to (2.428a) [(2.428b)]:

$$-c\nabla \wedge \bar{E} = \mu \dot{\bar{H}}, \quad c\nabla \wedge \bar{H} = \varepsilon \dot{\bar{E}} + \sigma_e \bar{E} + D\nabla T; \tag{2.428a, b}$$

(iii) of energy (2.425b) to (2.429):

$$\rho C_V \dot{T} = k\nabla^2 T - \rho T \alpha \left(\nabla \cdot \dot{\bar{u}} \right) - D\nabla \cdot \left(T\bar{E} \right); \tag{2.429}$$

(iv) of momentum (2.427) to (2.430):

$$\rho\ddot{\vec{u}} = \varepsilon\vec{E}(\nabla\cdot\vec{E}) + \frac{\mu}{c}\left(\sigma_e\vec{E} + D\nabla T\right)\wedge\vec{H} - \alpha\nabla T + \frac{E}{2(1+\sigma)}\left[\nabla^2\vec{u} + \frac{1}{1-2\sigma}\nabla(\nabla\cdot\vec{u})\right], (2.430)$$

where was used the divergence of the stress tensor (2.172d) in vector form (2.431a–c):

$$\partial_j T_{ij} = \frac{E}{2(1+\sigma)}\left[\partial_j\left(\partial_i u_j + \partial_j u_i\right) + \frac{2\sigma}{1-2\sigma}\partial_i\left(\partial_j u_j\right)\right]$$

$$= \frac{E}{2(1+\sigma)}\partial_j\partial_j u_i + \frac{E}{2(1+\sigma)}\left(1 + \frac{2\sigma}{1-2\sigma}\right)\partial_i\partial_j u_j \qquad (2.431a–c)$$

$$= \frac{E}{2(1+\sigma)}\left[\nabla^2\vec{u} + \frac{1}{1-2\sigma}\nabla(\nabla\cdot\vec{u})\right].$$

Thus *the electric $\vec{E}$ and magnetic $\vec{H}$ fields, displacement $\vec{u}$ and temperature T satisfy the fundamental equations of electricity (2.428a), magnetism (2.428b), energy (2.429) and momentum (2.430) for: (i) an isotropic elastic solid subject to electromagnetic forces; (ii) with coupled heat flux and electric current; (iii) assuming constant constitutive (2.148a–c; 2.134; 2.172d) [diffusion (2.363a, b)] coefficients; (iv) in unsteady conditions.*

In steady conditions (2.432a): (i) the electric field decouples (2.18a) ≡ (2.432b) considering (2.27b) only external electric charges (2.21b) ≡ (2.432c, d)

$$\frac{\partial}{\partial t} = 0: \quad \nabla\wedge\vec{E} = 0, \quad \nabla\cdot\vec{E} = \frac{\nabla\cdot\vec{D}}{\varepsilon} = \frac{q}{\varepsilon}; \qquad (2.432a–d)$$

(ii) the magnetic field also decouples (2.38a; 2.53b) ≡ (2.433a) considering only external electric currents (2.42b) ≡ (2.433b):

$$\nabla\cdot\vec{H} = 0, \quad c\nabla\wedge\vec{H} = \vec{j}; \quad D = 0 \neq k: \quad \nabla^2 T = 0, \qquad (2.433a–e)$$

(iii) the energy equation (2.430) considering only thermal conduction (2.433c, d) simplifies to Laplace's equation (2.433e) for temperature in the absence of heat sources, in agreement with (1.311d) with w = 0; (iv) the momentum equation (2.426) with external electric charges (2.432d) and currents (2.433b) and elastic stresses (2.431c) simplifies to (2.434):

$$\alpha\nabla T = \frac{E}{2(1+\sigma)}\left[\nabla^2\vec{u} + \frac{1}{1-2\sigma}\nabla(\nabla\cdot\vec{u})\right] + q\vec{E} + \frac{\mu}{c}\vec{j}\wedge\vec{H}. \qquad (2.434)$$

These equations are solved next in the one-dimensional case of a parallel-sided slab (subsection 2.4.23).

2.4.23 THERMOELASTIC ELECTROMAGNETISM IN A SLAB

Considering a parallel sided slab (2.435a) with Figure 2.10 electric field orthogonal to the sides and depending on the transverse coordinate x in (2.435b) then (2.432b) is satisfied and (2.432d) leads to (2.435c, d) where prime denotes derivative with regard to x:

$$0 \le x \le L: \quad \vec{E} = \vec{e}_x E(x), \quad \frac{q}{\varepsilon} = \frac{dE}{dx} = E'(x). \qquad (2.435\text{a–d})$$

In the case of constant electric charge (2.436a), the electric field on one side (2.436b) specifies the linear dependence of the electric field in the slab (2.436c) and the value on the other side (2.436d):

$$q = const, \quad E(0) = E_0: \quad E(x) = E_0 + \frac{q}{\varepsilon}x, \quad E(L) = E_0 + \frac{qL}{\varepsilon}. \qquad (2.436\text{a–d})$$

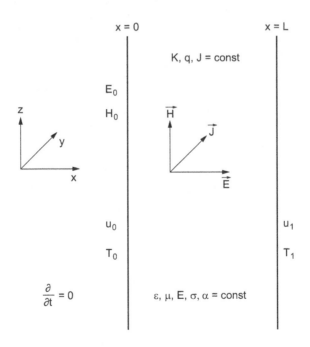

FIGURE 2.10 Steady thermoelastic piezoelectromagnetism in a parallel sided slab with: (i) all variables depending only on the coordinate x perpendicular to the slab; (ii) electric field normal to the slab and transverse magnetic field and electric current orthogonal to each other; (iii) constant electric charge q and current J densities; (iv) constant material properties, namely dielectric permittivity ε, magnetic permeability μ, Young modulus E, Poisson ratio σ and thermoelastic coefficient α.

Considering an electric current orthogonal to the electric field (2.437a) the Maxwell equation (2.433b) is satisfied by a magnetic field (2.437b) orthogonal to both leading to (2.437c, d):

$$\vec{J} = \vec{e}_z J, \quad \vec{H} = \vec{e}_y H(x): \quad \vec{e}_z \frac{J}{c} = \left(\vec{e}_x \frac{d}{dx} \right) \wedge \left[\vec{e}_y H(x) \right] = \vec{e}_z H'(x); \quad (2.437\text{a–d})$$

for a constant electric current (2.438a), the magnetic field on one side (2.438b) specifies (2.438c) the linear magnetic field inside the slab (2.438d) and the value on the other side (2.438e):

$$J = const, \quad H(0) = H_0: \quad H'(x) = \frac{J}{c}, \quad H(x) = H_0 + \frac{J}{c} x, \quad H(L) = H_0 + \frac{JL}{c}. \quad (2.438\text{a–e})$$

The temperature (2.433e) is also linear (2.439a) and is specified in the slab (2.439d) by the values at the two sides: (2.439b, c):

$$T''(x) = 0, \quad T_0 = T(0), \quad T_1 = T(L): \quad T(x) = T_0 + (T_1 - T_0) \frac{x}{L}. \quad (2.439\text{a–d})$$

For a displacement orthogonal to the sides (2.440a) the momentum equation (2.434) simplifies to (2.440b, c):

$$\vec{u} = \vec{e}_x u(x): \quad \frac{E}{1+\sigma} \frac{1-\sigma}{1-2\sigma} u''(x) = \alpha T'(x) - qE(x) - \frac{\mu J}{\sigma_e} H(x)$$

$$= \alpha \frac{T_1 - T_0}{L} - q\left(E_0 + \frac{q}{\varepsilon} x \right) - \frac{\mu J}{\sigma_e} \left(H_0 + \frac{J}{c} x \right), \quad (2.440\text{a–c})$$

The displacement satisfies (2.440c) ≡ (2.441a) involving two constants (2.441b, c):

$$u''(x) = 6ax + 2b:$$

$$\frac{E}{1+\sigma} \frac{1-\sigma}{1-2\sigma} \{2b, 6a\} \equiv \left\{ \alpha \frac{T_1 - T_0}{L} - qE_0 - \frac{\mu J}{\sigma_e} H_0, -\frac{q^2}{\varepsilon} - \frac{\mu J^2}{\sigma_e c} \right\}. \quad (2.441\text{a–c})$$

The displacement (2.441a) is given by (2.442a):

$$u(x) = ax^3 + bx^2 + C_1 x + C_2:$$

$$u_0 = u(0) = C_2, \quad u_1 \equiv u(L) = C_2 + C_1 L + bL^2 + aL^3, \quad (2.442\text{a–c})$$

with the two constants of integration determined by the values (2.442b, c) at the two sides. Substitution of (2.442b, c) in (2.442a) specifies the displacement (2.443):

$$u(x) = u_0 + (u_1 - u_0) \frac{x}{L} - (L - x)(bx + ax^2). \quad (2.443)$$

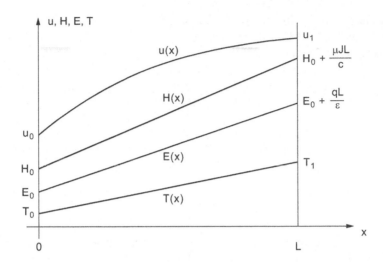

FIGURE 2.11 In the problem of the steady thermoelastic piezoelectromagnetism in a parallel sided slab (Figure 2.10) the solution is unique with six boundary conditions specifying: (i,ii) the electric E_0 and magnetic H_0 field on one side of the slab; (iii–vi) the temperature (T_0, T_1) and normal displacement (u_0, u_1) at the two sides of the slab. The solution as a function of the coordinate x normal to the sides of the slab is: (i–iii) linear for the temperature, electric and magnetic fields; (iv) cubic for the normal displacement.

Thus considering (Figure 2.10) a parallel-sided slab (2.435a) of width L, in steady conditions (2.432a) with one-dimensional dependence perpendicular to the sides: (i) a constant electric charge (2.436a) specifies a normal electric field (2.436c) with one boundary condition (2.436b); (ii) a constant (2.438a) tangential electric current (2.437a) specifies a magnetic field (2.438d) orthogonal to both (2.437b) with one boundary condition (2.438b); (iii) in the presence of thermal conduction (2.433d) without thermoelectric coupling (2.433c) and in the absence of heat sources (2.433e) the temperature (2.429a) is specified (2.439d) by two boundary conditions (2.429b, c); (iv) whereas all previous dependences (i–iii) are linear (Figure 2.11) the normal (2.440a) displacement (2.443; 2.441b, c) is a cubic with two boundary conditions (2.442b, c).

The structure of thermodynamics is indicated in the Diagram 2.1. The first (second and third) principles of thermodynamics [Section 2.1 (2.3)] lead to inequalities for the constitutive (diffusive) coefficients [Section 2.2 (2.4)] that apply to all matter, but do not specify the actual values for specific cases, like gases, liquids or solids. To complete the description of matter is needed an equation of state (Section 2.5) that enables the detailed description of: (i) thermodynamic processes such as adiabatic, isothermal, isobaric and isochoric; (ii) sequences of processes returning to the initial state and forming a thermodynamic cycle; (iii) specific thermodynamic cycles, like the Carnot, Atkinson, Stirling, Barber – Brayton; (iv) applications to a variety of engines and also heat pumps and refrigerators.

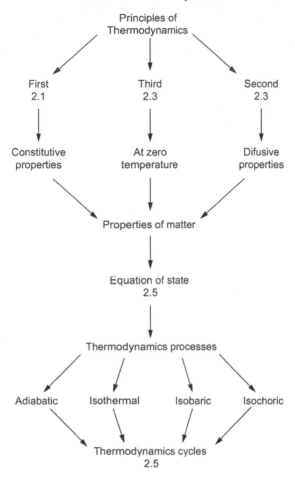

Structure of thermodynamics

DIAGRAM 2.1   The first (second) principle of thermodynamics relates to the constitutive (dissipative) properties of matter. A substance is specified by an equation of state, leading to the consideration of thermodynamic processes, for example, adiabatic (no heat exchanges), or isothermal/isobaric/isochoric, respectively, at constant temperature/pressure/volume. A sequence of processes leading back to the initial state forms a thermodynamic cycle.

References

Fourier, J. B. J. 1818. *"Theorie analytique de le chaleur"*, Paris, repr. Dover 1955.

Volta, A. 1821. "Nuova Memoria dull electricita animale", *Annali di chimica e Storie Naturale* **5**, 132–144.

Seebeck, T. J. 1822. "Magnetische Polarisation der Metalle und Erze durch Temperatur-Differenz", *Abbandlungen des Koniglichen Akademie des Wissenschaften Berlin*, 82, 265–373.

Navier, C. L. M. H. 1822. "Mémoires sur les Lois du Movement des Fluides", *Mémoires de l'Académie des Sciences de Paris* **6**, 389.

Ohm, G. 1827. *"Die Galvanische Kette, mathematisch bearbeitet"*, Berlin, Riemann.

Onsager, L. 1931. "Reciprocal Relations in Irreversible Processes", *Physical Review* **37**, 405–426.

Peltier, J. C. A. 1834. "Nouvelles Experiences sur la Caloricité des Courents Electriques", *Annale de Chimie et Physique* **56**, 371–386.

Stokes, G. G. 1845. "On the Theories of Internal Friction of Fluids in Moton", *Transactions of the Cambridge Philosophical Society* **7**, 287 (Papers 1, 75)

Joule, J. P. 1847. "On the Effects of Magnetism upon the Dimensions of Iron and Steel Bars", *The London, Edinburg and Dublin Philosophical Journal of Science* 30, 76–87.

Clausius, R. 1850. "Ueber die bewegende Kraft der Wärme und die Gesetze, welche sich darus für die Wärmelehre selbst ableiten lassen", *Annalen der Physik* **79**, 368–397.

Clausius, R. 1854. "Ueber eine veränderte Form der mechanischen Wärmetheorie", *Annalen der Physik* **93**, 481–506.

Thomson, W. 1857. "On a Mechanical Theory of Thermoelectric Currents", *Proceedings of the Royal Society Edinburg* 3, 91–98.

Clebsch, A. 1859. "Ueber der Integration der Hydrodynamische Gleichungen", *Journal fur reine und angewandte Mathematik* **56**, 1–10.

Maxwell, J. C. 1865. "A Dynamical Theory of the Electromagnetic Field", *Philosophical Transactions of the Royal Society* **155**, 459–512.

Maxwell, J. C. 1867. "On the Dynamical Theory of Gases", *Philosophical Transactions of the Royal Society* **157**, 49–88.

Gibbs, J. W. 1876–1878. *"Collected Works of J. Willard Gibbs"*, New York, Longmans Green, 1928, reprinted Dover, 1954.

Hall, E. 1879. "On a New Action of the Magnet on Electric Currents", *American Journal of Mathematics* **2**, 287–292.

Jacobi, C. G. J. 1891. *"Gesummelte Werke"*, Konigklische Presersiche Akademie der Wissenschaften, 8 vols., reprinted Berlin: Chelsea 1972.

Thomson, J. J. 1882. "On an Absolute Thermometric Scale Founded on Carnot's Theory", *Mathematical Papers* **1**, 302, Cambridge University Press.

Poyinting, J. H. 1884. "On the Transfer of Energy in the Electromagnetic Field", *Philosophical Transactions of the Royal Society* **155**, 343–361.

Duhem, P. M. M. 1886. *"Le potential thermodynamique et ses applications"*, Paris, Hermann.

Nerst, W. 1907. *"Experimental and Theoretical Applications of Thermodynamics to Chemistry"*, New York, Charles Scribner.

Caratheodory, C. 1909. "Untersuchungen über Grundlagen der Thermodynamik", *Mathematische Annalen.* **67**, 355–368.

Campos, L. M. B. C., Silva, M. J. S. & Moleiro, F.S. 2019. "Fundamental Equations of Unsteady Anisothermal Piezoelectromagnetism", *European Journal of Mechanics* **437**, 389–409.

Bibliography

The bibliography of the series "Mathematics and Physics for Science and Technology" is quite extensive since it covers a variety of subjects. In order to avoid overlaps, each volume contains only a part of the bibliography on the subjects most closely related to its content. The bibliography of earlier volumes is generally not repeated, and some of the bibliography may be relevant to earlier and future volumes. The bibliography covered in the four published volumes is:

A. General
  a. Overviews
    1. General mathematics: book 1
    2. Theoretical physics: book 2
    3. Engineering technology: book 3
  b. Reference
    4. Collected works: book 6
    5. Generic Encyclopedia: book 10
B. Mathematics
  c. Theory of functions
    6. Real functions: books 1 and 2
    7. Complex analysis: books 1 and 2
    8. Generalized functions: book 3
  d. Differential and integral equations
    9. Ordinary differential equations: book 4
    10. Partial differential equations: book 10
    11. Non-linear differential and integral equations: book 5
  e. Geometry
    12. Tensor calculus – book 3
  f. Higher analysis
    13. Special functions: book 8
C. Physics
  g. Classical mechanics
    14. Material particles: book 7
  h. Thermodynamics
    15. Thermostatics: book 2
    16. Heat: book 2
  i. Fluid mechanics
    17. Hydrodynamics: book 1
    18. Aerodynamics: book 1
  j. Solid mechanics
    19. Elasticity: book 2
    20. Structures: book 2
This choice of subjects is explained next.

The general bibliography consists of overviews and reference works. The overviews have been completed with mathematics, physics, and engineering, respectively in the volumes I, II and III corresponding to books 1, 2 and 3. The reference bibliography starts with the collected works of notable authors in book 6 of volume IV. Concerning mathematics, the bibliography on the theory of functions has appeared in the volumes I, II and III corresponding to books 1, 2 and 3. The bibliography on ordinary differential equations has appeared in book 4 and on non-linear differential equations in book 5, both in volume IV. The volume IV also contains in book 8 the bibliography on special functions. The bibliography on tensor calculus appears in book 3 that coincides with volume III. Concerning physics, the bibliography on material particles appears in book 7 of volume IV. The bibliography on thermodynamics and heat appears in book 2 that coincides with volume II. The bibliography on solid mechanics, including elasticity and structures, appears in book 2 that coincides with volume II. The bibliography on fluid mechanics, including hydrodynamics and aerodynamics appears in the book 1 that coincides with volume I. The present book 10 in volume V includes next the bibliography on:

1. Generic Encyclopedias
2. Partial differential equations

1. GENERIC ENCYCLOPEDIAS

Brockhaus Enzyklopedia. F. A. Brockaus, 1992, 22 vols., Wiesbaden.

D'Alembert, J. R. and Diderot, D. *Encyclopedie ou dictionaire raisonné des sciences, des arts et des métiers*. Briasson, 1751–1765, 17 vols., Paris.

D'Alembert, J. R. and Diderot, D. *Recueil de planches, sur les sciences, les arts liberaux et les arts méchaniques, avec leur explication*, Briasson, 1715–1765, 17 vols., Paris.

Encyclopedia Americana. Americana Corporation 1924, 26th ed. 1973, 30 vols., New York.

Encyclopedia Britanica. 1768, 15th ed. 1978, 30 vols., London.

Enciclpedia Luso-brasileira de Cultura. Editorial Verbo, 3rd ed. 2004, 22 vols., Lisbon.

Encyclopedia Publico. Rivadeneyra, 2004, 28 vols., Madrid.

Enciclopedia Salvat. 1970, 13 vols., Pamplona.

Encyclopedia Universal. Durccub, 2004, 29 vols., Madrid.

Encyclopedia Universalis. Encyclopedia Universalis, 1968, 20 vols., Paris.

Encyclopedia Universalis. 1987, 19 vols., Bertelsmann, Gütersloh.

Grande Enciclopédia do Conhecimento, Círculo dos Leitores. 2010, 24 vols., Lisbon.

Grande Enciclopédia Portuguesa e Brasileira. Editorial Enciclopedia, 1940, 2nd ed. 1990, 40 vols., Lisboa e Rio de Janeiro.

La Grande Encyclopedie Larousse. Librarie Larousse, 1971, 22 vols., Paris.

New Caxton Encyclopedia. 1973, 20 vols., London.

New Encyclopedia. 1992, 26 vols., Tammi Publishers, Helsinki.

Nova Enciclopedia Portuguesa. Ediclube, 1992, 22 vols., Lisbon.

2. PARTIAL DIFFERENTIAL EQUATIONS

Caratheodory, C. *Calculus of variations and partial differential equations of first order*. Repr. Chelsea: Chelsea Publishing Company 1982.

Constanda, C. *Solution techniques for elementary partial differential equations*. Boca Raton: Chapman & Hall 2020.

Epstein, B. *Partial differential equations: an introduction*. New York: McGraw-Hill 1962.

Garabedian, P. R. *Partial differential equations*. 1964, New York: Chelsea House 1986.

Greenspan, D. *Introduction to partial differential equations*. New York: McGraw-Hill 1961.

Godounov, S. *Equations de la physique mathematique*. Moscow: Editions Mir 1973.

Lie, S. *Differentialgleichungen*. Leipzig 1891, reprinted New York: Chelsea Publishing Co. Bronx 1967.

Mikhailov, V. *Equations aux derivées partielles*. Nauka 1976, Moscow: Editions Mir 1980.

Myint-U, T. and Debnath, L. *Partial differential equations for scientists and engineers*. New York: Prentice-Hall 1987.

Sneddon, I. N. *Elements of partial differential equations*. London: McGraw-Hill 1957.

Sobolev, S. L. *Partial Differential equations of mathematical physics*. Oxford: Pergamon Press 1964.

Taylor, M. E. *Partial differential equations*. New York: Springer 1996.

Tricomi, F. G. *Equazoni a derivate parziali*. Rome: Edizioni Cremonese 1957.

Webster, A. G. *Partial differential equations of mathematical physics*. 1927, 2nd edition 1933, reprinted New York: Dover 1955.

Young, E. C. *Partial differential equations: an introduction*. Boston: Aleyn & Bacon 1972.

Zachmanoglou, E. C. and Theoe, D. N. *Introduction to partial differential equations with applications*. New York: Williams & Wilkins 1976, reprinted Dover 1986.

Zauderer, E. *Partial differential equations of applied mathematics*. New York: Wiley 1989.

Index

Pages in *italics* refer figures and pages in **bold** refer tables.

A

Acceleration, 99
  of gravity, 100
Acoustic wave *see* sound wave
Activity, 97
Adiabatic
  process, 159–161
Affinity, 116–118
Angular velocity, 18–20
Anisothermal; *see also* heat
  piezoelectromagnetism, 219–225
Anisotropic, 126–137
  constitutive relations, 136–137
  diffusion relations, 200–205,
      208–211
  piezoelectromagnetism, 219–225
Avogadro number, 116–118

B

Balance of heat, 76–78
Biharmonic differential equation
  forced, 70–71
  unforced, 57–60
Bivector rotation, 115–116
Blow-up, 86–88
Bulk viscosity, 79–80
Burgers equation, 80–81

C

Characteristic
  curve, 10
  equations, 10
  polynomial, 82–83
  systems, 44–46, **45**
  variables, 10–11
Charge, electric, 101–102
Chemical potential, 116–118
Clebsch potentials, 31, 47–48, **47**
Coefficient
  of friction, 98
Coefficients of differential equation
  constant, 3–4
  variable, 2–4
Concentration, 117
Condition of integrability, 17
Conduction of heat *see* heat

Conductivity
  electrical
    Hall, 210–211
    Ohm, 201–204
    Thomson, 211–213
  thermal, 78
Conjugate thermodynamic variables, 121–123,
    127–128
Conservative vector field, 17, 22–24, 34, 36–37
Constant coefficients, differential equation with,
    2–4
Constant, gravitational, 100–101
Constitutive tensors, 129–150, 154–156
  dielectric permittivity, 103–104, 130, **131**
  elastic stiffness, 130, **131**, 132, 138
  electromagnetic coupling, 132–135, **133–134**,
    137
  magnetic permeability, 108–109
  piezoelectric, 135, 137–138
  piezomagnetic, **133–134**, 135, 137–138
  pyroelectric, 132, **133**, 137–138
  pyromagnetic, 132, **133**, 137–138
  specific heat, 130
  thermal stress, **133**, 135
Convective electric current, 211–212
Convention of summation, 97
Coupling
  electroviscous, 208
  thermoelectric, 202–204, 209
  thermoviscous, 208–209
Curves, characteristic, 10

D

Density
  of electric charge, 101–102
  of electric current, 105–107
  of force *see* force
  of heat, 76–78
Derivatives
  convective, 79
  local time, 79
  material, 79
  spatial, 79
Derivatives, thermodynamic
  first-order, 127–128
  second-order, 129–150, 155–156
Dielectric permittivity, 103–104, 130–131,
    149–150, 154